Modern Methods of Plant Analysis

New Series Volume 14

Editors
H. F. Linskens, Erlangen/Nijmegen/Amherst
J. F. Jackson, Adelaide

Volumes Already Published in this Series:

Forthcoming:

Seed Analysis

Edited by
H.-F. Linskens and J. F. Jackson

Contributors

L. A. Amberger   J.-C. Autran   D. Bullock   C. H. Fong
M. B. Forde   S. E. Gardiner   P. Gepts   M. W. Gray   S. Hasegawa
B. R. Hedges   Z. Herman   D. Hernández-López   A. H. C. Huang
J. F. Jackson   N. G. Larsen   V. Llaca   D. S. Luthe   S. de Meillon
K. J. Moore   W. R. Morrison   D. G. Myers   R. O. Nodari   T. R. Noel
R. G. Palmer   D. K. Pandey   L. Panella   O. Paredes-López
S. G. Ring   M. N. Schnare   D. F. Spencer   H. A. van de Venter
S. M. Wang   M. A. Whittam   C. W. Wrigley   M. Yamaguchi

With 72 Figures

Springer-Verlag
Berlin Heidelberg GmbH

Prof. Dr. HANS FERDINAND LINSKENS
Goldberglein 7
W-8520 Erlangen, FRG

Prof. Dr. JOHN F. JACKSON
Department of Viticulture, Enology and Horticulture
Waite Agricultural Research Institute
University of Adelaide
Glen Osmond, S.A. 5064
Australia

ISBN 978-3-662-01641-1   ISBN 978-3-662-01639-8 (eBook)
DOI 10.1007/978-3-662-01639-8

The Library of Congress Card Number 87-659239 (ISSN 0077-0183)

Production Editor: Herta Böning, Heidelberg
Reproduction of the figures: Gustav Dreher GmbH, Stuttgart
Typesetting: K+V Fotosatz GmbH, 6124 Beerfelden
31/3145-5 4 3 2 1 0 – Printed on acid-free paper

Introduction

Modern Methods of Plant Analysis

When the handbook *Modern Methods of Plant Analysis* was first introduced in 1954 the considerations were:
1. the dependence of scientific progress in biology on the improvement of existing and the introduction of new methods;
2. the difficulty in finding many new analytical methods in specialized journals which are normally not accessible to experimental plant biologists;
3. the fact that in the methods sections of papers the description of methods is frequently so compact, or even sometimes so incomplete that it is difficult to reproduce experiments.

These considerations still stand today.

The series was highly successful, seven volumes appearing between 1956 and 1964. Since there is still today a demand for the old series, the publisher has decided to resume publication of *Modern Methods of Plant Analysis*. It is hoped that the New Series will be just as acceptable to those working in plant sciences and related fields as the early volumes undoubtedly were. It is difficult to single out the major reasons for success of any publication, but we believe that the methods published in the first series were up-to-date at the time and presented in a way that made description, as applied to plant material, complete in itself with little need to consult other publications.

Contribution authors have attempted to follow these guidelines in this New Series of volumes.

Editorial

The earlier series *Modern Methods of Plant Analysis* was initiated by Michel V. Tracey, at that time in Rothamsted, later in Sydney, and by the late Karl Paech (1910–1955), at that time at Tübingen. The New Series will be edited by Paech's successor H. F. Linskens (Nijmegen, The Netherlands) and John F. Jackson (Adelaide, South Australia). As were the earlier editors, we are convinced "that there is a real need for a collection of reliable up-to-date methods for plant analysis in large areas of applied biology ranging from agriculture and horticultural experiment stations to pharmaceutical and technical institutes concerned with raw material of plant origin".

The recent developments in the fields of plant biotechnology and genetic engineering make it even more important for workers in the plant sciences to become acquainted with the more sophisticated methods, which sometimes come from biochemistry and biophysics, but which also have been developed in commercial firms, space science laboratories, non-university research institutes, and medical establishments.

Concept of the New Series

Many methods described in the biochemical, biophysical, and medical literature cannot be applied directly to plant material because of the special cell structure, surrounded by a tough cell wall, and the general lack of knowledge of the specific behavior of plant raw material during extraction procedures. Therefore all authors of this New Series have been chosen because of their special experience with handling plant material, resulting in the adaptation of methods to problems of plant metabolism.

Nevertheless, each particular material from a plant species may require some modification of described methods and usual techniques. The methods are described critically, with hints as to their limitations. In general it will be possible to adapt the methods described to the specific needs of the users of this series, but nevertheless references have been made to the original papers and authors. While the editors have worked to plan in this New Series and made efforts to ensure that the aims and general layout of the contributions are within the general guidelines indicated above, we have tried not to interfere too much with the personal style of each author.

There are several ways of classifying the methods used in modern plant analysis. The first is according to the technological and instrumental progress made over recent years. These aspects were used for the first five volumes in this series describing methods in a systematic way according to the basic principles of the methods.

A second classification is according to the plant material that has to undergo analysis. The specific application of the analytical method is determined by the special anatomical, physiological, and biochemical properties of the raw material and the technology used in processing. This classification will be used in Volumes 6 to 8, and for some later volumes in the series.

A third way of arranging a description of methods is according to the classes of substances present in the plant material and the subject of analytic methods. The latter will be used for later volumes of the series, which will describe modern analytical methods for alkaloids, drugs, hormones, etc.

Naturally, these three approaches to developments in analytical techniques for plant materials cannot exclude some small overlap and repetition; but careful selection of the authors of individual chapters, according to their expertise and experience with the specific methodological technique, the group of substances to be analyzed, or the plant material which is the subject of chemical and physical analysis, guarantees that recent developments in analytical methodology are described in an optimal way.

Volume Fourteen – Seed Analyses

This volume is devoted to various analyses relating to the commercially used seed of plants, which can be any of the cereal varieties, rice, and various legumes to citrus or almond. It is not our intention to cover the whole subject of seeds and their analysis, but rather selected methods have been chosen because they have been developed recently or are of typical interest in modern society and its commerce.

The range of methods described include protein analyses by electrophoresis, monoclonal antibodies and nuclear magnetic resonance, isozyme analyses, glycolipid analysis, conductivity, surface area, digital image analysis, proteinaceous inhibition of lipase, physicochemical analysis of starch, and mitochondrial DNA preparation.

Two chapters are included on the seed of fruit trees. The first deals with the detailed analysis of isozymes in the seed of almond, important in this crop since cross-pollination is essential for fruit production due to self-incompatibility in almond. Isozymes provide excellent molecular genetic markers and enable the scientist to trace pollen parents in commercial orchards so that the best juxtaposition of cross-pollinating pairs of almond cultivars can be determined under various conditions of climate, etc. The second fruit to be described is that of citrus. This chapter details the detection, identification, and quantitation of liminoids in citrus species.

The remainder of this volume deals with the seed of cereals, rice, and legumes. Thus, several chapters deal with the various cereals, using different approaches. One of these makes use of gel electrophoresis to identify proteins in wheat, barley, oats, rice, and maize. The various methods of protein separation are described, for each of these cereals, finishing with a section on the derivation of information about phenotype from these approaches. This is followed by a chapter on SDS-polyacrylamide gel electrophoresis of proteins from grasses and forage legumes, another on protein determination in corn, and then another dealing specifically with wheat using monoclonal antibodies and NMR. Another chapter on wheat deals with the isolation of mitochondrial DNA and RNA. The major storage proteins of rice, glutelin, is considered in another chapter together with in vivo labeling and electrophoresis, analyses of this and other rice storage proteins.

Starch analyses is the subject of two chapters in this volume, one concentrating on the physicochemical analyses of wheat starch, the other on a wider spread of analyses for the cereals in general. The latter covers the isolation of starch together with chemical analysis including total polysaccharide, amylase, lipid, protein, and phosphorus content, and physical analysis which involves the measurement of granule size, damaged starch, and gelatinization properties.

Such topics as differential scanning, calorimetry, bire fringence, end-point temperature, and others are included under gelatinization.

A general chapter on the analysis of lipids in seeds is presented, utilizing such methods as NMR, gas chromatography, and HPLC for total lipids, fatty acids, glyceride structure, and compounds such as phosphatides, sterols, and toxophenols. Fat determination in corn finds a place in this volume also.

The glycolipids in wheat are dealt with in a further chapter, using a large variety of analytical procedures while the proteinaceous inhibitors of lipase activities in soybean and other oil seeds are considered in an additional chapter. Conductivity testing of seeds finds a place in this volume, dealing with another method of evaluating seed quality, particularly pertaining to seed vigor and viability, cell membrane integrity, solute leakage, imbibition injury, and mechanical damage.

Two further chapters dealing with physical methods of approach to seed appraisal round off this volume. Image processing, e.g. weed in the discrimination of seeds (e.g. between wheat varieties), is one of these physical methods, the other is devoted to methods for the determination of total seed surface area, needed for various studies on water uptake and loss (as in drying operations) by seeds. If this volume deals with the cereals to a much greater extent than seeds, of other plants, then this is simply a reflection of the fact that cereals make up the major portion of the worldwide trade in seeds.

Acknowledgements. The editors express their thanks to all contributors for their efforts in keeping to production schedules, and to Dr. Dieter Czeschlik, Ms S. Mees and Ms H. Böning of Springer-Verlag for their cooperation in preparing this and other volumes in the Series *Modern Methods of Plant Analysis.*

Nijmegen/Siena and Adelaide, Summer 1992   H. F. LINSKENS
                                                           J. F. JACKSON

Contents

**Identification of Cultivars of Grasses and Forage Legumes
by SDS-PAGE of Seed Proteins**
S. E. GARDINER and M. B. FORDE (With 11 Figures)

**Analysis of Seed Proteins, Isozymes, and RFLPs for Genetic
and Evolutionary Studies in *Phaseolus***
P. Gepts, V. Llaca, R. O. Nodari, and L. Panella
(With 3 Figures)

**Determination of the Nitrogen-to-Protein Conversion Factor
in Cereals**
M. Yamaguchi (With 1 Figure)

**Protein Analysis of Wheat by Monoclonal Antibodies
and Nuclear Magnetic Resonance**
J.-C. AUTRAN (With 5 Figures)

Electrophoretic Analyses of Soybean Seed Proteins
B. R. HEDGES, R. G. PALMER, and L. A. AMBERGER

Analysis of Storage of Proteins in Rice Seeds
D. S. LUTHE (With 10 Figures)

Food Properties of Amaranth Seeds and Methods for Starch Isolation and Characterization
O. PAREDES-LÓPEZ and D. HERNÁNDEZ-LÓPEZ (With 5 Figures)

Glycolipid Analysis in Wheat Grains
N. G. LARSEN

Proteinaceous Inhibitors of Lipase Activities in Soybean and Other Oil Seeds

A. H. C. HUANG and S. M. WANG (With 4 Figures)

Conductivity Testing of Seeds
D. K. PANDEY (With 3 Figures)

Determination of the Surface Areas of Seed
H. A. VAN DE VENTER and S. DE MEILLON

The Discrimination of Seeds by Image Processing
D. G. MYERS (With 4 Figures)

Physicochemical Analysis of Wheat Starch
T. R. NOEL, S. G. RING, and M. A. WHITTAM (With 3 Figures)

The Isolation of Wheat Mitochondrial DNA and RNA
D. F. SPENCER, M. W. GRAY, and M. N. SCHNARE (With 3 Figures)

Analysis of Limonoids in Citrus Seeds
Z. HERMAN, C.H. FONG, and S. HASEGAWA (With 1 Figure)

List of Contributors

AMBERGER, LAURIE A., Iowa State University, G439 Agronomy, Ames, Iowa 50011, USA

AUTRAN, JEAN-CLAUDE, Laboratoire de Technologie des Céréales, INRA, 2 Place Viala, F-34060 Montpellier, France

BULLOCK, DONALD G., Department of Agronomy, University of Illinois, 11025 Goodwin Ave, Turner Hall Urbana, IL 61801, USA

FONG, CHI H., Citrus Products Technical Committee, 263 South Chester Avenue, Pasadena, CA 91106, USA

FORDE, MARGOT B., died in June 1992

GARDINER, SUSAN E., DSIR Fruit and Trees, Private Bag, Palmerston North, New Zealand

GEPTS PAUL, Department of Agronomy and Range Science, University of California, Davis, CA 95616-8515, USA

GRAY, MICHAEL W., Department of Biochemistry, Dalhousie University, Halifax, Nova Scotia B3H 4H, Canada

HASEGAWA, SHIN, USDA, ARS, PWA, Fruit and Vegetable Chemistry Laboratory, 263 S. Chester Avenue, Pasadena, CA 91106, USA

HEDGES, BRADLEY R., Agriculture Canada, Research Station, Harrow, Ontario, NOR 1GO, Canada

HERMAN, ZAREB, USDA, ARS, PWA, Fruit and Vegetable Chemistry Laboratory, 263 South Chester Avenue, Pasadena, CA 91106, USA

HERNÁNDEZ-LÓPEZ, DAVID, Departamento Ingeniería Bioguímica, Instituto Tecnológico de Celaya, Celaya

HUANG, ANTHONY H.C., Department of Botany and Plant Sciences, University of California, Riverside, CA 92521, USA

JACKSON, JOHN F., Department of Horticulture, Viticulture and Enology, Waite Agricultural Research Institute, University of Adelaide, Glen Osmond, S.A. 5064, Australia

LARSEN, NIGEL G., Grain Processing Laboratory, NZ Institute for Crop and Food Research, Private Bag, Christchurch, New Zealand

LLACA, VICTOR, Department of Agronomy and Range Science, University of California, Davis, CA 95616-8515, USA

LUTHE, DAWN S., Department of Biochemistry and Molecular Biology, P.O. Drawer BB, Mississippi State University, Mississippi State, MS 39762, USA

DE MEILLON, STEPHANUS, Margaretha Mes Institute for Seed Research, Department of Botany, University of Pretoria, Pretoria 0002, Republic of South Africa

MOORE, KENNETH J., USDA-ARS, University of Nebraska, 336 Keim Hall, Lincoln, NE 68583, USA

MORRISON, WILLIAM R., University of Strathclyde, Department of Bioscience and Biotechnology (Food Science), 131 Albion Street, Glasgow G1 1SD, Scotland, UK

MYERS, DOUGLAS G., Department of Computer Engineering, School of Electrical and Computer Engineering, Curtin University of Technology, P.O. Box U1987, GPO Perth, Western Australia 6001

NODARI, RUBENS O., Departamento de Fitotecnia, Universidade Federal de Santa Catarina, Caixa Postal 476, 88049-Florianopolis, SC, Brasil

NOEL, TIMOTHY R., AFRC Institute of Food Research, Norwich Research Park Colney, Norwich NR4 7UA, UK

PALMER, REID G., USDA, ARS, FCR, Iowa State University, G301 Agronomy, Ames, Iowa 50011, USA

PANDEY, DAYA K., Scientist (Plant Physiology), National Research Centre for Weed Science, 194 Ravindranagar Adhartal, Jabalpur (M. P.)-482004, India

PANELLA, LEONARD, Department of Agronomy and Range Science, University of California, Davis, DA 95616-8515

PAREDES-LÓPEZ, OCTAVIO, Lab. Biotecnología de Alimentos, Unidad Irapuato, Centro de Investigación y de Estudios Avanzados del IPN, Apdo Postal 629, 36500 Irapuato, Gto., México

RING, STEPHEN G., AFRC Institute of Food Research, Norwich Research Park Colney, Norwich NR4 7UA, UK

SCHNARE, MURRAY, N., Department of Biochemistry, Dalhousie University, Halifax, Nova Scotia B3H 4H7, Canada

SPENCER, DAVID F., Department of Biochemistry, Dalhousie University, Halifax, Nova Scotia B3H 4H7, Canada

VAN DE VENTER, HENDRIK A., Margaretha Mes Institute for Seed Research, Department of Botany, University of Pretoria, Pretoria 0002, Republic of South Africa

WANG, SHUE MEI, Department of Botany, National Taiwan University, Taipei, Taiwan

WHITTAM, MARY A., AFRC Institute of Food Research, Norwich Research Park Colney, Norwich NR4 7UA, UK

WRIGLEY, COLIN W., CSIRO Grain Quality Research Laboratory, P.O. Box 7, North Ryde (Sydney) NSW 2113, Australia

YAMAGUCHI, MICHO, Laboratory of Nutritional Chemistry, Department of Food Science, Jissen Women's University, 4-1-1 Osakaue, Hino, Tokyo, 191 Japan

Genotype Determination in Almond Nuts for Paternity Analysis

J. F. JACKSON

1 Introduction

Almond (*Prunus dulcis* (Mill.) D.A. Webb syn *P. amygdalus* Batsch), as grown commercially around the world, is essentially self-incompatible and therefore cross-compatible varieties of almonds need to be interplanted within commercial orchards to achieve cross-fertilization and nut set (Jackson and Clarke 1991 a). Much guesswork has gone into the manner of interplanting, as little is understood of the way in which pollen movement (or "gene flow") takes place so as to achieve pollination of the most valued almond cultivar, Nonpareil, by the other almond cultivars planted in the orchards in California and Australia.

The vector for pollen movement is the honeybee, introduced into the orchards at flowering time in large numbers. The fact that the honeybee is used has led to some assumptions about the scope and distance which viable and compatible pollen can be carried about the orchard. The honeybee can and does travel great distances (up to several kilometers per flight) (Levin and Waller 1989), and so one can believe it is possible that viable and compatible pollen can also move in such a way over great distances and lead to successful pollination, fertilization, and nut set. In other words, gene flow (by pollen) within the orchard could well take place over great distances. The "Almopol" model for almond pollination (DeGrandi-Hoffmann et al. 1989) takes advantage of this possibility, pollen from different sources being brushed off one bee to another within the hive in the model, so that chance contact within the hive allows one cultivar to be pollinated by pollen collected by another bee from a distant source.

On the other hand, trial and error over the years has led the modern Californian orchardist to interplant pollenizers and Nonpareil such that each row of Nonpareil is separated by one row of pollenizer cultivar (i.e., a 1:1 planting). Often two or three or more pollenizer cultivars are used within the orchard, perhaps in different parts of the orchard, or alternately in rows, still preserving the 1:1 planting of Nonpareil and pollenizers. The reason often given for the use of several pollenizers in this way is that each will have a slightly different flowering time to Nonpareil and so pollenizers can be chosen to overlap early or later portions of the Nonpareil flowering time and thus apparently maximizing nut set. Factual information is needed, however, to determine whether these practices are efficient and the assumptions correct. It would thus be useful to map actual gene flow by pollen about commercial orchards, and answer the question just how far does viable pollen travel about the orchard, and how does this vary in different geographic locations with orchards planted out in different geometric patterns and with beehives set out in different ways and at varying densities. Examining these questions over several seasons would provide information to guide growers as to which horticultural practices are most efficient for particular situations. In particular, the information could be used to decide which cultivars are the best pollenizers under given conditions and how they should be set out in the orchard for replanting or when establishing a new orchard.

The best way to determine gene flow by pollen is to examine nuts (or embryos) collected from various parts of the orchards for contained genotype (and therefore contained pollen genes), making sure that sufficient embryos are examined to ensure that a truly representative picture of gene flow is reconstructed. The method used needs to be relatively cheap, rapidly carried out and able to distinguish the various almond cultivar genotypes from one another.

2 Methods for Determining Genotype of Almond Embryos

In general, it can be said that the nuts borne by particular cultivars are characteristic of that cultivar (i.e., characteristic of the female genotype) and that the pollen genes contained within the nut have only minor effects on nut size and shape. There is not short cut, therefore, in determining contained pollen genes within the embryo. The effect of female genotype on nut size and shape is considered further below. The genotype of the embryos can be determined by fingerprinting each embryo DNA by restriction fragment length polymorphism (RFLP), for example; the method is, however, relatively expensive and somewhat lengthy. Since thousands of nuts need to be analyzed in any one orchard for gene flow determination, this method may not be the one of choice for the present purposes.

Another method, which is considerably cheaper than RFLP to carry out, is one which depends on isozyme polymorphism. This has the advantage that several isozymes have already been shown to exhibit polymorphism in almond and to yield useful genetic markers for almond cultivar identification (Hauagge et al. 1987a, b; Cerezo et al. 1989). Furthermore, Jackson and Clarke (1991a) have identified several more polymorphic isozymes in almond, so as to provide enough isozyme markers for many situations to determine embryo genotype and thus the contribution of pollen genes to that genotype.

In what follows, the visual influence of female genotype on nut shape, size, etc. will be considered first, and then the total genotype of almond nuts (embryos) by isozyme polymorphism, including a brief outline of the theory involved, description of methodology, and interpretation of results.

3 Effect of Female Genotype on Nut Characteristics

As can be seen in Fig. 1a−i, there is considerable variation in the size and shape of almond nuts, depending mostly on the female genotype. The cultivars Fritz, Nonpareil, Peerless, Carmel, Ne Plus Ultra, and Mission are cultivars developed in California, and grown commercially in both California and Australia. The cultivars Keanes, Chellaston, and Johnston were bred South Australia, Keanes only recently; Chellaston and Johnston more than 50 years ago. The latter is a large nut, sought after for specialist purposes. Chellaston is a blanching variety. Of the Californian cultivars, the best known is Nonpareil (also known as Californian papershell). It is a sweet nut, the most desirable for its eating qualities. The other American cultivars are largely growth for cross-pollination purposes (on Nonpareil), each having specialist uses for blanching, cooking, inclusion in chocolates, and so on. Nonpareil is the most widely grown cultivar in both California and Australia. While it is obvious that female genotype has the greatest influence on nut appearance, size, etc., the effect, if any, that the pollen parent genotype has on the nut is unknown. Even if it is a small effect, for example, on nut weight, even a 1% increase in weight translates to a relatively large increase in tonnage of almond kernels over the whole orchard. Experiments are underway in the author's laboratory to determine the effect of pollen parent genotype.

It is emphasized that only the most common of the almond cultivars are considered here. Many other varieties are grown in California and Australia, including Price, Jordalano, Saute, and many more. Nor are the many other cultivars considered that are grown in Italy, Spain, or Iran, all areas producing a considerable tonnage of almond meats. In these latter countries, however, there is considerable variation in tree genotype (especially where they have been raised from seed), compared with California and Australia where most lines have been reproduced clonally for decades, and in the case of Nonpareil, for about 100 years.

4 Isozyme Polymorphism as Used to Determine Almond Kernel Genotype

The genes coding for proteins (enzymes) are polymorphic, i.e., they exist as one or more alleles. These alleles are generally codominant, so that they are expressed at the same time in a heterozygous organism (e.g., embryo of the almond kernel). This allows the early determination of genotype from the observed phenotype. The product of the alleles (i.e., enzyme protein) can be separated by gel electrophoresis and detected on the gels by staining for enzyme activity with substrate and dyes which interact with enzymatically changed substrate. The separated allele products are known as isozymes. The isozymes may be separated in this way due to differences in electrical charge between the isozymes because of one or more amino acid differences between them. This has usually arisen due to one or more nucleotide base changes in the gene, giving rise to the one or more alleles at that locus, when the enzyme is said to exhibit polymorphism. The reader is referred to Pasteur et al. (1988) and Richardson et al. (1986) for further general principles of polymorphism as it is related to isozymes. As will be discussed further, the separated isozymes after staining may well be the result of expression of more than one gene or locus, and often it is difficult to decide if this is so or not. However, there are approaches which can help decide whether you have alleles from one or more locus, again the references quoted above deal at length with this question. The scoring of gels to give a phenotype to the isozyme pattern obtained and interpretation to yield the genotype will be dealt with below using a typical gel "run" involving the analysis of kernel extracts for isocitrate dehydrogenase isozymes. Interpretation, of course, is only possible if each of the cultivars has first been "typed" for the polymorphic isozymes by running leaf and pollen extracts from each cultivar. We found that by using eight polymorphic isozymes, we had enough markers to give a good estimate of the paternal genotype contribution to the genetic makeup of diploid kernel embryo in most cases. The isozymes used were IDH (isocitrate dehydrogenase, EC 1.1.1.42),

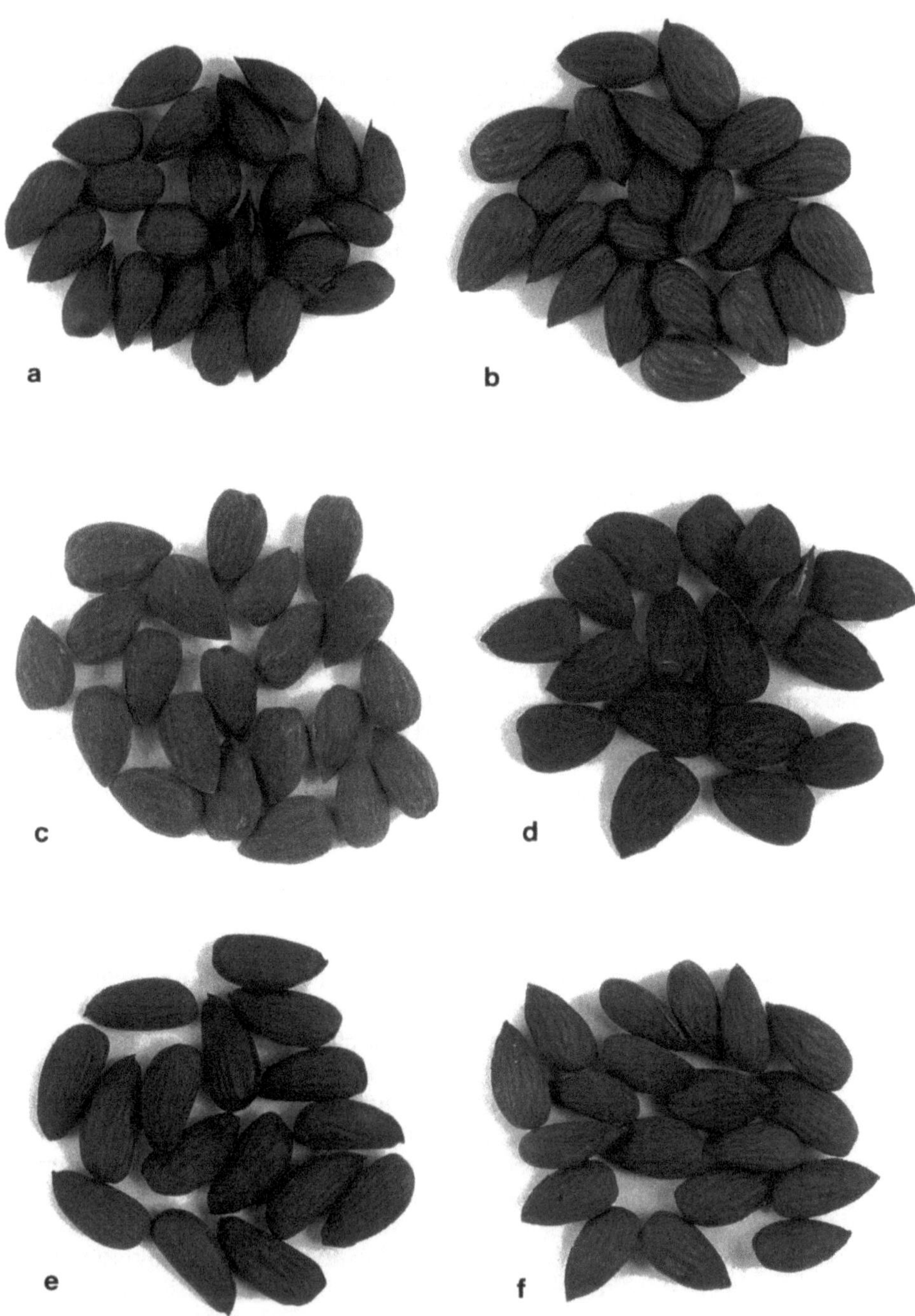

Fig. 1a–f

Fig. 1a–i. Effect of female genotype on kernel appearance. Each cultivar of almond bears kernels which have a characteristic size, shape, color, etc., so that the female genotype is more important than the pollen parent in determining these characteristics. The photographs in **a–i** show a sample of nuts from each of nine cultivars, reproduced to approximately 0.4×life size. **a** Fritz kernels; **b** Nonpareil kernels; **c** Keanes kernels; **d** Peerless kernels; **e** Chellaston kernels; **f** Carmel kernels; **g** Ne Plus Ultra kernels; **h** Johnston kernels; **i** Mission kernels

G6PD (glucose-6-phosphate dehydrogenase, EC 1.1.1.49), SKDH (shikimate dehydrogenase, EC 1.1.1.25), GPI-2 (glucose phosphate isomerase, locus 2, EC 5.3.1.9), LAP-1 (leucine aminopeptidase, locus 1, EC 3.4.1.1), AAT-1 (aspartate amino transferase, locus 1, EC 2.6.1.1), PGM-1 and PGM-2 (phosphoglucomutase, locus 1 and 2, EC 2.7.5.1). Procedures for the first three mentioned (IDH, G6PD, and SKDH) were developed for almond by Jackson and Clarke (1991), and the others (GPI-2, LAP-1, AAT, PGM-1, and PGM-2) by Hauagge et al. (1987a, b).

5 Isozyme Analysis in Practice

5.1 Extraction of Almond Tissue

Leaf samples should be obtained from fresh new growth, preferably in spring (Fig. 2). Problems can be met with by using older leaves (e.g., as in Fig. 3). Pollen in sufficient quantities can be obtained by collecting flowers just before opening, the anthers excised, and allowed to reach anthesis in the laboratory. In a laboratory at 20 °C, leaving the anthers freely exposed to the air, anthesis can be achieved in 24 h. Extraneous anther tissue is removed with a sieve, as described by Jackson (1989), and the pollen sieved through and used immediately for isozyme analysis. Nuts should be fully mature, and before extraction the embryo should be excised (Hawker and Buttrose 1980) for isozyme analysis. We have found that nuts can be stored (as free kernels) at room temperature (18 – 24 °C) for at least 1 year before analysis without any obvious effect on isozyme patterns of the contained embryo.

Leaf material was extracted by grinding 0.3 g leaf with 0.15 g polyvinylpyrrolidone and 2 ml of extraction buffer containing 0.05 M Tris-0.15% citric acid-0.12% cysteine HCI-0.1% ascorbic acid, pH 8. Grinding was carried out in a pestle and mortar, the mixture was centrifuged at 3000 x g in an Eppendorf 5414S centrifuge, and the supernatant used for gel electrophoresis. Pollen was extracted in the same way, except that 2 mg was extracted with 0.25 ml extraction buffer, no polyvinylpyrrolidone being used in this case. Embryos were extracted as described above for leaf material except that 0.4 g of embryo was used. In all cases, gel electrophoresis of the supernatant for isozyme analysis should be carried out within 1 h of extraction.

Fig. 2. Early spring growth of Nonpareil almond cultivar; the new growth with two or three bronze-green leaves are suitable for isozyme analysis

Fig. 3. Summer growth of almond with leaves that are generally unsuitable for isozyme analysis

5.2 Gel Electrophoresis

Although earlier isozyme studies on almond leaves and embryos (Hauagge et al. 1987a, b) and almond pollen (Cerezo et al., 1989) made use of starch gel, we prefer to use Cellogel. This product, a cellulose acetate gel, is made by Chemtron (Milan, Italy). We find Cellogel to be more predictable in its behavior from day to day, and in our hands it gave sharper bands than starch gel. The Cellogel is supplied in airtight plastic bags containing aqueous methanol, and should be stored as such at 4 °C. Several sheets can be kept in a covered tank containing 30% methanol shortly before use, and should be cut to the size needed with a scalpel and ruler. Dry gels should not be exposed to the air for more than a few seconds. Since one side of the gel is porous and the other has a plastic-coated nonporous surface, care should be taken to apply the sample to the porous side of the gel.

After cutting to size the gels are soaked in buffer for a least 10 min, blotted, and positioned in the electrophoresis tank so as to be "stretched" between the anode and cathode compartments containing buffer. Three long bar magnets can be positioned at the inner wall of each buffer compartment (but above the

Fig. 4. Gel electrophoresis apparatus ready for application of the kernel embryo extracts. The seven tanks holding the Cellogel in place are placed in banks supported by wooden blocks. Plastic rulers are fixed in place to ensure that application of samples occurs on a straight line across the cathodic side of the gel. The operator can be seen holding an architect's type draftsman's pen, which is used to apply each sample

level of liquid), in order to hold the Cellogel in a "stretched" manner across the space between the anode and cathode compartments (Fig. 7). The magnets are held in this position by attraction to a metal strip on the other side of the wall. Buffer is placed in the compartments, the buffer used depending on the isozymes to be detected on that gel. For GPI, LAP, AAT, IDH, and SKDH, 0.05 M Tris-maleate, pH 7.8 is used; for PGM, 0.025 M Tris-glycine, pH 8.5; for G6PD, 0.02 M sodium phosphate, pH 7 is best.

For gene flow studies, we routinely extract 50 kernel embryos at a time, and apply 50 extract supernatants to each of seven gels in seven different electrophoresis tanks. Seven gels are needed so that the supernatants can be used to analyze for all the isozymes IDH, G6PD, SKDH, GPI, LAP, AAT, and PGM (1 and 2). To facilitate loading of approximately 1 µl of supernatants, a bank of wooden blocks was set up so as to place each of the seven tanks containing seven gels for easy access for loading (Fig. 4). As shown in Fig. 4, a plastic ruler is held in place over each gel to act as a guide to loading of the samples, a draftsman's pen (architect's type) being used to apply the supernatant. When loading, sufficient pressure should be exerted to lightly mark the surface of the gel (Fig. 5), the length of each mark being monitored on the ruler to allow for 50 different samples in line across the gel. For almond isozymes in the buffer

Fig. 5. Application of samples to Cellogel. An ice bucket containing almond embryo extract supernatants can be seen in the *background*, together with the centrifuge used to produce the supernatants. The process of loading samples is best carried out in a cold room, adjacent to the area where electric current will be applied across the gels. Carrying the electrophoresis tanks filled with buffer electrolyte and loaded gels any distance can bring disaster to the experiment

systems indicated, we apply the samples in a line (called the origin) 0.5 to 1.5 cm from the cathode edge of the flat porous surface of the gel.

Each tank is then connected to the power pack, so that all tanks receive 200 V (Fig. 6), and the electrophoresis "run" is carried out at 4 °C for approximately 1.75 h. At the end of the run, the gels, which so far have been treated the same, apart from some variations in buffer, are ready for enzyme-specific staining.

5.3 Staining for Isozymes

Staining solutions for each enzyme were made up immediately before use by mixing stock solutions (contained in the racks in Fig. 7) and spreading the final mix on a disposable plastic sheet which was stretched over cardboard with the aid of clips (Fig. 7). After turning off the electric current, the gel is then taken out of the electrophoresis tank, holding it by the portions that were in contact with cathodal and anodal buffer liquid. The gel was dipped into the stain spread on the plastic sheet and rocked gently back and forth (Fig. 8). The stain and gel are left in contact with occasional rocking motion for 30 to 60 s. The gel was then blotted to remove excess stain, and placed between two plastic sheets. The "wrapped" gel is now incubated at 37 °C to allow isozyme reaction,

Fig. 6. Electrophoresis in progress, a constant 200 V is applied across each Cellogel to ensure that each gel receives the same treatment. Safety plugs are employed, a notice should be evident to warn of the danger of high voltage

some may take a few seconds, others up to 1 h, to show discrete visible bands reflecting positions of isozymes. A record of band positions can be obtained by photocopying wrapped gels.

Staining ingredients for each of the isozymes and notes on the enzyme-specific staining are given below.

Isocitrate Dehydrogenase (IDH)

Stain 10 mg DL-isocitric acid, 2 ml 0.1 M Tris-HCl pH 8, 0.1 ml 25 mM NADP, 0.1 ml 0.2 mM $MgCl_2$, 0.1 ml 14.5 mM methyl thiazolyl blue (MTT), 0.1 ml 6.5 mM phenazine methosulfate (PMS).

Notes Where isocitrate dehydrogenase is located on the gel, it catalyzes the conversion of isocitrate to α-ketoglutarate in the presence of $MgCl_2$ and at the same time the reduction of NADP to NADPH. The NADPH produced enzymatically reacts chemically with the dye MTT, using PMS as intermediate, yielding an insoluble purple formazan. Purple areas therefore correspond to the location of isocitrate dehydrogenase.

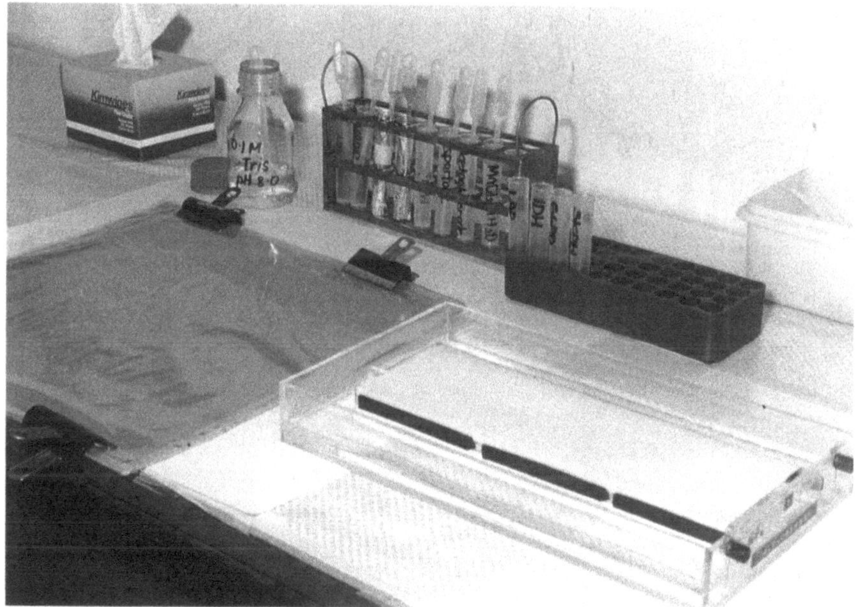

Fig. 7. Preparation to stain for an isozyme. The various components of the isozyme staining mixture are shown in the racks in the *background*, a plastic surface is shown at the *left* where the Cellogel will be brought into contact with the staining solution. The Cellogel is shown in the electrophoresis tank with the magnetic bars still holding it in place, after the electric current has been passed for a sufficient time

Glucose-6-Phosphate Dehydrogenase (G6PD)

Stain 6 mg glucose-6-phosphate, 2 ml 0.1 M Tris-HCl pH 8, 0.1 ml 25 mM NADP, 0.1 ml 1 M MgCl$_2$, 0.1 ml 14.5 mM MTT, 0.1 ml 6.5 mM PMS.

Notes Where G6PD lies on the gel, it catalyzes the conversion of glucose-6-phosphate to 6-phosphogluconate in the presence of MgCl$_2$, and at the same time reduces NADP to NADPH. As for IDH above, the NADPH so produced reacts chemically with the dye MTT (in the presence of PMS) to give a purple formazan.

Shikimate Dehydrogenase (SKDH)

Stain 6 mg shikimic acid, 0.1 ml 25 mM NADP, 0.1 ml 14.5 mM MTT, 0.1 ml 6.5 mM PMS, 2 ml 0.1 M Tris-HCl pH 8.5.

Notes In locations where SKDH is on the gel, shikimate is oxidized to 3-dehydroshikimate and NADP is reduced to NADPH. As for IDH and G6PD above, purple formazan is precipitated where bands of SKDH in this case are located.

Glucose Phosphate Isomerase (GPI)

Stain 5 mg fructose-6-phosphate, 2 ml 0.1 M Tris-HCl pH 8, 0.1 ml 25 mM NADP, 0.1 ml 1 M MgCl$_2$, 0.1 ml 14.5 mM MTT, 0.1 ml 6.5 mM PMS, 2 international units of glucose-6-phosphate dehydrogenase.

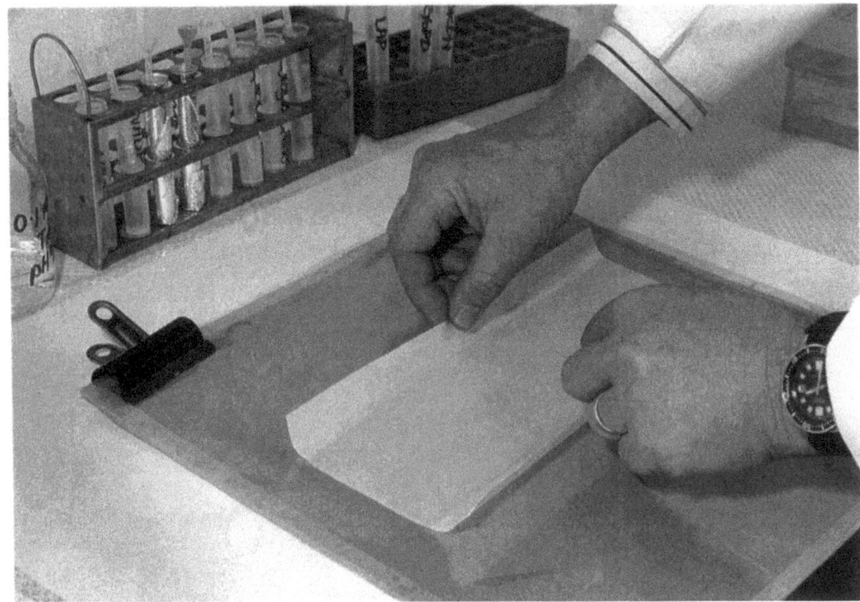

Fig. 8. The staining mixture being passed over the Cellogel with gentle agitation from the operator's hands, shown holding the Cellogel by the portions that had been dipping into buffer electrolyte during the run

Notes Wherever GPI appears on the gel, fructose-6-phosphate is converted to glucose-6-phosphate. The so-called linking enzyme glucose-6-phosphate dehydrogenase added to the stain converts glucose-6-phosphate in turn to 6-phosphogluconate while also reducing NADP to NADPH in the presence of $MgCl_2$. As for other enzymes above, the NADPH reacts chemically with MTT (+PMS) to give purple formazan which is precipitated onto bands where GPI is located on the gel.

Leucine Aminopeptidase (LAP)

Stain 5 mg L-leucyl-β-naphthylamide, 1.5 mg Black K salt, 0.25 ml 0.5 M $MgCl_2$, 2 ml 0.2 M, Tris-maleate pH 5.5.

Notes Where LAP is located on the gel, it catalyzes the hydrolysis of leucyl-β-naphthylamide to leucine and naphthol. The naphthol so formed reacts chemically with Black K to give a violet precipitate overlaying areas of peptidase activity.

Aspartate Aminotransferase (AAT)

Stain 6 mg Fast Garnet GBC salt, 0.2 ml 50 mg/ml L-aspartate pH 8, 0.2 ml 50 mg/ml α-ketoglutarate pH 8, 2 ml 0.1 M Tris-HCl pH 8.

Notes Where AAT lies on the gel, it catalyzes the transferase reaction between α-ketoglutarate and L-aspartate to yield glutamate and oxaloacetate.

The oxaloacetate so formed reacts chemically with the diazonium salt Fast Garnet GBC to give a colored precipitate which becomes occluded in the gel where AAT is located. AAT is also known as glutamate-oxaloacetate transaminase (GOT).

Phosphoglucomutase (PGM)

Stain 10 mg glucose-1-phosphate containing 1% glucose-1,6-diphosphate, 0.1 ml 25 mM NADP, 0.1 ml 1 M $MgCl_2$, 0.1 ml 14.5 mM MTT, 0.1 ml 6.5 mM PMS, 2 international units of glucose-6-phosphate dehydrogenase (also used in the GPI stain mixture above).

Notes Where PGM is located on the gel, glucose-1-phosphate is enzymatically converted to glucose-6-phosphate in the presence of glucose-1,6-diphosphate. The glucose-6-phosphate so formed is in turn converted to 6-phosphogluconate in the presence of glucose-6-phosphate dehydrogenase and $MgCl_2$, at the same time NADP is reduced to NADPH. As for the above staining reactions yielding NADPH, this compound reacts chemically with MTT in the presence of PMS to give an insoluble purple formazan which thus becomes occluded in areas of the gel containing active PGM.

5.4 Interpretation of Stained Gels

First, one needs to know how many polypeptides make up the active enzyme involved. For almond, only LAP and PGM of the enzymes used are monomers (one polypeptide), the others are dimers. For those enzymes with a monomer structure, the interpretation of gels is simplest. Thus, in homozygotes (i.e., in homzygote diploid embryo or leaf) only one kind of polypeptide is synthesized and there is only one band. In another embryo or leaf, which is homozygous for a second allele, there is also one band, but it may migrate a different distance from the origin compared to the first mentioned. The genotype of each diploid embryo or leaf could be designated aa and bb. In the heterozygote, genotype ab, the cells of the embryo or leaf will produce both "a" polypeptide and "b" polypeptide, and so there will be two bands. We see two alleles in almond for LAP-1 and for PGM-1 and PGM-2. The numbers after the isozyme name refer to the particular locus (or gene) involved. Usually, the bands for each locus are well apart and there is no difficulty in telling one from the other. In the case of PGM-1 and PGM-2, where both are heterozygotes (as in Nonpareil almond cultivar, for example), the genotype is ab for PGM-1 and ab for PGM-2, and four clear enzyme bands are seen on the gel, stained for PGM.

One of the most useful isozymes for almond identification is IDH. This is a dimer and three alleles have been found in almond. The reader is referred to the publication by Jackson and Clarke (1991a) for demonstration of the fact that in almond one obtains three alleles at the one locus for IDH. Proof of this involves the use of pollen which is haploid and which therefore cannot have hybrid dimers in stained gels. An almond cultivar which is heterozygous

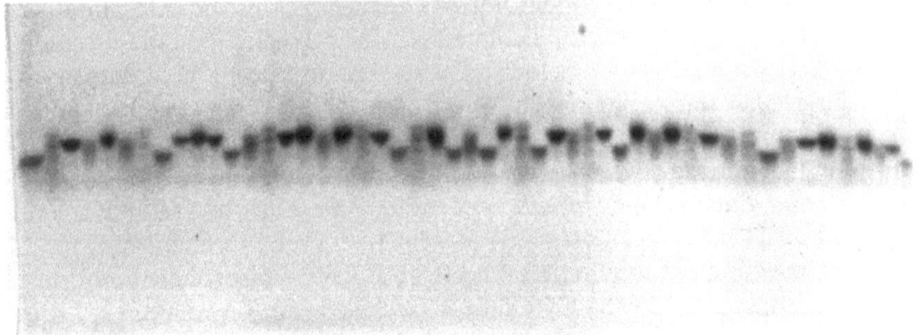

Fig. 9. Cellogel stained for IDH after electrophoresis following loading with 50 almond embryo extract supernatants. Interpretation of the IDH staining patterns is given in the text. Reading from *left to right*, IDH genotype is scored as aa, ac, bb, ab, cc, ab, bc, aa, bb, cc, bb, aa, ab, ac, bb, cc, ab, cc, ac, bb, aa, ac, bc, aa, ab, aa, bc, ac, aa, bb, ab, ac, bb, aa, bc, ab, bc, ab, bb, ab, ac, aa, ab, bb, bc, ac, bc, ab, bb, aa

"ab" for IDH yields three bands for diploid cells, dimer (a protein + a protein), dimer (b protein + b protein), and hybrid dimer (a protein + b protein), and two bands for haploid pollen, dimer (a protein + a protein) in "a" genotype pollen and dimer (b protein + b protein) in "b" genotype pollen and no hybrid dimer as there is only one gene per cell in pollen from these heterozygous plants. In Fig. 9 a typical pattern obtained by electrophoresis of kernel embryo extracts is shown which is subsequently stained for IDH isozymes. In this gel 50 different embryo extracts are examined; the results show all of the possible genotypes resulting from the three alleles. Thus, examination of Fig. 9 suggests that as one reads from left to right, the first extract yields one band and is thus homozygous. It has low mobility, so we give it a genotype aa. The second extract (lane 2 from the left) shows three bands and is thus heterozygous. One band migrates the same as observed in lane 1, and is thus the a allele product; the fastest band moves as fast as any band on the whole gel and can be the product of the c allele. The middle band is thus the hybrid ac dimer, and the genotype of this embryo is ac. The third lane from the left shows one band which is not as fast as the fastest in the second embryo and cannot be cc. It must therefore be bb genotype. The fourth lane shows three bands and is therefore heterozygous; the fastest band is the same as the single band in lane 3, and corresponds therefore to the product of the b allele. It can only be that this embryo has the genotype ab. The fifth lane from the left has one fast band and is homozygous, corresponding to cc genotype. The sixth is ab, while the seventh lane has three bands and is interpreted as having the genotype bc. In summary, reading from left to right, the first seven kernel embryos in Fig. 9 correspond to genotypes aa, ac, bb, ab, cc, ab, and bc. Similarly, the remaining 43 lanes in Fig. 9 can be assigned genotypes. With practice, the genotype can be readily inferred from the freshly stained gel and recorded in the laboratory notebook for all 50 embryos in a few minutes. In fact, the time taken from the start of extraction of each of the 50 kernel embryos to the end of recording genotype for all eight isozymes (IDH, G6PD, SKDH, GPI, LAP, PGM-1, and PGM-2), i.e., 400 entries, can be with practice 5 to 6 h. By dovetailing the operations, two experienced laboratory workers can determine the genotype for each of the eight isozyme loci for 100 and occasionally 150 nuts per working day. This rate of progress makes it possible to carry out gene flow reconstruction from almond kernels collected in appropriate locations in the orchard.

6 Isozyme Genotype for Some Almond Cultivars

The results obtained above can then be interpreted in terms of the pollen parent that gave rise to nuts collected from particular trees after the genotype for each almond cultivar is determined for each of the eight isozyme loci. Simple Mendelian inheritance has been found to be the rule in all cases so far examined in almond. Where a reconstruction of gene flow is not readily apparent and a more refined mathematical treatment of the results is needed, a general Monte Carlo method for estimating gene flow in angiosperms is given by Devlin and Ellstrand (1990). The genotype for several commonly grown almond cultivars is listed in Table 1 for each of the eight loci discussed above. Our results differ in some respects from those obtained by Hauagge et al. (1987a); however, we have repeated them many times and the use of Cellogel has made our determinations more precise.

Table 1. Almond genotype

Cultivar	GPI-2	LAP-1	AAT-1	PGM-1	PGM-2	IDH	G6PD	SKDH[3]
Carmel	aa	cc	ab	ab	bb	ab	aa	bb
Fritz	ab	nb[1]	na[1]	ab	bb	ab	ab	bb
Ne Plus Ultra	aa	bc	ab	bb	ab	ac	ab	bb
Nonpareil	aa	bc	ab	ab	ab	ab	ab	bb
Mission	ab	nc[1]	ab	aa	bb	aa	[2]	[2]
Peerless	aa	bb	nn[1]	ab	ab	bb	ab	bb
Price	ab	cc	aa	aa	bb	ab	ab	bb
Thompson	aa	bc	bb	ab	ab	aa	ab	bb

[1] n denotes a null gene. Some genes have alleles which have no product that can be detected by gel electrophoresis followed by staining. These are called "silent" or "null" genes; they are recessive.

[2] Not as yet determined.

[3] All of the cultivars listed are homozygous "bb" for SKDH. Some Australian varieties are heterozygous at this locus (Jackson and Clarke 1991b).

7 Conclusions

Genotypes for isozymes exhibiting polymorphism at eight loci in almond can readily be determined in almond kernel embryos; the necessary methods are described herein. The eight loci are GPI-2, LAP-1, AAT-1, PGM-1, PGM-2, IDH, G6PD, and SKDH. This analysis can be used in many cases to determine the pollen parents giving rise to nuts on particular almond trees, given that the genotype at the eight loci is known for the almond cultivars in common use. A table showing some of these genotypes is presented. Gene flow by pollen can then be reconstructed for the previous flowering season in the orchard situation.

Acknowledgments. The author acknowledges the excellent technical work of Mr. Geoff Clarke, who, with the assistance of Emily Telfer and Narelle Davidson, was responsible for many of the advances reported here. A grant to JFJ from the Horticultural Research and Development Corporation, the Almond Cooperative Ltd., and Excell Almonds made the work possible.

References

Cerezo M, Socias i Company R, Arus P (1989) Identification of almond cultivars by pollen isozymes. J Am Soc Hortic Sci 114:164–169

DeGrandi-Hoffman A, Roth SA, Loter GM (1989) Almopol: a cross-pollination and nut set simulation model for almond. J Am Soc Hortic Sci 114:170–176

Devlin B, Ellstrand NC (1990) The development and application of a refined method for estimating gene flow from angiosperm paternity analysis. Evolution 44:248–259

Hauagge R, Kester DE, Dooy RA (1987a) Isozyme variation among Californian almond cultivars. I. Inheritance. J Am Soc Hortic Sci 112:687–693

Hauagge R, Kester DE, Arulsekar S, Parfitt DE, Liu L (1987b) Isozyme variation among Californian almond cultivars. II. Cultivar characterization and origins. J Am Soc Hortic Sci 112:693–698

Hawker JS, Buttrose MS (1980) Development of the almond nut [*Prunus dulcis* (Mill.) D.A. Webb]. Anatomy and chemical composition of fruit parts from antheses to maturity. Ann Bot 46:313–321

Jackson JF (1989) Borate control of protein secretion from petunia pollen exhibits critical temperature discontinuities. Sex Plant Reprod 2:11–14

Jackson JF, Clarke GR (1991a) Gene flow in an almond orchard. Theor Appl Genet 82:169–173

Jackson JF, Clarke GR (1991b) Patterns of bee visitations to flowers of almond cultivars in an orchard as determined by molecular genetic marker analysis of pollen in the "pollen basket". Int Symp on Angiosperm pollen and ovules: basic and applied aspects. Lake Como, Italy, June 27

Levin MD, Waller GD (1989) The role of pollinating insects in future world food production. Apiacta 24:18–21

Pasteur N, Pasteur G, Bonhomme F, Catalan J, Britton-Davidian J (1988) Practical isozyme genetics. Ellis Horwood, Publ, Chichester

Richardson BJ, Baverstock PR, Adams M (1986) Allozyme electrophoresis. Acad Press, Sydney

Identification of Cereal Varieties
by Gel Electrophoresis of the Grain Proteins

C. W. WRIGLEY

1 Introduction

1.1 Grain Proteins as Documents of Identity

If information about the identity, or even genetic history, of a grain sample is sought, it should be possible to read this information from the grain-protein composition. As direct products of gene transcription and translation, proteins contain a wealth of genetic information, ready to be read off, given the appropriate techniques. Zuckerkandl and Pauling (1965) have described the relationship between chemical composition and genotype (or even progenitors) as occurring at several different levels. Semantides (sense-carrying molecules, such as DNA, RNA and proteins) provide reliable information about genotype; this is not so for episemantic molecules (the products of enzymes) or for asemantic molecules (not specifically produced by the organism), since in these latter cases, environmental or extraneous influences might be expected to predominate.

The most reliable information about identity will thus be provided by primary or secondary semantides (DNA or RNA), e.g. by restriction fragment length polymorphism (RFLP) analysis. Examination of proteins (classed as tertiary semantides) is a reasonable compromise between the direct study of the genotype and ease of accessibility to chemical constitutents. Analysis of protein composition would thus be expected to provide a better basis for varietal identification than the study of pigmentation or even morphology.

1.2 Reading Identity from Protein Composition

Gel electrophoresis has been the traditional technique for "reading off" information about identity from plant proteins. This involves four basic steps (used below in describing the methods):

1) Gel preparation.
2) Sample preparation: protein extraction, yielding a solution representative of all the proteins, or possibly of only a particular class of proteins.
3) Electrophoresis: fractionation of the proteins in a gel medium according to a combination of their size and charge, yielding a "fingerprint" of banding patterns.

4) Gel staining/interpretation: comparison of this arrangement of protein bands with those of authentic samples, fractionated under the same conditions.

For the identification of wheat varieties, gel electrophoresis of the gliadin class of proteins at pH 3 has been the most popular procedure. In addition, gel electrophoresis in the presence of the detergent sodium dodecyl sulfate (SDS) has provided complementary information about identity, because this method usually involves extraction of all grain proteins (not only gliadins), fully reduced as polypeptides. SDS gel electrophoresis has thus proved to be more generally applicable to a wider range of grains.

If the seed technologist looks more widely than at just the grain proteins, the choice of tissue or protein class can determine the information provided. For example, anodic gel electrophoresis of the leaf proteins has provided useful distinctions at the genus level for the various cereals, but not at a lower level (Wrigley 1982). Electrophoretic analysis of the water-soluble albumins of cereal grains has elucidated relationships at the species level, without (confusing) additional information about subspecies distinctions, such as might be provided by gliadin composition (used primarily for varietal distinction).

Various review articles are available to give fuller information about the relative advantages of variations of the electrophoretic technique to the specific crop varieties (Konarev et al. 1979; Wrigley et al. 1982; Cooke 1984, 1988; Konarev and Gavriljuk 1988). Figure 1 summarizes a range of wheat-identifi-

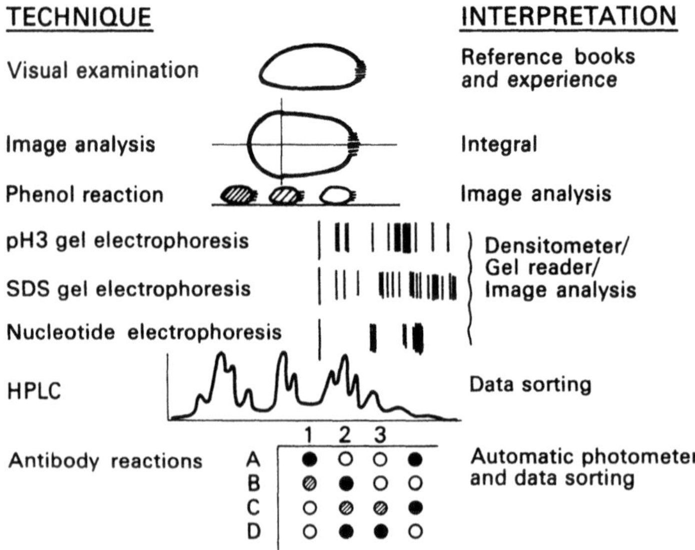

Fig. 1. Approaches to identifying wheat varieties. (Wrigley and McMaster 1989)

cation methods, including procedures for gel electrophoresis (details in Wrigley et al. 1989). Cooke's review (1988) includes a tabulation of all experimental conditions used for over 40 published methods for electrophoretic identification of wheat and barley varieties.

In recent years, high performance liquid chromatography (HPLC) (Bietz 1986) has invaded the previously exclusive province of gel electrophoresis. This technique substitutes, in step 3 above, a tightly packed column of adsorbant material on which fractionation of the proteins takes place, yielding an elution profile (generally based on ultraviolet absorbance) of peaks and valleys to serve as the "fingerprint" of identity. HPLC has the advantages of speed (generally less than 1 h) and automatic sorting of the output to provide matching of the profile for an unknown sample against those of a library of authentic samples (Bekes et al. 1991). Despite the advantages, for HPLC, of automatic loading of samples and overnight operation, it provides analysis of only one sample at a time. Its throughput is thus limited, in relation to capital outlay, when compared to electrophoretic methods.

1.3 Extending the Specificity of Gel Electrophoresis

Routine gel electrophoresis involves staining for the presence of all proteins, but this approach can provide too much information for easy interpretation. For example, many of the protein bands may be common to all the varieties that are to be distinguished. To overcome this problem, specific staining of only certain proteins can be achieved by the use, for example, of enzyme staining thereby indicating the presence of isozymes. At present, none of the routine identification methods uses enzyme staining, and none is described in detail in this chapter, but there are many reports of successful applications of this approach (reviewed by Wrigley et al. 1982; Cooke 1984, 1988); it should be considered as an alternative when devising a new identification strategy.

An alternative version of this approach is the use of specific antibodies to reveal the presence of homologous proteins, generally after transferring protein zones from the electrophoretic gel to a membrane medium such as nitrocellulose. Alternatively, if antibody specificity is sufficient, it may be possible to avoid the electrophoretic step and examine the reaction of the antibody with the whole-grain extract. The application of this group of methods for obtaining more efficient/better distinction between varieties of wheat and barley has been described by Konarev et al. (1981), Wrigley et al. (1987), and by Skerritt et al. (pp. 110−123 in Konarev and Gavriljuk 1988).

1.4 The Need for Varietal Identification

The need for initial identification or for verification of varietal identity arises throughout the sequence of events from breeding, through variety release, pure-seed propagation and sowing, at harvest, and into marketing and processing of the harvested grain. The introduction of Plant Breeders' Rights has brought even more exacting requirements for genotype identification and distinctness testing in seed certification (Cooke 1988; see also reviews for specific countries in Plant Varieties and Seeds, Vol. 3, No. 3, 1990). More recently, the use of specific proteins as markers of grain quality attributes, or other complex aspects of phenotype, takes gel electrophoresis beyond the function of genotype fingerprinting into the realm of selection for these attributes (reviewed by Cooke 1988; MacRitchie et al. 1990).

Specific requirements of identification differ at these various stages, as reviewed by Cooke (1988). The breeder needs to be sure, for example, that yield and quality evaluations have been performed on the same genotype. Electrophoretic identification can provide assurance that seed is true to label for sowing, but can also indicate the nature of off-type plants or strangers during propagation (Appleyard et al. 1979; Cooke and Draper 1986).

1.5 Complications in Electrophoretic Identification

An important aim of genotype identification is the elimination of irrelevant factors, such as the effects on phenotype of growing conditions. Such factors, which complicate the task of visual identification by grain morphology, have generally not affected the results of electrophoretic techniques, and this has been a great advantage of this general approach. For example, neither severe frosting of immature wheat nor germination of mature wheat seed for up to 44 h caused gliadin electrophoretograms to be significantly changed (Lookhart and Finney 1984). However, this potential complication must be borne in mind, especially for samples of dubious origins, and when exploring identification possibilities for a new species. Analysis of F 1 hybrid seed presents particular problems and opportunities for electrophoretic identification, as reviewed by Cooke (1988) and Konarev and Gavriljuk (1988, p. 14).

Another complication frequently encountered is that a variety is polymorphic for protein composition as indicated by the electrophoretic method used; that is, analysis of individual grains of the same variety gives more than one electrophoretic pattern, due to the presence of multiple biotypes within the variety. All these biotypes may be legitimate parts of the variety, merely reflecting a degree of inhomogeneity (e.g. sister lines) in the original selection(s) used to make up the variety (Appleyard et al. 1979). A solution to this problem may lie in changing to a less discriminating method of electrophoresis, but more likely, the complication must be accommodated in routine analysis, by taking the main biotypes to define the variety.

2 Other Methods to Complement Gel Electrophoresis

2.1 Making Electrophoresis More Efficient

As electrophoretic identification is a reasonably labour-intensive task, it should not be used if simpler methods are adequate for a particular job of identification. Furthermore, its use may be rendered more efficient by combining it with other methods of identification. Figure 1 lists a range of methods suited to wheat-grain varietal identification, starting with the simplest, i.e. the visual study of grain morphology. Figure 2 shows how such complementary methods can be used to increase the efficiency of electrophoretic analyses, firstly by the application of pre-sorting tests, and then taking into account the specific information needed.

2.2 Pre-Sorting Methods

Visual examination of grain samples (first method in Figs. 1 and 2) has served the seed industry well. It requires a high degree of skill and experience, but reference handbooks, available for many national sets of varieties, can accelerate the process of acquiring such skills. Image analysis promises to take the subjective element out of this approach, adding rapid computation of grain outline-shapes and dimensions (Neuman et al. 1987; Draper and Keefe, Konarev and Gavriljuk 1988, pp. 27–35; Myers and Edsall 1989). With further development, image analysis may provide on-the-spot identification of many grains in a sample, leaving a lab test such as electrophoresis to check the abnormal or questionable grains.

The phenol reaction for wheat is another example of a simple test that can be applied to a large number of grains to pre-select, for example, grains, that do not conform to the declared identity. Likewise, the sodium hydroxide test can be used to distinguish red- from white-grained wheat varieties (see Wrigley et al. 1982 for details of both tests). Electrophoretic analysis of grains from both treatments is possible, after water-washing. Pre-sorting tests may be applied to grains of other cereals in a similar manner. For example, blue or white aleurone coloration is a useful distinguishing character for varieties of barley, particularly after a light pearling treatment to remove the outer husks.

A software program, called WhatWheat, has been devised by Bekes et al. (1991) to increase the efficiency of electrophoretic identification by combining it with other methods. The program provides for the recording of data for a large number of varieties according to a range of techniques, e.g. for wheat: morphological descriptions of grain (possibly also heads and plants), grain hardness, phenol reaction, and pattern groupings for other tests such as gel electrophoresis, HPLC and RFLP analyses. The program serves as a source of this information and as a means of identifying a sample after determining its characteristics, but more importantly, the program can be interrogated to indi-

C. W. Wrigley

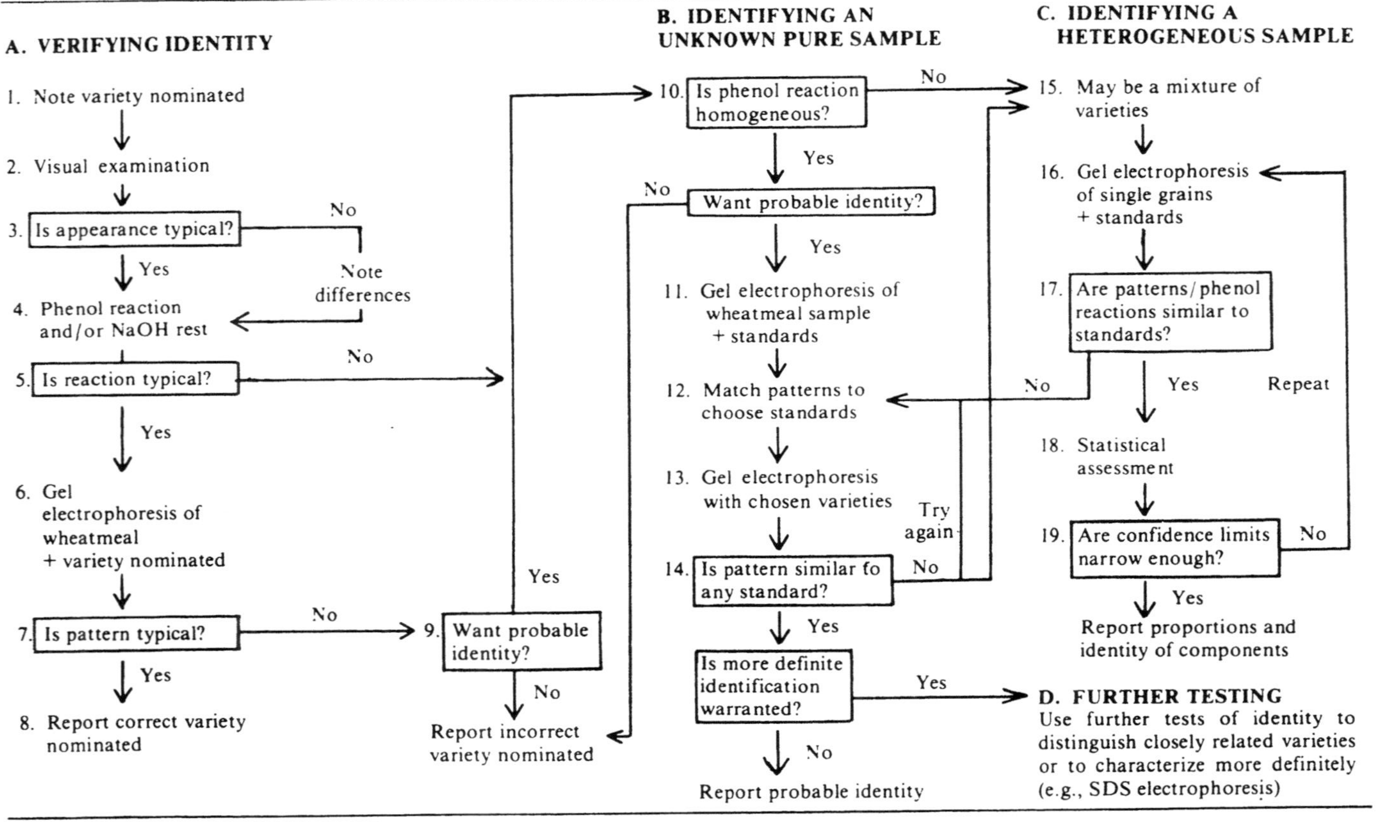

Fig. 2. Procedure for systematic identification of wheat varieties. (Wrigley et al. 1982)

cate the simplest strategy for distinguishing between a particular combination of the varieties recorded.

2.3 Milled vs Single-Grain Samples

The procedure in Fig. 2 (Step 6) suggests that the first step in electrophoretic analysis should be to examine an extract of the milled grain, in which all the component genotypes are represented in their proportions. Comparison of the resulting pattern with that of an authentic (pure) sample of the nominated variety will indicate firstly if the sample is correctly labelled, and secondly if there is significant contamination with a variety of different patterns. Whether further analysis is warranted depends on the type of information requested, as set out in the boxed questions in Fig. 2. Detailed information about the degree of contamination in an impure sample (or about a sample known to be a mixture) requires the further tedious task of analyzing grains one by one, bearing in mind the degree of statistical significance required.

3 Interpretation of Results

3.1 Statistical Analysis of Results

The degree of contamination is generally indicated as confidence limits, e.g. based on a confidence coefficient of 0.95 (see Wrigley et al. 1982 for more detail). For total grain numbers of less than 100, confidence limits are determined by reference to tables of binomial distribution (e.g. Beyer 1968). For example, one grain of different identity, in a total of 20 grains analyzed, represents a degree of contamination of 5%, with confidence limits of 0.1% to 20.35% (or 0.02% − 26.18% for a 0.99 confidence coefficient). This is interpreted to mean that this contaminating grain may have been virtually the only contaminant in a large cargo (represented by the extreme lows of 0.1% or 0.02%), or that it may indicate the presence of a much more significant contamination (to over 20%).

As a normal distribution can be assumed for total grain numbers of over 100, the following formula may be used to calculate confidence limits. The proportion (P%) of a component in a mixture is given by

$$P \pm \sqrt[2]{\frac{P(100-P)}{n}} \, ,$$

where n is the total number of grains analyzed (if n < 100, use statistical tables).

Table 1. Confidence limits for the proportion of a variety in a sample, based on the proportion found in a subsample consisting of a certain number of grains. Confidence coefficient of 95% used. Tables (Beyer 1968) used for grain numbers under 100

Proportion found (%)	Confidence limits (%), depending on number of grains examined				
	20 Grains	50 Grains	100 Grains	200 Grains	500 Grains
1	–	–	0 – 5	0 – 3	0 – 2
2	–	0 – 10	0 – 7	0 – 4	1 – 3
3	–	–	1 – 8	1 – 5	2 – 4
4	–	1 – 13	1 – 10	2 – 7	2 – 6
5	0 – 25	–	2 – 11	2 – 9	3 – 7
6	–	1 – 16	2 – 12	3 – 10	4 – 8
8	–	2 – 19	3 – 13	4 – 12	6 – 10
10	1 – 32	3 – 22	5 – 17	6 – 14	7 – 13
20	6 – 44	10 – 34	13 – 29	15 – 25	17 – 23
30	12 – 54	18 – 45	21 – 40	24 – 36	26 – 34
40	19 – 64	26 – 55	30 – 50	33 – 47	36 – 44
50	27 – 73	35 – 65	40 – 60	43 – 57	46 – 54

According to this formula, a single-grain contaminant in 500 grains represents 0.2% ±0.4%, that is, between 0.0% and 0.6%. Detection of 1 grain in 50 represents a proportion of 2%, with confidence limits of 0.05% and 10.65% (from statistical tables), indicating that over 50 grains must be analyzed to be 95% sure of detecting one grain of a 10% contaminant with distinguishable characteristics. The range of examples of confidence limits shown in Table 1 indicates the obvious principle that the limits narrow as the number of grains increases. Table 1 provides a rough guide for interpreting the results of grain-by-grain analyses.

3.2 Recording and Comparison of Results

The stained gel will keep indefinitely sealed in a plastic bag; alternatively, it may be dried onto a glass plate for storage after soaking in 10% – 20% glycerol and wrapping with cellophane film onto the glass sheet. Photography is the usual method of recording the results (e.g. the gel in Fig. 3), using uniformly transmitted light from below the gel and an orange-red filter on the camera, mounted above the gel. A quantitative record of results can be obtained by the further step of densitometry or image analysis, thereby scanning the optical absorbance of each lane to produce a series of peaks, each corresponding to a band on the original gel pattern.

This approach offers the advantage that the data output can be matched automatically against a library of scans for authentic samples. Several studies have provided means of pattern matching (e.g. Jones et al. 1982; Sapirstein and Bushuk 1986; Autran and Abbal 1988; Bekes et al. 1991), all with a similar aim – to provide an immediate identification of a pattern of unknown identity.

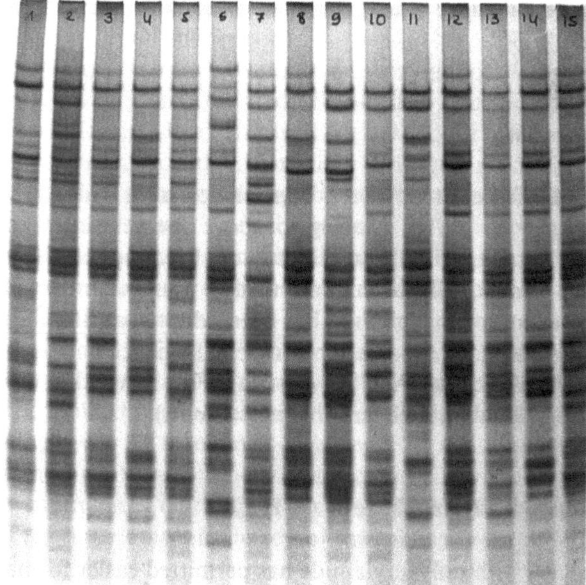

Fig. 3. Acidic polyacrylamide gel electrophoresis patterns for gliadin proteins from European wheat varieties, obtained by a procedure similar to the Draft ISO method (see Sect. 4.3): *1* Festival; *2* Longbow; *3* Camp-Remy; *4* Capitaine; *5* Recital; *6* Carolus; *7* Slejpner; *8* Soissons; *9* Florence-Aurore; *10* Centauro; *11* Goelent; *12* Drakkar; *13* Monopol; *14* Hugo; *15* Thesee. (The figure was provided by Mrs. J. Le Brun, Institut Technique des Cereales et des Fourrages, Paris, France)

Such an approach is obviously limited by the extent of the library of reference patterns and the degree of reproducibility of the analytical procedure, combined with the ability of the pattern-matching method to compensate for pattern differences due to variations in procedure.

In its most basic form, electrophoretic identification is a comparative method, involving the side-by-side comparison of a sample of declared identity with a sample known to be of that identity. This comparison should produce a yes/no answer, as set out in Fig. 2 (Step 7). The further stage of determining identity after a "no" answer is much more difficult, requiring a combination of the approaches described in the preceding paragraph and in Fig. 2.

4 Wheat

The need for varietal identification is probably greater for wheat than for any other cereal grain because it is grown so widely, and because of the need to distinguish between varieties of differing quality type after harvest. Methodological details will thus be given for wheat, both for their own sake and as a basis for other cereals.

4.1 Starch Gel Electrophoresis

One of the earliest gel media for electrophoresis was formed from hydrolyzed potato starch, after mixing with buffer, heating to above the gelatinization temperature, and pouring into a suitable mould to set. Electrophoresis of wheat gliadin proteins has traditionally been performed in aluminium lactate buffer, pH 3.1. A French version of this method has been adapted as Standard Method No. 142 of the International Association for Cereal Science and Technology (ICC) (P.O. Box 77, Schwechat, A-2320, Austria). This method is now largely replaced by polyacrylamide gel electrophoresis (PAGE).

4.2 Acidic Polyacrylamide Gel Electrophoresis (A-PAGE)

The replacement of starch gel with polyacrylamide has permitted easier formation of a more reproducible gel with much wider variation in molecular sieving of proteins, providing a combination of charge- and size-based separation. However, acrylamide monomer, used for gel formation, is neurotoxic, and extreme caution should be exercised in handling it before it is polymerized (which renders it harmless). An alternative procedure that avoids this risk is the use of pre-formed gels, also described below.

4.3 A-PAGE-ISO and ISTA Methods

A pH of 3.1 is retained for routine gliadin electrophoresis, generally using either aluminium or a sodium lactate buffer. The most recent version of the ICC Standard Method (No. 143, revised 1987) for A-PAGE uses a vertical gel slab with aluminium lactate buffer (pH 3.1). It is similar to the ISO Draft Method (ISO/TC 34/SC 4 N 527 E, dated April 1990), outlined below. An alternative (also below) is the Standard Reference Method adopted by the International Seed Testing Association (ISTA; Draper 1987). It involves the use of a 10% polyacrylamide gel, a buffer of glycine-acetic acid (pH 3.2), and 2-chloroethanol as the extractant for gliadins (or other prolamins, such as the hordeins of barley grain, with other additions, see Sect. 5.1).

4.3.1 ISO Draft Method

Gel Preparation. Hoefer SE 600 ($14 \times 12 \times 1.5$ mm vertical slab) or LKB Model 2001 equipment is suggested, regulated to 12 °C with 3- or 4-mm wide sample slots. The gel/electrode buffer contains 15 g aluminium lactate (e.g. Fluka

Product No. 69812), dissolved in 5.5 l deionized water, adjusted to pH 3.1 with lactic acid and made to 6 l.

The gel solution contains 6 g acrylamide, 0.3 g N,N-methylene bisacrylamide, 0.02 g ascorbic acid, dissolved in electrode buffer and made to 100 ml. This solution (30 g) is mixed with 0.075 ml 1% ferrous sulfate heptahydrate (made freshly) and 0.12 ml 1% hydrogen peroxide solution. Degassing, by evacuation prior to adding the last reagent, helps to avoid the formation of bubbles in the gel. The gel mixture is introduced into the apparatus according to the manufacturer's instructions, and the gel is left to polymerize with a slot-forming "comb" in place; on removal of the comb, pockets are left to receive the sample extracts.

Sample Preparation. In a small tube (e.g. 75×10 mm), extract the sample of flour, milled whole meal or crushed single grains with 70% ethanol (in water), using about 0.2 ml extractant for every 50 mg sample. Vortex (agitate vigorously) for about 10 s, and allow the extracting mixture to stand for at least 1 h at 20 °C, or overnight. Centrifuge to clarify the extract. Apply 2 µl of extract into each of the sample slots in the top of the gel, ready for the electrophoretic separation.

Electrophoresis. Assemble apparatus with the cathode (negative electrode) at the bottom, connect cooling water (10 °C), and switch on the current. The appropriate voltage and time depend on the geometry of the apparatus; the standard method suggests 70 min at 600 V. When the volt/time combination has been established to provide good resolution of the gliadin proteins, with the water-soluble albumin proteins run off the lower end of the gel, these conditions should be kept constant to provide reproducible results. WARNING: Disconnect power before manipulating the apparatus. Ensure that electrical connections are in good repair, with no electrical leaks.

Gel Staining/Interpretation. Disconnect the power and remove the gel. Take care not to touch it with the hands, to avoid the appearance of fingerprints on the gel after staining. Place the gel in a flat trough, or in a plastic bag, and cover it with about 100 ml staining solution, made up by mixing 95 ml 10% trichloroacetic acid with 5 ml of a 0.5% solution of Coomassie Brilliant Blue R 250 in ethanol. Cover the trough and agitate it gently/occasionally for some hours (4 to 18 h is suggested).

Band patterns will appear after a few hours, and become gradually stronger. Destaining is not needed with this method as the solvent does not allow dissolution of much of the dye. Nevertheless, excess precipitate of dye should be wiped from the surface of the gel before examination, photography or densitometry. Record results by one of the methods suggested above. Compare patterns of the various samples, bearing in mind the strategy outlined in Fig. 2, namely, checking the pattern of a sample for analysis against those of authentic samples, preferably run on the same gel.

The main reason for international agreement on a standard method is to permit comparison of results. To further assist to this end, an agreed form of nomenclature has been established for the individual gliadin bands, based on the publication of Bushuk and Zillman (1978). It involves calculating the relative mobility of each band with respect to that of a reference band, designated as 50, present in the patterns of certain standard varieties, particularly the variety Marquis. As a result, it is possible to represent the electrophoretic pattern by a series of numbers, each representing the mobility of a gliadin band, the low numbers (say, 20 to 40) being those of low mobility, with numbers extending on up to 100 or so. In a later version (Sapirstein and Bushuk 1985a), three reference bands are used. Translation of patterns into numerical form also permits rapid sorting or matching of the pattern of an unknown against those established for many other varieties, possibly using an international data bank. To help in work towards such a goal, Bekes et al. (1987) have compared different approaches to the quantitative evaluation of gliadin patterns. Catalogues have been prepared for various national sets of wheats, e.g. for Canada (Sapirstein and Bushuk 1985b; Ng et al. 1988), for Germany (Quaite et al. 1987), for Italy (Dal Belin Peruffo et al. 1981), for US wheats (Jones et al. 1982; Mecham et al. 1985), and for Yugoslavia (Jost 1989).

4.3.2 ISTA Standard Reference Method

The ISTA method (Draper 1987) is similar to the above with the following modifications. Figure 4 shows examples of the resulting patterns, in this case for durum wheats.

Gel Preparation. The electrode buffer consists of 0.4 g glycine and 4 ml glacial acetic acid made to 1 l. The gel mixture contains 10 g acrylamide, 0.4 g bisacrylamide, 0.1 g ascorbic acid, and 0.005 g ferrous sulfate dissolved in a total volume of 100 ml. Polymerization is initiated by quickly stirring in 0.35 ml 0.6% hydrogen peroxide.

Sample Preparation. Chloroethanol (20%) is used as extractant.

Electrophoresis and *Gel Staining.* These are similar to the ISO method.

4.4 Acid Gradient (AG)-PAGE

The polyacrylamide gel used in the method above (about 6%) provides only modest sieving of the proteins, thereby accentuating differences between the gliadins based on their charge properties, but not taking advantage of the potential of higher gel concentrations to sharpen bands and to provide further fractionation. Such advantages may be obtained with a gradient of gel concentration (increasing from low at the top to higher further down) thereby providing graded pore sizes: larger near the start, appropriate to the larger proteins, grading down the gel to smaller pores with better sieving of the smaller proteins of higher mobility.

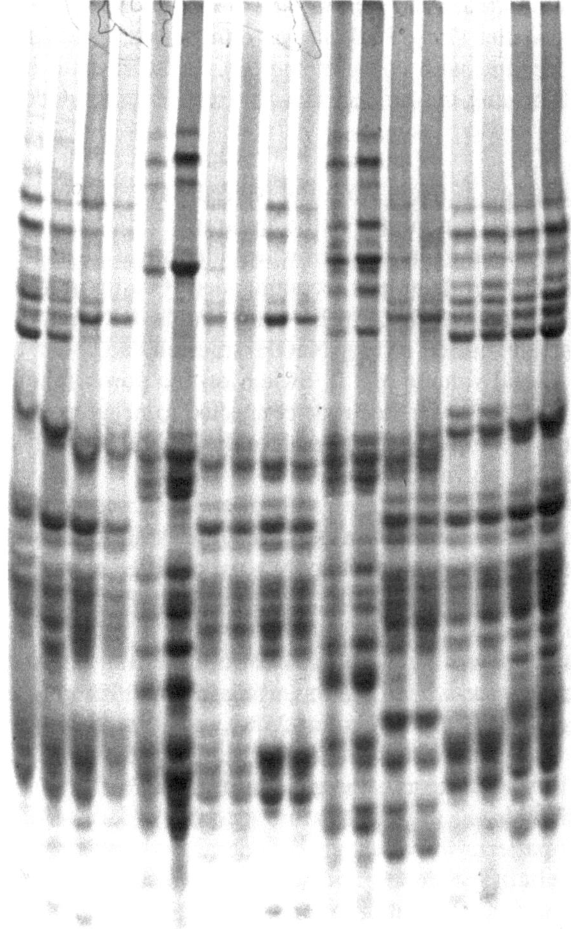

Fig. 4. Electrophoretic patterns for a selection of durum wheats, whose gliadin proteins have been fractionated by the ISTA acid PAGE method (Sect. 4.3.2). (Cooke, in Konarev and Gavriljuk 1988, p. 14)

Gradient gel electrophoresis (du Cros and Wrigley 1979) thus offers advantages for identification, but the gel is more difficult to make in the lab. This disadvantage has been overcome by the mass production and sale of gradient gels of appropriate pore ranges, so that gels of reproducible sieving properties may be purchased. This approach to electrophoretic identification has the added advantage of avoiding the hazards of handling toxic acrylamide monomer and the convenience of having gels ready to use without waiting for gel preparation, bringing the method within the scope of inexperienced staff.

This method has been officially adopted for routine use in the Official Testing Methods (Cereal Chemistry) of the Royal Australian Chemical Institute (RACI 1988). The basic steps are similar to those given above (Sect. 4.3), but gel preparation is almost eliminated and times are much shorter as the commercially available gels are smaller ($75 \times 75 \times 3$ mm gel dimensions, in a glass cassette measuring $80 \times 80 \times 5$ mm, with space at the top of the cassette for sample application). The following description must be read in conjunction with Section 4.3.

Gel Preparation. The gradient gel is taken from the packed (take care not to freeze the gels during storage) and is mounted in the apparatus, after inserting the manufacturer's device for spacing samples. For wheat identification, a gradient of about 3% to 13% polyacrylamide is recommended (Gradipore Cat. No. GS-313 from Amicon, Boston MA, USA or E. Merck, Poole, UK). Fill the apparatus with sodium lactate buffer (0.17 g NaOH solution, adjusted to pH 3.1 with lactic acid in a total volume of 1 l). Switch on the power (about 200 V and 40 mA per gel) with the lower electrode connected to negative, and run for about 1 h to allow the electrode buffer to replace the general electrolyte in which the gel was originally polymerized.

Sample Preparation. This method specifies the use of 6% urea solution in water as extractant, using about 6 µl/mg sample. Apply 10 µl clarified extract to each sample position.

Electrophoresis. Switch on the power and continue electrophoresis at 200 V; a period of 2.5 h has proved best for wheat gliadins using 3% to 13% gels at 25 °C. Alternatively, run at 400 V for 1 h or so at 25 °C.

Gel Staining. Switch off the power, remove the gel cassette, peel off the tape and remove the gel for staining in the staining mixture specified in Section 4.3. Figure 5 shows examples of patterns for wheat gliadins obtained by this procedure. Table 2 lists variations in the procedure above, required to suit it to other cereal grains.

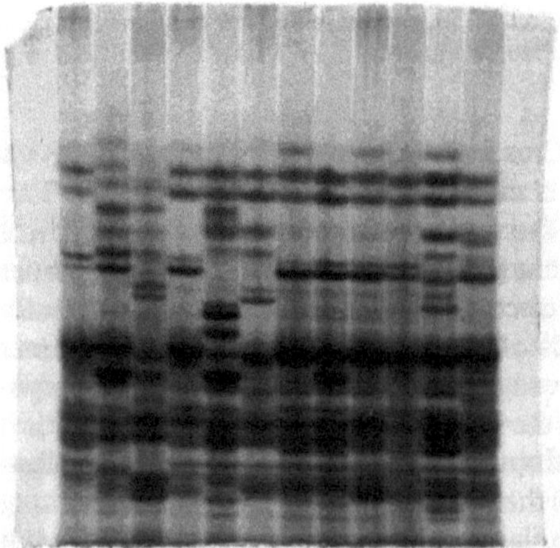

Fig. 5. Gradient-gel electrophoretograms of gliadin proteins, fractionated at pH 3.1 in a Gradipore gel, 3% to 13% polyacrylamide gradient, according to the method in Section 4.4 (RACI 1988)

Table 2. Variation of the conditions in the RACI acid gradient PAGE method suiting it to various cereals

	Wheat, rye	Barley	Oats	Rice
Extracting solution	6% Urea	6% Urea + 1% ME[a]	15% Urea	18% Urea + 1% ME[a]
Extracting solution volume/mg grain	6 µl	4 µl	6 µl	4 µl
Gel type (gradient range)	3 – 13%	3 – 27%	3 – 27%	3 – 27%
Electrophoresis time at 200 V	150 min	120 min	140 min	90 min

[a] Mercaptoethanol.

4.5 Rapid AG-PAGE

The procedures described in the two sections above are hardly rapid enough to permit identification appropriate to some requirements, e.g. checking samples within 1 h or so to determine acceptability of grain cargoes during harvest or on delivery at a flour mill. For such requirements, or even to ensure efficient throughput of analyses and use of equipment, the following rapid method has been devised (Wrigley et al. 1991 a), taking advantage of very small, pre-formed gradient gels, measuring only 25 mm in migration length, 1 mm in gel thickness, but the same gel/cassette width as for the gels in Section 4.4. These Gradipore Micrograd gels are available from Amicon (Boston MA, USA) or E. Merck (Poole, UK); best results for wheat identification have obtained with a 3% to 15% gradient (Gradipore Cat. No. MG-315).

Gel Preparation. Take the gel from the packet and mount it in the apparatus, including a sample slot. Add sodium lactate buffer (pH 3.1), as for Section 4.4. Pre-run for 5 min or so at 200 V.

Sample Preparation. As in Section 4.4, but apply only 3 to 5 µl extract. Extraction with ethylene glycol (< 5% water; as suggested by Clements 1988; Wrigley et al. 1991), in place of urea solution for example, provides efficient extraction of gliadins, without the need for the step of centrifugation to clarify the extracts (because this solvent allows the flour particles to settle out quickly).

Electrophoresis. Continue electrophoresis for only 8 to 10 min at 300 V (25 °C).

Gel Staining. Use the staining mixture of Section 4.3. Staining is more rapid due to the thinness of the gel. Staining can be further accelerated (to a matter of minutes) by increasing the temperature to about 50 °C. Figure 6 shows an example of patterns provided by this procedure. The patterns are compressed, compared to those of more conventional gel formats (e.g. Fig. 3), but resolu-

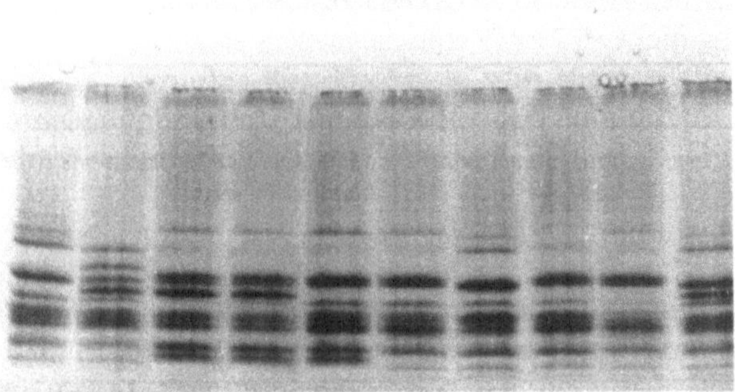

Fig. 6. Micro-grad electrophoretograms of gliadin proteins, fractionated at pH 3.1 in a Gradipore mini-gel, according to the method in Section 4.5 of Wrigley et al. (1991a)

tion is great due to the gradient in the gel, and the pattern is well suited to screening samples to check that they are true to label (see Step 7 in Fig. 2).

4.6 SDS-PAGE

All the methods in Sections 4.4 to 4.5 relate to the gliadin class of proteins of the wheat grain. If the resulting patterns for a combination of varieties does not provide sufficient discrimination (as indicated in Fig. 2D), a useful difference may be provided by examining another aspect of protein composition, e.g. the glutenin polypeptides (see review by Wrigley and Bietz 1988). This class of grain proteins, and all others, is released into solution as the reduced polypeptides by extraction with a detergent solution (e.g. sodium dodecyl sulfate, SDS) and reducing agent (e.g. mercaptoethanol). The resulting extract is fractionated by gel electrophoresis in the presence of SDS, generally in a discontinuous buffer system at about neutral pH (e.g. King and Laemmli 1971; Shewry et al. 1978b; and Wrigley et al. 1982).

This procedure has the advantage that it is more generally applicable to seed source, irrespective of species, than acid-PAGE (Sects. 4.1 to 4.5). On the other hand, however, the SDS gel pattern is generally complex with many bands (e.g. see the patterns in Fig. 7), often making comparison difficult.

4.6.1 ISTA SDS Method

The following procedure is based on that being considered for adoption into the Rules of the International Seed Testing Association.

Gel Preparation. The discontinuous gel system requires the formation of two gel layers: the main (resolving) gel, in which band separation takes place, and the short upper (stacking) gel, on which samples are applied and in which the

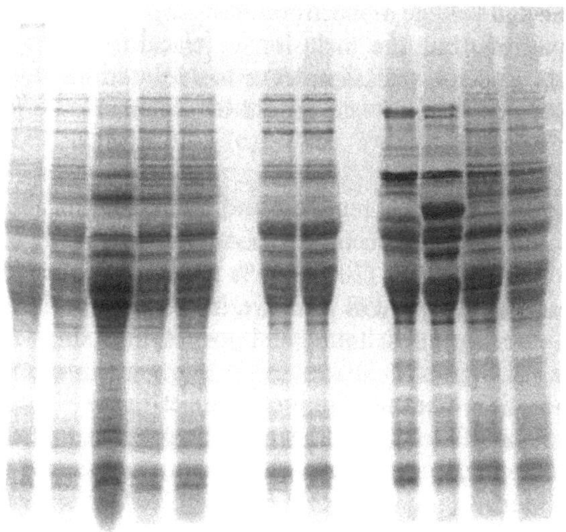

Fig. 7. SDS gel electrophoretograms, showing patterns of SDS-extracted wheat-grain polypeptides (after reduction of SSD bonds). (Wrigley et al. 1987)

protein zones are concentrated to give very thin starting zones. Otherwise, the general procedure is similar to that described above in Section 4.3.

The main (resolving) gel, occupying the main space in the gel slab, is formed by mixing 56.4 ml Tris buffer (121.1 g Tris, plus HCl to pH 8.8 in 1 l) with 86.25 ml gel solution (19.6 g acrylamide + 0.26 g bisacrylamide, made to 90 ml). Degas, and mix in 3.75 ml 1% ammonium persulfate solution, 1.5 ml 10% SDS solution, and 0.075 ml TEMED.

On top of this gel is set a short layer (1 or 2 cm) of the stacking gel into which the sample-loading positions are set. This gel consists of 10 ml 1 M Tris buffer (30.3 g Tris in water adjusted with HCl to pH 6.8 and made to 250 ml), 67.2 ml gel solution (49 g acrylamide + 0.07 g bisacrylamide, made to 67.2 ml). Degas, and mix in 3 ml 1% ammonium persulfate solution, 0.8 ml 10% SDS solution and 0.080 ml TEMED.

The electrode solution is of yet a different composition, containing 3 g Tris, 14.1 g glycine, and 1 g SDS dissolved in water to a volume of 1 l.

Sample Preparation. Mix the sample (crushed grain, milled wheat meal or flour) with extraction buffer in the ratio of about 8 µl extractant/mg sample. Stock extract buffer consists of 12.5 ml Tris (1 M, pH 6.8) + 20 ml glycerol + 24.1 ml water + 4 g SDS. On the day of extraction, 17 parts of stock buffer (by volume) are mixed with 3 parts mercaptoethanol and 40 parts water. After standing for about 1 h at 25 °C, the extracting mixture is heated for 10 min in a boiling water bath, and then centrifuged to give a clear solution ready for application on top of the upper gel layer.

Electrophoresis. The time and voltage of electrophoresis depend on the geometry of the gel and apparatus, but the inclusion of tracking dye (5 µl 1% bromophenol blue in 10% glycerol, with samples or in a few sample slots) provides an indication of progress; the current should be switched off when the dye reaches the bottom of the gel.

Gel Staining/Interpretation. The gel is removed and immersed for about 1 h in fixing solution (methanol − glacial acetic acid − water, 4:1:5, by volume), and then overnight in staining solution (200 ml 15% trichloroacetic acid solution + 10 ml 1% Coomassie Blue, or PAGE Blue dye, in methanol). The stained gel is rinsed in distilled water for a few hours, and possibly in fixing solution, before examination and photography.

As SDS gel electrophoresis is often used to provide an estimate of molecular weight (based on comparison of mobility with standard proteins), it is possible to attempt a listing of zones according to size, but the large number of closely spaced bands usually obtained (e.g. see Fig. 7) makes this exercise difficult.

4.6.2 Rapid SDS Methods

The complex procedure required for standard SDS gel electrophoresis, such as that above, suits it poorly for routine or rapid identification. For these reasons, modifications have been devised. A major change has been the elimination of the need for two separate gel layers. This, together with the provision of stable pre-cast gels, offers the possibility for more widespread use of SDS gel electrophoresis. The Gradipore Micrograd gels, described above in Section 4.5, are also available ready for use in SDS electrophoresis, thus providing speed, convenience and simplicity (Wrigley et al. 1991 b). In addition, an automated system employing horizontal pre-cast gels ($50 \times 43 \times 0.45$ mm) for the SDS method is available from Pharmacia. This Phast system has been reported by Marchylo et al. (1989) to be applicable to wheat identification. They recommended the use of an 8% to 25% gel gradient.

4.7 Isoelectric Focusing (IEF)

Another approach to electrophoretic fractionation is gel isoelectric focusing, which involves separation of protein species according to differences in their isoelectric points in a pH gradient (reviewed by Wrigley and Bietz 1988). This method has been used to fractionate a range of cereal-grain proteins, but more for research purposes than for routine analysis.

4.8 Two-Dimensional Gel Electrophoresis

Gel IEF has been particularly useful as the first dimension of two-dimensional (2-D) fractionation, the second generally being acid or SDS-PAGE (see reviews, such as Wrigley and Bietz 1988). An example of 2-D fractionation of wheat-grain proteins for identification is given by Dunbar et al. (1985), who fractionated "hundred of proteins" from a single seed. However, the extra work involved in such procedures means that they are reserved for specialized aspects of identification (e.g. Plant Variety Rights registration).

4.9 Capillary Electrophoresis

Many other forms and variations of electrophoresis have been applied to cereal grain proteins, and there are potential advantages to be gained in many cases. Of these, a distinctly different one is capillary electrophoresis, which rivals HPLC (see Sect. 1.2) in its resolution, format (capital expense), and potential value for automatic loading and data interpretation. See the Volume 11, Number 9 (1990) of the journal *Electrophoresis* for a current review of developments in the use of this promising technique.

5 Other Cereals

The SDS gel electrophoresis procedure described above (Sect. 4.6) is applicable to all cereal grains. The acid PAGE methods are not so generally applicable as they depend on the composition of the gliadin proteins that are specific to wheat grain. This class of proteins, generally known as prolamins, is paralleled to a degree in some other cereals, e.g. secalins in rye, hordeins in barley, and avenins in oats. Modifications of the acid PAGE procedures are thus also likely to suit such cereal grains, as summarized in Table 2, from the Standard RACI method (Sect. 4.4), although the proteins being examined in these procedures are not exclusively the relevant prolamins.

5.1 Barley

Electrophoretic methods adapted to barley identification include most of those described in Section 4, particularly acid PAGE, SDS-PAGE, and IEF. The details of 16 methods are tubulated by Cooke (1988); see also Cooke (1984). Cooke and Morgan (1986) were able to divide 191 European barley varieties into 41 groups on the basis of electrophoretic patterns obtained by acid PAGE in a 10% polyacrylamide gel using the glycine/acetic acid buffer of the ISTA Standard Reference Method (see Sect. 4.3). In general, successful extraction of hordeins requires more stringent conditions than for the gliadins, such as the use of mercaptoethanol (1% specified in Table 2, and in the ISTA method which also requires 20% chloroethanol and 18% urea). Figure 8 shows example patterns, obtained by the ISTA method (according to Cooke, in Konarev

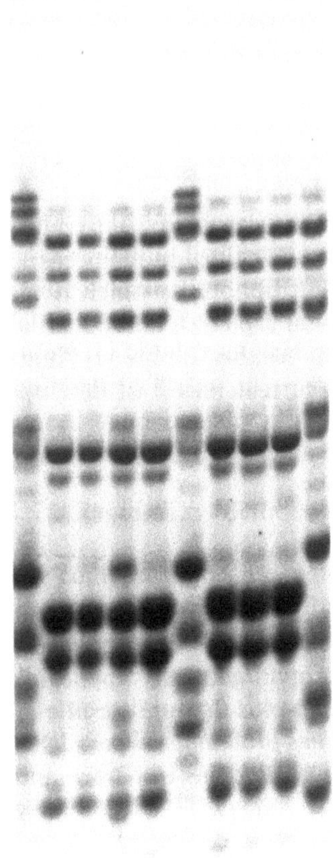

Fig. 8. Electrophoretic patterns (ISTA acid PAGE) for barley-grain proteins, extracted from Triumph (*lanes 1*, *left*, and *6*), an authentic sample of Corniche (*extreme right*) and off-type seeds of Corniche. (Courtesy of Dr. R. J. Cooke, NIAB, Cambridge, UK)

and Gavriljuk 1988, p. 14), for Triumph barley and for Corniche (one authentic and seven off-type grains).

In view of the added difficulty of extracting barley-grain proteins (compared to the gliadins), it is not surprising that SDS gel electrophoresis has proved more popular than acid PAGE involving, as it usually does, extraction with reducing agent as well as detergent. The method specified in Section 4.6 is suitable. Shewry et al. (1979) were able to divide 164 European barley varieties into 32 groups on the basis of SDS-PAGE, indicating that it is generally suitable, but that similar varieties may not be distinguished (possibly, in comparison to wheat, because barley is not polyploid). As IEF has also been used effectively for barley identification (Cooke 1988), two-dimensional combinations of IEF with PAGE have been discriminating for certain cases (Shewry et al. 1978a).

5.2 Oats

Extraction of oat prolamins (avenins) for acid PAGE has been achieved with concentrated urea solution (RACI 1988; Table 2) or with 25% chloroethanol (according to a modification of the ISTA acid PAGE method reported by Cooke 1988). In both these cases, use of smaller pore polyacrylamide gels is recommended (3% to 27% gradient or 12% uniform gel, respectively). Figure 9 illustrates patterns obtained for extracts of oat grains by the procedure of Cooke (in Konarev and Gavriljuk 1988, p. 14), showing both distinction between varieties of cultivated oats and identification of wild oat species. Characterization of species of wild oats and distinction between wild and cultivated oats have been a valuable application of gel electrophoresis (Cooke and Draper 1986). Similar studies have been pursued by Sanchez de la Hoz and Fominaya (1989) using isoenzyme staining methods. Rapid SDS gel electrophoresis has also been applied to oats, using the Pharmacia Phast system, by Hansen et al. (1988).

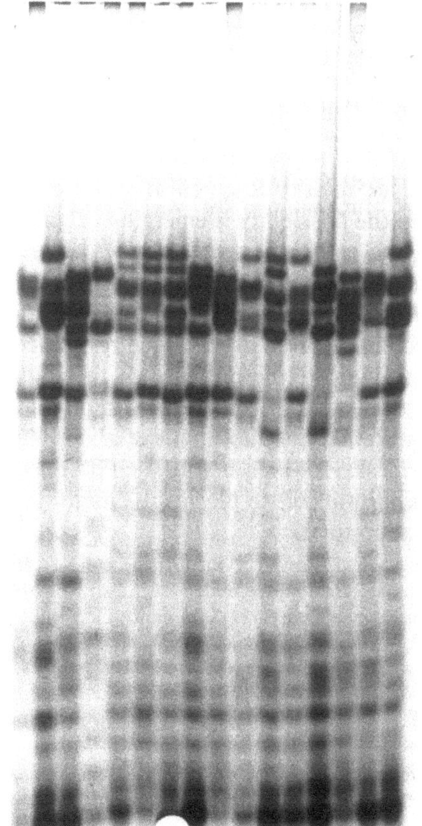

Fig. 9. Electrophoretic patterns for oat-grain proteins, fractionated by the ISTA acid PAGE method (Sect. 5.2). The first four patterns (*left*) are for wild-oat samples; the remainder are for varieties of cultivated oat. (Courtesy of Dr. R.J. Cooke, NIAB, Cambridge, UK)

5.3 Rice

Varietal identification by electrophoresis has been reported less for rice than for other major cereals. Reasonably harsh extraction conditions are generally used, combined with smaller pore gel composition, e.g. see Table 2 (conditions based on du Cros et al. 1979). Hussain et al. (1989) have adapted the ISO/ICC method (Sect. 4.3.1) to rice by advocating 5 M acetic acid (containing 1% sucrose) as extractant. Isoenzyme analysis has also proved to be useful for rice identification (Glaszmann 1987).

5.4 Maize

As maize (corn) is often sown as F1 hybrid seed, identification in this case may present the additional challenge of the seed being heterozygous. Cooke (1988) provides as discussion of the implications of this aspect of identification. The prolamin of maize (zein) contains a range of polypeptides of similar size, but different isoelectric points. IEF has thus been applied effectively for this cereal, e.g. Righetti and Bosisio (1981). So, too, has SDS electrophoresis (Wilson et al. 1981) and two-dimensional IEF-acid PAGE (Wall et al. 1984).

6 Derivation of Information About Phenotype

Electrophoretic methods of identification are mainly designed to indicate the genotype of the seed − no more. However, the possibility is emerging of going further, as information accumulates on the associations of specific protein components as genetic markers (or possibly even causal agents) of certain aspects of phenotype. If, for example, there is a close linkage between the gene for that protein and the gene for a key protein involved in pathogen resistance, the electrophoretic pattern may be able to indicate the presence of the resistance gene. The best prospect for provision of such information currently lies with specific aspects of grain quality, as reviewed by MacRitchie et al. (1990), including dough properties and grain hardness. However, before such promises are fulfilled, gel electrophoresis has an important role to play as the major approach to varietal identification in the laboratory.

References

Appleyard DB, McCausland J, Wrigley CW (1979) Checking the identity and origin of off-types in the propagation of pedigreed seed. Seed Sci Technol 7:459−466
Autran JC, Abbal P (1988) Wheat cultivar identification by a totally automatic soft-laser scanning densitometry and computer-aided analysis of protein electropherograms. Electrophoresis 9:205−213

Bekes F, Kemeny A, Meresz P, Varga J, Demeter L (1987) Comparison of different quantitative evaluation methods of gliadin. In: Lasztity R, Bekes F (eds) Proc 3rd Int Workshop Gluten Proteins, World Scientific Singapore, pp 598–603

Bekes F, Batey IL, Wrigley CW, Gore PJ (1991) Rapid electrophoresis of gliadin proteins: integration of lab tests to efficiently identify wheat varieties. In: Bushuk W, Tkachuk R (eds) Gluten Proteins 1990. Am Assoc Cereal Chemist, St Paul, MN, pp 467–475

Beyer WH (1968) Handbook of tables of probability and statistics, 2nd edn. Chemical Rubber Co, Ohio

Bietz JA (1986) High-performance liquid chromatography of cereal proteins. Adv Cereal Sci Technol 8:105–170

Bushuk W, Zillman RR (1978) Wheat cultivar identification. I. Apparatus, method and nomenclature. Can J Plant Sci 58:505–515

Clements RL (1988) A continuous acetic acid system for polyacrylamide gel electrophoresis of gliadins and other prolamines. Electrophoresis 9:90–93

Cooke RJ (1984) The characterisation and identification of crop cultivars by electrophoresis. Electrophoresis 5:59–72

Cooke RJ (1988) Electrophoresis in plant testing and breeding. In: Chrambach A, Dunn MJ, Radola BJ (eds) Advances in electrophoresis, vol 2. VCH Verlagsgesellschaft, Weinheim, Germany, pp 171–261

Cooke RJ, Draper SR (1986) The identification of wild oat species by electrophoresis. Seed Sci Technol 14:157–167

Cooke RJ, Morgan AG (1986) A revised classification of barley cultivars using a standard reference electrophoresis method. J Natl Inst Agric Bot 17:169–178

Dal Belin Peruffo, Pallavicini AC, Varanini Z, Pogna NE (1981) Analysis of wheat varieties by gliadin electrophoresis. 1. Catalogue of electrophoregram formulas of 29 common wheat cultivars grown in Italy. Genet Agrar 14:195–208

Draper SR (1987) ISTA variety committee. Report of the working ground for biochemical tests for cultivar identification 1983–1986. Seed Sci Technol 15:431–434

du Cros DL, Wrigley CW (1979) Improved electrophoretic methods for identifying cereal varieties. J Sci Food Agric 30:785–794

du Cros DL, Wrigley CW, Blakeney AB (1979) Fractionation of rice-grain proteins by gradient gel electrophoresis and gel isoelectric focusing: characterisation of rice genotypes. Riso 28:275–284

Dunbar BD, Bundman DS, Dunbar BS (1985) Identification of cultivar-specific proteins of winter wheat (*T. aestivum* L.) by high resolution two-dimensional polyacrylamide gel electrophoresis and color-based silver stain. Electrophoresis 6:39–43

Glaszman JC (1987) Isoenzymes and classification of Asian rice varieties. Theor Appl Genet 74:21–30

Hansen AE, Nassuth A, Altosaar I (1988) Rapid electrophoresis of oat (*Avena sativa* L.) prolamins from single seeds for cultivar identification. Cereal Chem 65:153–154

Hussain A, Scanlon MG, Juliano BO, Bushuk W (1989) Discrimination of rice cultivars by polyacrylamide gel electrophoresis and high-performance liquid chromatography. Cereal Chem 66:353–356

Jones BL, Lookhart GL, Hall SB, Finney KF (1982) Identification of wheat cultivars by gliadin electrophoresis: electrophoregrams of the 88 wheat cultivars most commonly grown in the United States in 1979. Cereal Chem 59:181–188

Jost M (ed) (1989) Handbook of Yugoslav winter wheat cultivars. Podravka 7:1–115

King J, Laemmli UK (1971) Polypeptides of the tail fibres of bacteriophage T4. J Mol Biol 62:465–477

Konarev VG, Gavriljuk IP (eds) (1988) Biochemical identification of varieties. Int Seed Testing Assoc and NI Vavilov All-Union Inst Plant Ind, Leningrad

Konarev VG, Gavriljuk IP, Gubareva NK, Peneva TI (1979) Seed proteins in genome analysis, cultivar identification and documentation of cereal genetic resources: a review. Cereal Chem 56:272–278

Konarev VG, Gavriljuk IP, Gubareva NK, Choroshajlov HG (1981) Electrophoretic and serological methods in seed testing. Seed Sci Technol 9:807–817

Lookhart GL, Finney KF (1984) Polyacrylamide gel electrophoresis of wheat gliadins: the effect of environment and germination. Cereal Chem 61:496–499

MacRitchie F, du Cros DL, Wrigley CW (1990) Flour polypeptides related to wheat quality. Adv Cereal Sci Technol 10:79–145

Marchylo BA, Handel KA, Mellish VJ (1989) Fast horizontal sodium dodecyl sulfate gradient polyacrylamide gel electrophoresis for rapid wheat cultivar identification and analysis of high molecular weight glutenin subunits. Cereal Chem 66:186–192

Mecham DK, Kasarda DD, Qualset CA (1985) Identification of western U.S. wheat varieties by polyacrylamide gel electrophoresis of gliadin proteins. Hilgardia 53 (7):1–32

Myers DG, Edsall KJ (1989) The application of image processing techniques of the identification of Australian wheat varieties. Plant Varieties Seeds 2:109–116

Neuman M, Sapirstein HD, Shwedyk E, Bushuk W (1987) Discrimination of wheat class and variety by digital image analysis of whole grain samples. J Cereal Sci 6:125–132

Ng PKW, Scanlon MG, Bushuk W (1988) A catalog of biochemical fingerprints of registered Canadian wheat cultivars by electrophoresis and high-performance liquid chromatography. Publ No 139, Food Science Dept, Univ Manitoba, Canada

Quaite E, Schildbach R, Burbidge M (1987) Protein-elektrophoretische Identifikation der in der Bundesrepublik Deutschland zugelassenen Weizensorten. Getreide Mehl Brot 41:259–264

Righetti PG, Bosisio AB (1981) Applications of isoelectric focusing to the analysis of plant and food proteins. Electrophoresis 2:65–75

Royal Australian Chemical Institute (RACI) (1988) Electrophoretic identification of cereal varieties. In: Official Testing Methods of the Cereal Chemical Division. RACI Melbourne, pp 24–27

Sanchez de la Hoz P, Fominaya A (1989) Studies of isozymes in oat species. Theor Appl Genet 77:735–741

Sapirstein HD, Bushuk W (1985a) Computer-aided analysis of gliadin electrophoregrams. I. Improvement of precision of relative mobility determination by using a three reference band standardization. Cereal Chem 62:372–377

Sapirstein HD, Bushuk W (1985b) Computer-aided analysis of gliadin electrophoregrams. III. Characterization of the heterogeneity in gliadin composition for a population of 98 common wheats. Cereal Chem 62:392–398

Sapirstein HD, Bushuk W (1986) Computer-aided wheat cultivar identification and analysis by densitometric scanning profiles of gliadin electrophoregrams. Seed Sci Technol 14:489–517

Shewry PR, Ellis JRS, Pratt HM, Miflin BJ (1978a) Comparison of methods for the extraction and separation of hordein fractions from 29 barley varieties. J Sci Food Agric 29:433–441

Shewry PR, Falks AJ, Pratt HM, Miflin BJ (1978b) The varietal identification of single seeds of wheat by sodium dodecyl sulfate polyacrylamide gel electrophoresis of gliadin. J Sci Food Agric 29:847–849

Shewry PR, Pratt HM, Faulks AJ, Parmer S, Miflin BJ (1979) The storage protein (hordein) polypeptide pattern of barley (*Hordeum vulgare* L.) in relation to varietal identification and disease resistance. J Nat Inst Agric Bot 15:35–40

Wall JS, Fey DA, Paulis JW (1984) Improved two-dimensional electrophoretic separation of zein proteins: application to study of zein inheritance in corn genotypes. Cereal Chem 61:141–146

Wilson CM, Shewry PR, Miflin BJ (1981) Maize endosperm proteins compared by sodium dodecyl sulfate gel electrophoresis and isoelectric focusing. Cereal Chem 58:275–281

Wrigley CW (1982) The use of genetics in understanding protein composition and grain quality. Qual Plant Plant Foods Hum Nutr 31:205–227

Wrigley CW, Bietz JA (1988) Proteins and amino acids. In: Pomeranz Y (ed) Wheat chemistry and technology, vol I. Am Assoc Cereal Chem, St Paul, MN, pp 159–275

Wrigley CW, McMaster GJ (1989) New approaches to testing the product suitability and processing of wheat. In: Logan LA (ed) Proc 1989 Cereal Sci Conf, DSIR Crop Research Division, New Zealand, pp 88–91

Wrigley CW, Autran JC, Bushuk W (1982) Identification of cereal varieties by gel electrophoresis of the grain proteins. Adv Cereal Sci Technol 5:211–259

Wrigley CW, Batey IL, Campbell WP, Skerritt JH (1987) Complementing traditional methods of identifying cereal varieties with novel procedures. Seed Sci Technol 15:679–688

Wrigley CW, Tomlinson JD, Skerritt JH, Batey IL, Sing W (1989) Efficient identification of wheat varieties by established and novel procedures. Cereal Foods World 34:629–632
Wrigley CW, Gore PJ, Manusu HP (1991 a) Rapid (< 10 min) electrophoresis for identification of wheat varieties. Electrophoresis 12:384–385
Wrigley CW, Batey IL, Bekes F, Gore PJ, Margolis J (1991 b) Rapid and automated characterization of seed genotype using Micrograd electrophoresis and pattern matching software. Appl Theor Electrophoresis (in press)
Zuckerkandl E, Pauling L (1965) Molecules as documents of evolutionary history. J Theor Biol 8:357–366

Identification of Cultivars of Grasses and Forage Legumes by SDS-PAGE of Seed Proteins

S. E. Gardiner and M. B. Forde†

1 Introduction

The banding patterns produced following polyacrylamide gel electrophoresis of total seed storage proteins in the presence of sodium dodecylsulphate (SDS-PAGE) have proved an effective laboratory method for distinguishing cultivars of the largely cross-fertilized pasture grasses and legumes despite their high innate genetic variability (Ferguson and Grabe 1984, 1986; Gardiner et al. 1986; Gardiner and Forde 1987, 1988a; Clark et al. 1989). The positions of the stained bands visible in the gel following electrophoresis are determined by the molecular weight of the seed proteins or their constituent polypeptides and differences between cultivars arise because of the polymorphic nature of seed storage proteins within each species. Similar techniques have been used extensively by workers concerned with cultivar identification of cultivated inbreeding crops (Cooke 1984, 1989) but to a lesser extent for differentiating cultivars of outbreeding species (Gilliland 1989).

Some advantages which this type of electrophoretic descriptor offers over the traditional morphological and other descriptive criteria derived from field trials include freedom from the influence of environmental or management practices. No land or plant growth facilities are required, as the analyses are performed on dry seed. Labour inputs are reduced as there is no need to examine individual genotypes because the analysis of a ground meal from a bulked sample of seed produces stable results which can differentiate cultivars which are not readily distinguishable morphologically, especially as single plants (Ferguson and Grabe 1986; Gardiner et al. 1986; Gardiner and Forde 1987, 1988a, b). Electrophoretic analysis of allozymes (isoenzymes at a single locus) has been demonstrated to be a powerful method for distinguishing cultivars of *Lolium perenne* L. (Gilliland et al. 1982; Gilliland 1989) but inputs are higher than for the seed protein technique as individual seedlings from the population comprising each cultivar must be analyzed in order to determine genotype frequencies.

The technique for electrophoresis of seed proteins as described here has been developed to give optimum resolution of the often numerous bands of varying intensities obtained when bulked samples of ground seed from cross-fertilized forage species are analyzed. Variations on the basic technique giving best results for a number of genera and species are outlined. Methods for ryegrass and peas, differing only in detail, are in the process of being included in a handbook (ISTA Electrophoresis Handbook) for the use of seed testing authorities worldwide. Rules for the testing procedures will be submitted to the Rules Committee of the International Seed Testing Association for consideration at the 1992 ISTA Congress. It is crucial to note that profiles produced by different methods will differ to some extent and should not be compared directly for the purposes of cultivar identification. Indeed, detailed comparisons between gel profiles should best be made between samples analyzed on the same gel, as slight differences between electrophoretic runs are unavoidable (see Sect. 2.5).

2 Procedures

All chemicals for electrophoresis should be of analytical grade, as impurities, particularly in the polyacrylamide, can give to poor resolution of protein bands and lack of reproducibility (Biorad 1984).

2.1 Sample Preparation

A finely ground seed meal should be prepared from a sample of dry seeds, using a hammer mill (1-mm screen) or an electric motor-type coffee grinder with a lid modified by the addition of flanges to direct the seeds on to the rotor blades. A minimum of 200 seeds are required to give a reproducible sample of the genetic variation in seed proteins present in the cultivar population of an outbreeding species and a larger sample is preferable. As most pasture species have very small seeds, a sample of 1 g is usually very adequate, but greater quantities will be required for the larger-seeded legumes. Seed should be of normal harvestable quality. Immature seed will not give clear results as the proportion of different seed proteins changes during seed development (Luthe 1987). The ground seed is stable for about 1–2 years at room temperature.

If the species is inbreeding and the cultivars are known to be pure lines containing only a single genotype, then single seeds may be analyzed to characterize a cultivar (Sect. 3). However, initially a number should be analyzed individually to confirm this supposition. Single seeds can be crushed using flat-nose pliers with a smooth gripping surface and the amount of buffer to be added is calculated from the weight of an average seed.

The extraction buffer is that of Smith and Payne (1984). The buffer stock solution which is stable at room temperature consists of: 12.5 ml of 1 M Tris-HCL, pH 6.8; 24 ml water; 20 ml glycerol (or 11.5 g Ficoll, see Sect. 4.1); 4 g sodium dodecylsulphate; and 12 mg bromophenol blue.

Immediately before use, the stock solution is mixed with 2-mercaptoethanol, dimethylformamide and water in the ratio 3:1.06:1.76:3 and added

Table 1. Summary of technical parameters and utility of SDS-PAGE for cultivar identification in different genera and species

Genus/species	Polyacrylamide concentration (% w/v)	Extraction buffer per 20 mg (ml)	Volume loaded		Rating[a] (1 − 4)
			with glycerol (µl)	with Ficoll (µl)	
A. Grasses					
Agrostis capillaris	15	0.25	–	6 − 8	3
	15	0.33	10	–	
Bromus sect. *Ceratochloa*	12.5	0.25	6	–	3
Cynosurus cristatus	11, 17.8	0.25	6	–	4
Dactylis glomerata	15	0.25	–	14	3
Festuca spp.	11	0.25	–	15	1
	11	0.25	8	–	
Lolium spp.	11	0.25	–	–	2
	11	0.25	8	–	
Paspalum dilatatum	12.5	0.125	40	–	4
Poa pratensis	15	0.25	–	15	1
B. Legumes					
Astragalus spp.	12.5	0.25	–	6 − 9	2
Chamaecytisus palmensis	12.5	0.25	6	–	3
Lotus spp.	12.5	0.5	4 − 6	–	3
Medicago sativa	12.5	0.5	6	–	4
Ornithopus spp.	12.5	0.5	5 − 8	–	3
Trifolium fragiferum	12.5	0.25	–	6	3
Trifolium michelianum	12.5	0.5	5	–	2
Trifolium pratense	12.5	0.25	–	6	1
	12.5	0.5	6 − 8	–	
Trifolium repens	12.5	0.25	–	6 − 8	3
	12.5	0.5	10	–	
Trifolium subterraneum	12.5	0.25	–	10	2
	12.5	0.5	3 − 4	–	
Vicia spp.	12.5	0.25	–	6	1

[a] 1 = Very good; 2 = good; 3 = satisfactory; 4 = unsatisfactory.

to weighed samples in 1.5-ml stoppered plastic microcentrifuge tubes (preferably in a fumehood). Routinely, 20 mg of seed meal is extracted with 0.25 ml of the buffer (see Table 1 for exceptions). After standing for 1 h at room temperature, the samples are mixed thoroughly with a motor-driven pestle constructed to fit the centrifuge tubes closely. [Construct by adding a small volume (300 − 400 µl) of Araldite glue to a tube of the same brand to be used, and clamping a metal rod vertically in the glue until it hardens.] The samples are left at room temperature overnight, resuspended using a vortex mixer, heated for 10 min at 85 °C in a water bath (a small slit is made in the caps to prevent build-up of pressure), and then remixed with a vortex mixer.

We have found that this through extraction procedure gives very reproducible results. Other methods may result in uneven extraction of proteins.

The extracts are cleared by centrifuging (5 min) in a microcentrifuge. We have found it occasionally necessary to recentrifuge an aliquot of the supernatant in order to remove all particulate matter (see Sect. 4.2). Samples are then taken for electrophoresis. (A list of volumes for different species and genera is given in Table 1.)

2.2 Electrophoresis

Discontinuous SDS-polyacrylamide gel electrophoresis is performed using a system based on that of Laemmli (1970), but with increased buffer concentrations to give better resolution (see Fling and Gregerson 1986). The running gel routinely contains 0.56 M Tris-HCL (pH 8.8), 0.1% SDS and 11% − 15% acrylamide (Table 1). The stacking gel contains 0.19 M Tris-PO$_4$ (pH 6.8), 0.1% SDS and 5% acrylamide (Gardiner and Forde 1988a).

It is convenient to prepare stock solutions for preparation of both gels. Gloves should be worn when handling acrylamide solid or solutions as it is a neurotoxin (Biorad 1984).

Solutions

1. Stacking gel buffer: 0.75 M Tris-PO$_4$, pH 6.8. Store at 4 °C.
2. Running gel buffer: 1.5 M Tris-HCL, pH 8.8. Store at 4 °C.
3. Stacking gel acrylamide stock: 20% acrylamide, 0.8% N,N^1-methylene-bis-acrylamide (filtered through Whatman No. 1 paper after preparation). Store at 4 °C for up to 1 month.
4. Resolving gel acrylamide stock: 30% acrylamide, 0.39% N,N^1 methylene-bis-acrylamide (filtered as above). Store at 4 °C for up to 1 month.
5. 10% SDS. Store at room temperature.
6. TEMED (N,N,N^1,N^1-tetramethylethylenediamine). Store at −20 °C.
7. 2% ammonium persulphate, *prepared fresh* each day.

The resolving gel is prepared first. For 100 ml of solution (sufficient for two gels); 37.5 ml of stock buffer, 1 ml 10% SDS and 50 µl TEMED are added to appropriate volumes of stock acrylamide (30%) and water (e.g. 36.6 ml and 25.9 ml respectively for 11% gels). The solution is mixed by *gentle* swirling. (If air bubbles are not introduced during mixing, degassing is unnecessary). Ammonium persulphate (2.5 ml) is added, the solution gently mixed, and the gel poured immediately and without causing foaming into prepared clean dry gel cassettes. Space for the stacking gel should be left, and the top of the solution overlaid with 0.5 − 1 ml of water-saturated isobutanol to exclude air, using a syringe. Polymerization typically occurs in 20 − 30 min. If it is too fast or too slow, the amount of ammonium persulphate should be adjusted for subsequent gels. The isobutanol is then poured off (or removed with a syringe) and the top of the gel rinsed thoroughly with distilled water and surplus water drained off. A paper towel may be used to absorb the last drops.

A 5% stacking gel solution (25 ml) is prepared by combining 6.2 ml of stock buffer, 6.2 ml acrylamide stock (20%), 12.2 ml water, 0.26 ml 10% SDS, 20 µl

TEMED and 1 ml ammonium persulphate, in the same fashion as for the resolving gel. This is poured onto the top of the resolving gel and the well-former ("comb") inserted to within $1-1.5$ cm of the top of the resolving gel. The gel may be left overnight with the comb in place provided the top is covered with Parafilm or plastic wrap to prevent it from drying out. The slots in the gel should be rinsed with the upper buffer prior to assembling the apparatus and filling the upper buffer reservoir. Aliquots of the protein extracts are loaded into the wells under the buffer using a pipetting device or syringe (Table 1).

The upper electrode buffer contains 0.038 M Tris, 0.29 M glycine and 0.1% SDS and is diluted $2:1$ (approximately) for the lower tank buffer. We use a Biorad Protean II electrophoresis apparatus. The 1.5-mm-thick running gels are 13 cm long and 16.2 cm wide with 15 7-mm-wide wells (or 20 5-mm-wide wells in gels to be used for screening). Although thinner gels (1 mm) often give improved resolution of bands, they are more difficult to handle without tearing. Water at $4 \pm 1\,°C$ is circulated through the central core of the apparatus and for best resolution the lower buffer level is extended as far as possible up the gel (i.e. to the top of the running gel). This buffer is stirred by use of a magnetic stirrer and, in addition, the whole apparatus is placed in an ice/water bath. This cooling system enables a relatively short run time and we find a suitable protocol to be 25 mA/gel for 1 h; then 15 W/gel until the tracking dye reaches the bottom of the gel (about 4 h). The lower buffer may be reused about four times but the upper buffer becomes exhausted after one use.

2.3 Staining

Following electrophoresis, the gels are immersed in methanol:water:acetic acid $(5:5:1, v/v/v)$ for 30 min to precipitate proteins, then washed with three changes of distilled water for a total of 30 min (both steps with slow agitation). Proteins are then stained with a solution containing 0.02% Coomassie Blue R dissolved in 5% ethanol, 6% trichloroacetic acid and 25% methanol (reagent grade methanol or less is adequate). Bands are clearly visible after overnight staining and maximum staining of proteins is attained after 3 days. To reduce the background colour and to intensify the staining of some bands not readily visible after the first staining, a second staining prior to photographing the gels is normally performed using the Coomassie Blue G stain described by Blakesley and Boezi (1977). This is conveniently prepared as follows:

Two g Coomassie Blue G-250 is suspended in 1.95 l water and 54 ml of conc. H_2SO_4 is added carefully and left stirring overnight. The solution is then filtered through Whatman No. 1 filter paper. KOH (123.4 g) made up to 220 ml with water is added to the filtrate with stirring, followed by 300 g trichloroacetic acid made up to 300 ml with water.

The resulting green-blue solution is filtered through Whatman paper No. 1 and stored in the dark until use. The gel stained with Coomassie Blue G250 should be rinsed in water (two changes) for $1-2$ h before soaking in this second stain for about 2 days in the dark. The blue-stained bands fade if the gels are left in daylight.

2.4 Photographs

We photograph the double-stained gels on a light box, using a red filter and Ilford Pan F film. The film is developed under standard conditions with Ilford ID 11 developer and printed on Ilford Ilfospeed paper (3.1 M grade).

2.5 Evaluation

We evaluate the protein profiles visually by comparing samples with each other and those of authentic reference samples run on the same gel and see little advantage in using currently available densitometer systems for evaluation of these profiles. Protein molecular weight references may be run on the gels to define the molecular weights of bands. Assessment is on the basis of qualitative banding differences (i.e. presence/absence of a band at a particular position) and quantitative differences (differences in relative intensity of specific bands). Because cultivars of forage species have high genetic variability, and are often quite closely related, quantitative differences in band intensity (representing different frequencies of particular common genes in the two populations) rather than complete absence/presence of bands must be expected to be the most common differentiating features and this makes any calculation of similarity indices difficult.

Because samples with apparently identical seed protein banding profiles could still have differences in other characters, it is not possible to state that an unknown cultivar is cultivar X, only "that the seed protein banding pattern of the sample following analysis by SDS-PAGE cannot be distinguished from that of cultivar X" (or "is consistent with that of variety X"). In practice, however, we have found the method extremely useful and reliable as a cultivar descriptor.

However, the value of SDS-PAGE profiles of seed proteins as a cultivar descriptor does vary between and within different genera. Some genera, such as *Festuca* show a great deal of variation at the variety or ecotype (i.e. population) level but few species-specific bands (Gardiner et al. 1986; Gardiner and Forde 1987). Others, such as *Lotus* and *Ornithopus* have very little variation in banding profiles at the variety level but strong species identifiers, making the technique more useful in taxonomic studies than in cultivar description (Gardiner and Forde 1988a; Forde and Gardiner 1991). Similar differences occur between species within the same genus, e.g. cultivars of *Trifolium pratense* L. and *T. subterraneum* L. are much easier to identify than those of *T. repens* L. (Fig. 7 and Gardiner and Forde 1988a).

3 Strategies for Cultivars of Cross-Fertilized and Self-Fertilized Species

SDS-PAGE analysis of proteins extracted from individual seeds of a cultivar of a cross-fertilized species such as *Trifolium pratense* produces distinct but very diverse banding patterns. A few of these are demonstrated for cv. Grasslands Turoa (Fig. 1, lanes 1–6). However, a seed protein banding profile characteristic of the interbreeding population which comprises the cultivar can readily be obtained by analyzing an extract from a subsample of a seed meal produced by grinding at least 200 seeds (Fig. 1, lane 7).

In contrast, individual seeds of a cultivar of a self-fertilized species such as *Trifolium subterraneum* ssp. *subterraneum* give rise to identical banding profiles (Fig. 1, lanes 8–14), provided the cultivar is a pure line. Where these circumstances occur, single seeds may be used for cultivar identification of such species, but caution is necessary as some cultivars comprise two or more seed protein genotypes (Fig. 1, lanes 15–20) or may have become contaminated during multiplication. In such cases either a statistically significant number of single seeds should be analyzed in order to characterize the cultivar, or more conveniently, a bulked sample should be analyzed, as for outbreeding species (Fig. 1, lane 21).

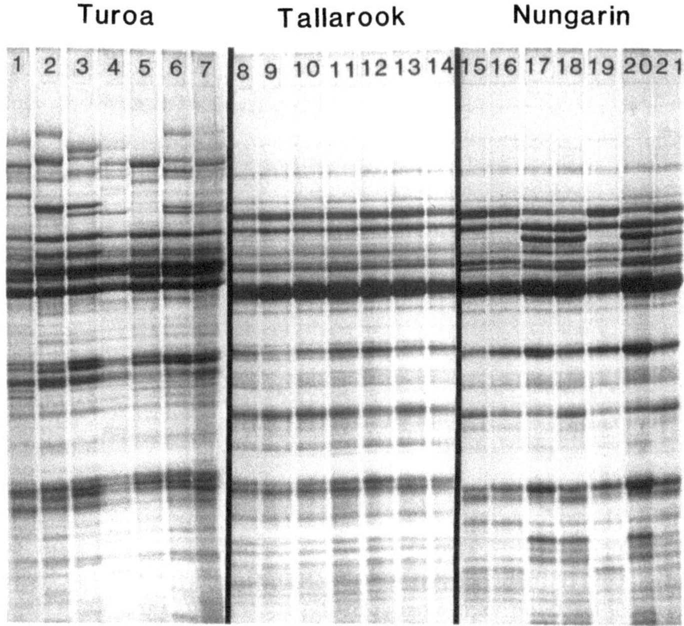

Fig. 1. The banding patterns of proteins extracted from single seeds of *Trifolium pratense* cv. Grasslands Turoa (*lanes 1–6*) and of *T. subterraneum* cvs. Tallarook (*lanes 8–14*) and Nungarin (*lanes 15–20*). *Lanes 7* and *21* are from extracts of bulk seed meals of Turoa and Nungarin respectively

4 Factors Affecting Band Resolution

4.1 Extraction Buffer

The standard extraction buffer originally described by Smith and Payne (1984)
for barley has proved more satisfactory than the buffer described by Laemmli
(1970). The replacement of glycerol by Ficoll in the extraction buffer (see
Sec. 2.1 above) can give finer resolution of a complex profile. It is normally
necessary when using Ficoll to carefully adjust the volume of sample loaded
as it is very easy in its presence to produce the band distortions characteristic
of overloading.

4.2 Sample Volume and Quality

The volumes of extract to be loaded specified in Table 1 should be regarded
as an indication only. In practice, it is necessary to run several lanes at different
loadings when establishing an electrophoretic system for a particular species.
As an aid to diagnosis, Fig. 2 demonstrates the different types of band distor-

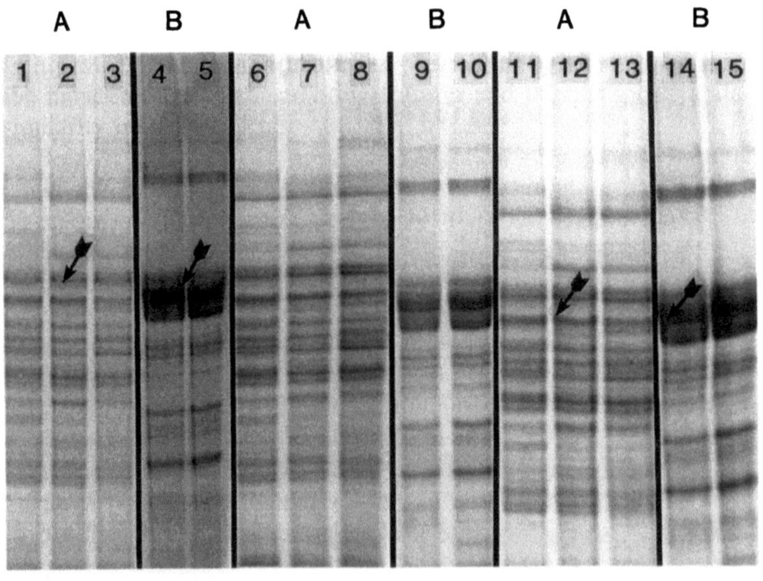

Fig. 2. *A* composite figure demonstrating the effect on seed protein profiles of under- and overloading of protein extracts. SDS-PAGE was on 11% gels and the volumes indicated are of extracts
produced from 20 mg of seed meal and 0.25 ml of buffer containing Ficoll. **A** *Festuca arundinacea*; *B Lolium perenne*. Characteristic distortions are indicated by *arrows*

tions obtained for under- and overloading on profiles for two representative grasses. For each type, the distortions are more severe for *L. perenne* than for *F. arundinacea* Schreb. Underloading of extract leads to spikiness of some bands which run slightly ahead of their normal position. It is unusual for all bands of the same intensity to be affected. Overloading leads to a characteristic dumb-bell shape of the more intense bands or the more serious distortions in Fig. 10. Occasionally, the protein profiles produced by a particular seed line are very faint, possible due to a low nitrogen supply during production of the seed crop. In such cases the volume loaded should be increased so that adjacent profiles are of comparable intensity of staining. A streaking of stain from the origin, which has the effect of blurring the profile (e.g. Fig. 9, lane 11), is caused by insufficient centrifugation during preparation of the protein extract.

4.3 Buffer Concentration

Higher ($\times 1.5$) buffer concentration for the upper tank and gels are routinely used in this system compared with that of Laemmli (1970). This practice has been found to improve resolution and is based on Fling and Gregerson (1986). If the appearance of the bands indicates that some are overloaded but not others, and the problem is not relieved by decreasing the sample volume, a 33% decrease in buffer concentration may assist. This problem can occur in genera where there is a wide variation in band intensity (e.g. *Lotus, Vicia, Astragalus*; see Fig. 10).

4.4 Acrylamide Concentration

The concentration of acrylamide in the running gel should be adjusted so that the bands exhibiting variability between cultivars fall in the top half of the gel, to minimize band "fuzziness". Suitable concentrations range from 11 to 15% according to genus (Table 1).

4.5 Cooling

The seemingly elaborate cooling system described in Section 2.2 has been found to increase resolution in experiments with ryegrass extracts (unpub. data). In the absence of a commercially available electrophoresis apparatus with a cooling core, electrophoresis runs may be extended over a longer time period (e.g. overnight) in a refrigerator or a cold room. Our early experiments were performed using laboratory-constructed equipment (Slack et al. 1985) under such conditions.

5 Stability of Profiles

The consistency in seed protein profiles between generations is demonstrated in Fig. 3 for several cultivars of white clover including two New Zealand multiplied lines of each of the British cultivars Menna and Olwen as well as five generations of Grasslands Pitau. The lack of influence of country of origin or generation on banding profiles observed here is consistent with results reported previously for *L. perenne* (Ferguson and Grabe 1986; Gardiner et al. 1986) and

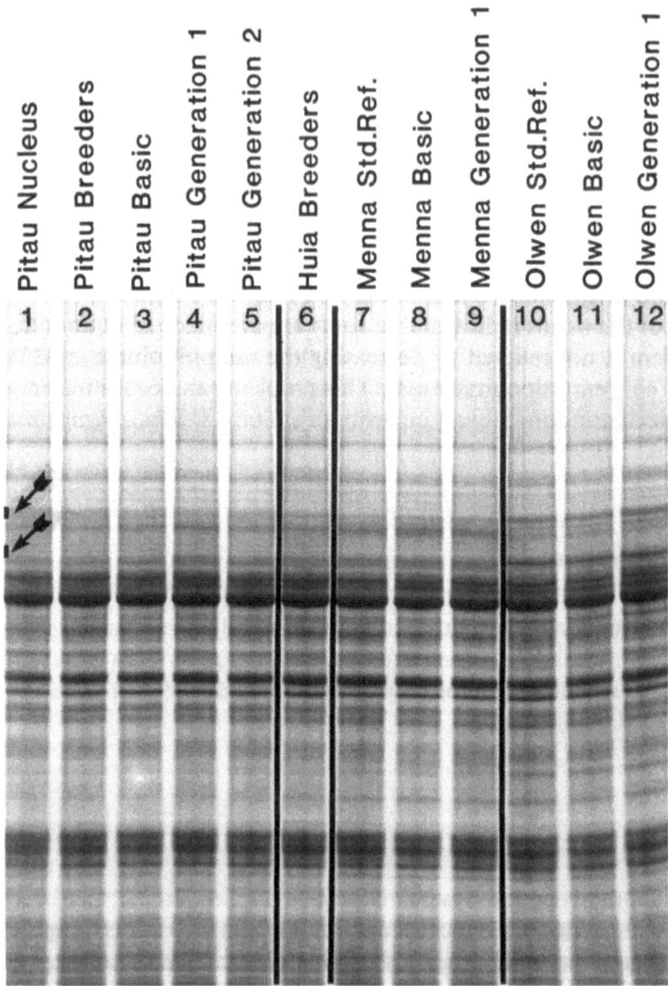

Fig. 3. Stability of banding profiles of seed proteins of cultivars of *T. repens* over several generations and in different countries of production. The seed in *lanes 7* and *10* was produced in Great Britain, the rest in New Zealand. The *arrows* indicate differences in the profiles between Grasslands Pitau and Grasslands Huia

Festuca arundinacea (Gardiner and Forde 1987) and the profiles in Fig. 5 for Grasslands Apanui. Some of the samples in Fig. 3 had shown unusually great variability in plot trials, though not enough to exclude the seed from certification. The profiles confirmed the identity of the samples, i.e. seed not excluded from certification by plot trials would not be excluded by SDS-PAGE. Note that although Grasslands Huia and Grasslands Pitau are very closely related (Pitau is a backcross progeny of Huia), their profiles are distinguishable (see arrows indicating bands of interest).

6 Examples of Typical Results

6.1 Grasses

A "best-case" example of the utility of SDS-PAGE for cultivar identification (Gardiner et al. 1986; Gardiner and Forde 1987) is given by Fig. 4 which dem-

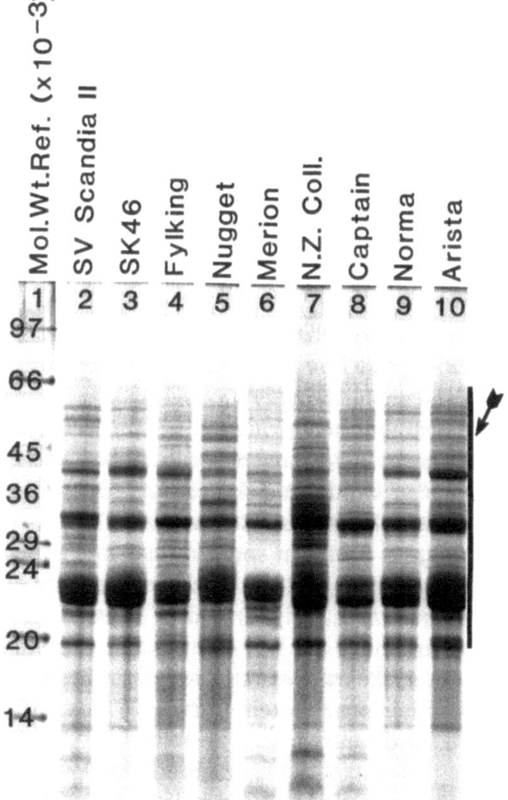

Fig. 4. Profiles of seed protein bands following SDS-PAGE of proteins extracted from a range of cultivars and one wild collection of *Poa pratensis*. The countries of origin are as follows: *lane 2* Sweden; *3* Poland; *4* Canada; *5, 6* USA; *7* New Zealand; *8, 10* Netherlands; *9* Denmark. The region of variable banding is indicated

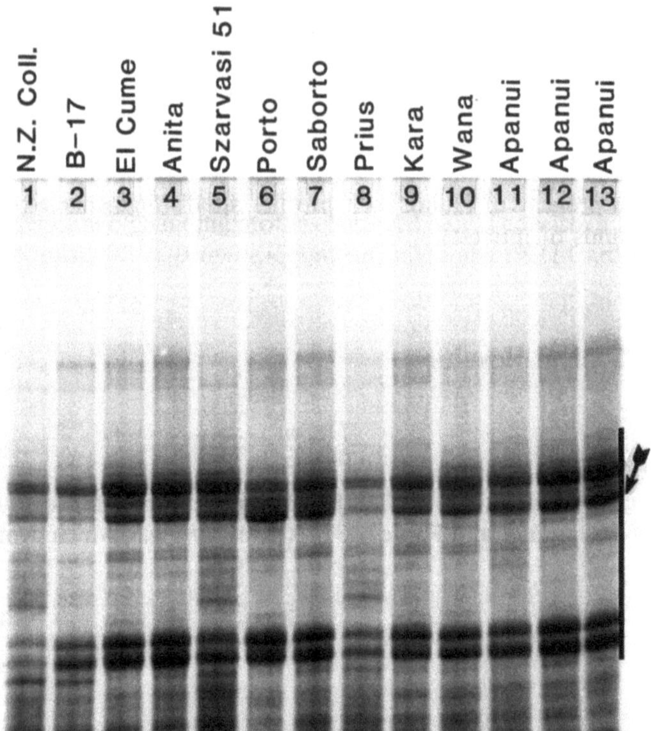

Fig. 5. The protein banding patterns given by extracts of bulk seed samples of 11 different cultivars and accessions of *Dactylis glomerata*. The three lines of Grasslands Apanui in *lanes 11*, *12* and *13* are respectively Breeders, 1979, Nucleus, 1981 and Breeders, 1981. The region of variable banding is indicated

onstrates the typical range of variation between banding patterns for a world selection of cultivars of the popular turf grass *Poa pratensis* L. Differences are detectable over a large proportion of the profile MW 19000 – 66000) and allow such cultivars to be unequivocally distinguished from the others, with the greatest differences evident between the variety in lane 6, a wild New Zealand collection displaying very large number of bands, and the bred cultivars from Europe, USA and Scandinavia. Similar cultivar distinctness with differences exhibited over an even wider MW range is found in some species of *Festuca* (Gardiner et al 1986; Gardiner and Forde 1987, 1988b). Such profiles may be contrasted with those of *Dactylis glomerata* L. (Fig. 5) and of *Agrostis* ssp. (Fig. 6), where the bands exhibiting variability are less distinct and clustered between MW 20000 and 45000. However, the cultivars and selections illustrated are still readily distinguishable by their seed protein profiles.

We have found the occasional species of grass (e.g. *Cynosurus cristatus* L. and *Paspalum dilatatum* L.) which is intractable to differentiation by this technique. *Cynosurus cristatus* seed produces profiles with clear bands but there

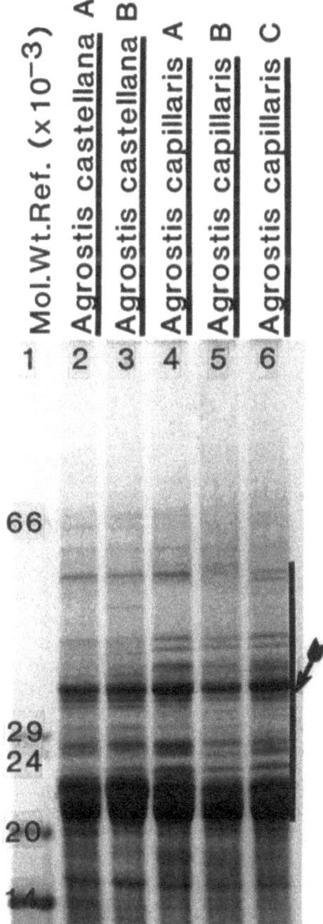

Fig. 6. SDS-PAGE seed protein banding profiles for two selections of *Agrostis castellana* Boiss. et Reut. and three cultivars of *Agrostis capillaris* L (*A, B, C* are respectively Grasslands Sefton, Bardot and Grasslands Egmont). The region of variable banding is indicated

are few observable differences. In the panicoid species *P. dilatatum* the problem appears to be different and due to the small amount of stored seed protein relative to the husk, giving rise to very faint and inconsistent banding profiles following SDS-PAGE analysis of seeds, even with high loadings (Table 1). It would be interesting to investigate whether poor banding of seed proteins using SDS-PAGE is a general feature of panicoid grasses, which usually have lower viability as seed crops compared with the temperate festucoid grasses.

6.2 Legumes

Figure 7 demonstrates banding profiles for a number of pasture legumes. Cultivars of *Trifolium subterraneum* and *T. pratense* show the most intercultivar variation in seed protein banding profiles. For *T. pratense* cultivars, proteins exhibiting variability are clustered near the top of the profile, but these proteins are more evenly distributed in the *T. subterraneum* profiles.

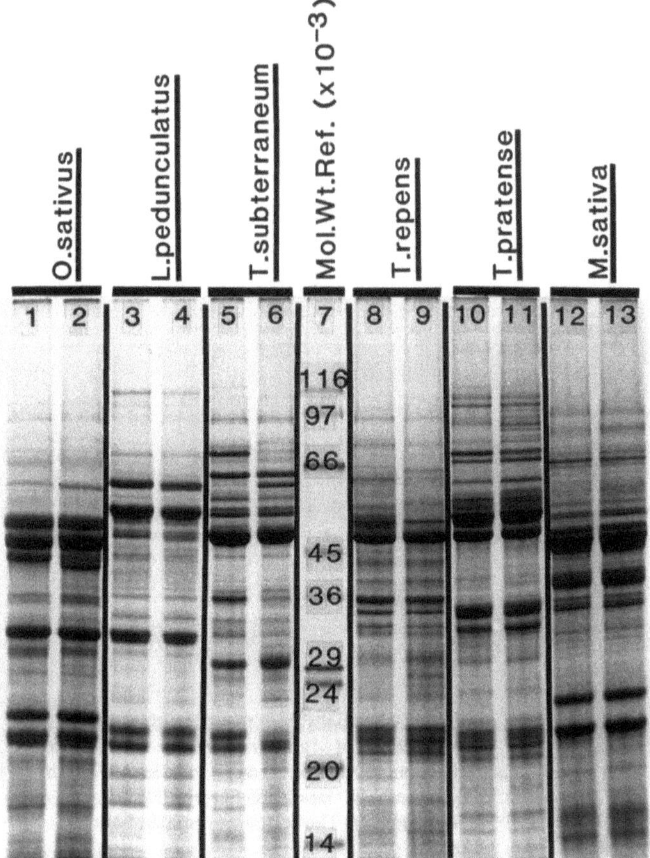

Fig. 7. Comparison of SDS-PAGE banding patterns of seed proteins from a range of pasture legumes. Cultivars are as follows: *Ornithopus sativus*, Grasslands Koha, Vinar; *Lotus pedunculatus*, Grasslands Maku, a line ex Chile; *Trifolium subterraneum*, Dalkeith, Dwalganup; *T. repens*, Grasslands Huia, Ross; *T. pratense*, Redwest, Grasslands Hamua, *Medicago sativa*, Boreal, Pioneer

Cultivars of *T. repens, Lotus pedunculatus* Cav. and *Ornithopus sativus* Brot. exhibit less variability in seed protein banding pattern between cultivars, but they are still distinguishable when clear profiles such as these are obtained. The variable bands are of different sizes in different genera and hence located in different regions of the gel. Profiles for two cultivars of *Medicago sativa* L. are included to illustrate the case of a legume for which cultivar differentiation by SDS-PAGE of seed proteins does not seem feasible. Although the two cultivars demonstrated exhibit a difference in banding of proteins around MW 45000 daltons, a range of other cultivars were not distinguishable (unpubl. results). A comparison of profiles from seed of two cultivars of *Stylosanthes hamata* (L.) Taub. (A. Vieritz pers. comm., Gardiner, unpubl.) suggest that

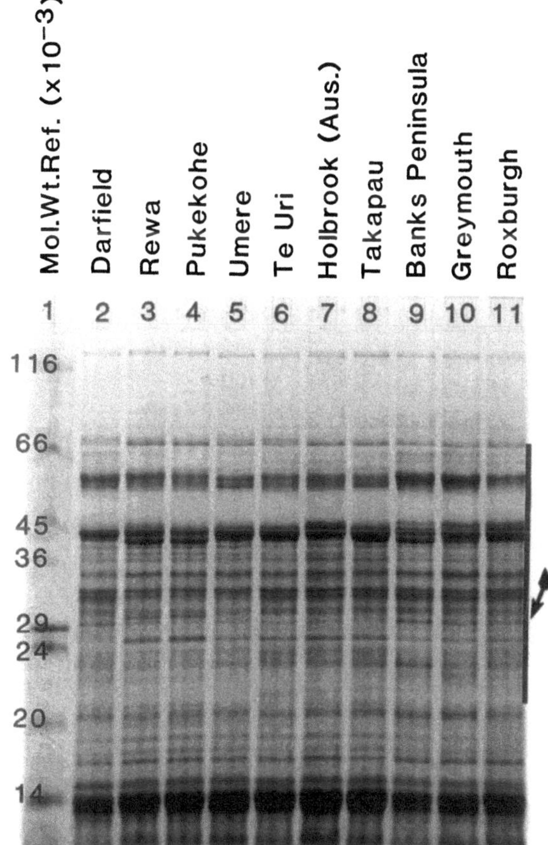

Fig. 8. A survey of SDS-PAGE profiles of seed extracts from different accessions of *Chamaecytisus palmensis*. Accessions were derived from collections in different parts of New Zealand, except that in *lane 7* which was from Australia. The region of banding variation between the profiles is indicated

there may be a similar problem with this tropical legume. It is possible that DNA-based techniques such as fingerprinting with the M13 probe (Zimmerman et al. 1989) may prove more suitable for such recalcitrant subjects where by the present technique polymorphism is barely evident.

We have recently extended our scope to include the forage shrub *Chamaecytisus palmensis* (Christ) Hutch. Figure 8, which demonstrates the results from a survey of different accessions from New Zealand and Australia, suggests that SDS-PAGE of seed proteins will be useful for identifying cultivars of this species.

7 Practical Applications

The technique was developed primarily as a further descriptor for use in supplementing morphological descriptors for the purposes of Plant Variety Rights

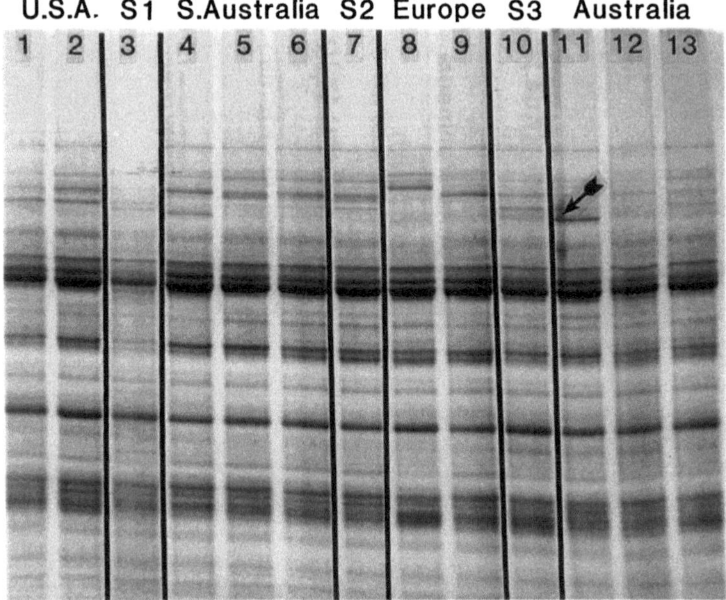

Fig. 9. Variation in seed protein profiles for *Trifolium michelianum* selections (*S1*, *S2*, *S3*) and accessions obtained from sources in USA, Australia and Europe. The *arrow* indicates an artefact caused by insufficient clearing of the protein extract by centrifugation. (Forde and Gardiner 1990)

applications. We have found it of great use for this purpose and routinely supply photographs of SDS-PAGE analyses with such applications.

Banding profiles may also be used by breeders for confirming hybridity following interspecific crosses. Examples are illustrated in Gardiner and Forde (1988a) for such crosses within *Lotus* and *Ornithopus*, and the breeding of a hybrid ryegrass and its backcross progeny is tracked in Gardiner et al. (1986).

We have also used SDS-PAGE to estimate the diversity between accessions of species in our genetic resource collection. An example is *Trifolium michelianum* Savi, a comparatively rare wild annual forage legume also known as *T. balansae* Boiss. in part of its range. Figure 9 is an attempt to survey variation in seed protein banding patterns in a small number of wild collections and match these up with each other, and with selections made by breeders in New Zealand (lanes 3 and 10) and an Australian cultivar (lane 7). There are clear similarities between some profiles for accessions from the wild, indicating some duplications within the genetic resource collection, e.g. lanes 1 and 2, lanes 5 and 6 and also lanes 12 and 13. The selection in lane 7 appears closely related to the accessions in lanes 5 and 6 and the selection in lane 10 to the accession in lane 11.

We have recently investigated the technique for its utility in classifying accessions of the genera *Vicia* and *Astragalus* (Forde and Gardiner 1990). Accessions of *Astragalus* collected in the Caucasus (Fig. 10) were identified as

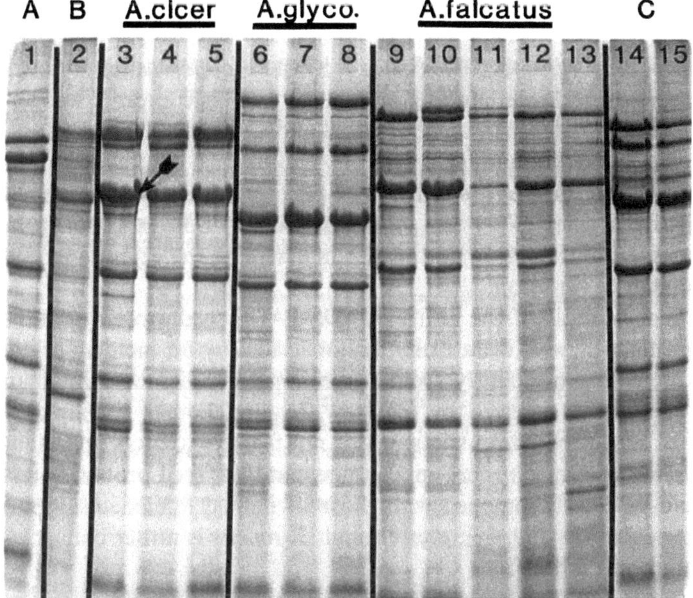

Fig. 10. Identification of accessions of *Astragalus* spp. using SDS-PAGE profiles of seed proteins. *Lanes 3, 6* and *9* contain seed proteins from standard reference accessions for *A. cicer*, *A. glycophyllos* and *A. falcatus* respectively, and the other lanes proteins from collections in the Caucasus. The species labelled *A*, *B* and *C* were not identified. The *arrow* indicates band distortion indicative of considerable overloading. (Forde and Gardiner 1990)

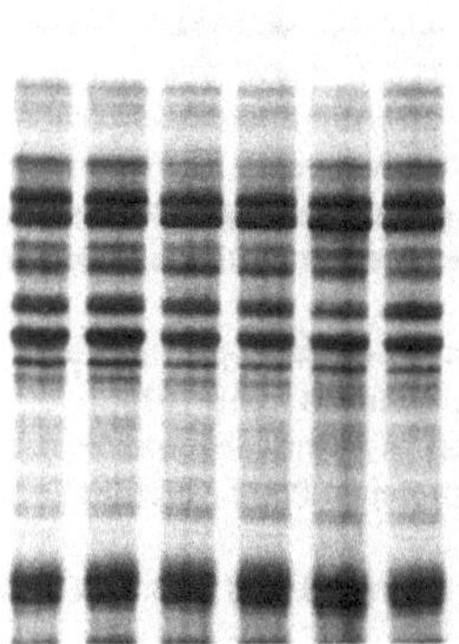

Fig. 11. Use of SDS-PAGE profiles of seed proteins for identification of three samples of *Trifolium repens* seed. *Lanes 1, 3,* and *5* contain extracts of reference samples for the three named cultivars, and *lanes 2, 4* and *6* extracts from unlabelled samples. The profiles match in pairs, 1 with 2, 3 with 4, and 5 with 6

A. cicer L. (lanes 4, 5), *A. glycophyllos* (lanes 7, 8) and *A. falcatus* Lam. (lanes 10–13) by comparison with the banding profiles of reference lines. It was apparent that three other unidentified species (A, B and C; lanes 1, 2, and 14 and 15) were also represented. Individual accessions within each species were distinguishable. Population studies have also been performed of the endangered New Zealand native species *Clianthus puniceus* (D.Don) Sol. ex Lindl. (unpubl. results) to assist conservation authorities in assessing the extent and distribution of its genetic variation.

A final practical example of the utility of SDS-PAGE seed protein profiles for cultivar identification was beautifully demonstrated by an incident when varietal names were lost accidently from three large samples of American Ladino clover (*T. repens*) received by the New Zealand Forage Germplasm Centre in 1990 for seed production trials. It was not feasible to grow and identify these similar varieties by physical appearance, but their seed protein banding profiles could be clearly matched with those of named reference samples of the same three cultivars (Fig. 11), even though *T. repens* is rather conservative in band variation (Table 1).

Acknowledgments: We thank Linda Muir and Lisa Cambridge for capable technical assistance.

References

Biorad (1984) Acrylamide polymerisation – a practical approach. Bulletin 1156:1–7

Blakesley RW, Boezi JA (1977) A new staining technique for proteins in polyacrylamide gels using Coomassie Brilliant Blue G250. Anal Biochem 82:580–582

Clark KW, Hussain A, Bamford K, Bushuk W (1989) Identification of cultivars of *Agrostis* species by polyacrylamide gel electrophoresis of seed proteins. In: Takatoh H (ed) Proc 6th Int Turfgrass Research Conference. Tokyo, Japan, pp 121–125

Cooke RJ (1984) The characterisation and identification of crop cultivars by electrophoresis. Electrophoresis 5:59–72

Cooke RJ (1989) The use of electrophoresis for the distinctness testing of varieties of autogamous species. Plant Varieties Seeds 2:3–13

Ferguson JM, Grabe DF (1984) Separation of annual and perennial species of ryegrass by gel elctrophoresis of seed proteins. J Seed Technol 9:137–149

Ferguson JM, Grabe DF (1986) Identification of cultivars of perennial ryegrass by SDS-PAGE of seed proteins. Crop Sci 26:170–176

Fling SF, Gregerson DS (1986) Peptide and protein molecular weight determination by electrophoresis using a high-molarity Tris buffer system without urea. Anal Biochem 155:83–88

Forde MB, Gardiner SE (1991) Electrophoretic seed protein banding patterns as genetic resource descriptors. In: Zakri AH et al (eds) Conservation of plant genetic resources through in vitro methods. FRIM/MNCPGR, Kuala Lumpur, pp 95–108

Gardiner SE, Forde MB (1987) SDS polyacrylamide gel electrophoresis of grass seed proteins: a method for cultivar identification of pasture grasses. Seed Sci Technol 15:663–674

Gardiner SE, Forde MB (1988a) Identification of cultivars and species of pasture legumes by sodium dodecylsulphate polyacrylamide gel electrophoresis of seed proteins. Plant Varieties Seeds 1:13–26

Gardiner SE, Forde MB (1988b) Fingerprinting pasture species. In: Zakri AH (ed) Plant breeding and genetic engineering. SABRAO, Malaysia, pp 353–359

Gardiner SE, Forde MB, Slack CR (1986) Grass cultivar identification by sodium dodecylsulphate polyacrylamide gel electrophoresis. NZ J Agric Res 29:193–206

Gilliland TJ (1989) Electrophoresis of sexually and vegetatively propagated cultivars of allogamous species. Plant Varieties Seeds 2:15–25

Gilliland TJ, Camlin MS, Wright CE (1982) Evaluation of phosphoglucosomerase allozyme electrophoresis for the identification and registration of cultivars of perennial ryegrass (*Lolium perenne*) Seed Sci Technol 10:415–430

Laemmli UK (1970) Cleavage of structural proteins during the assembly of the head of bacteriophage T4. Nature 227:680–685

Luthe DS (1987) Storage protein synthesis during oat (*Avena sativa* L.) seed development. Plant Physiol 84:339–340

Slack CR, Hancock DA, Griffin WB, McEwan JM (1985) Separation of proteins from grain of New Zealand grown wheat and barley varieties by sodium dodecylsulphate polyacrylamide gel electrophoresis (SDS-PAGE). Plant Physiology Division, DSIR, Tech Rep 21:1–19

Smith DB, Payne PI (1984) A procedure for the routine determination of electrophoretic band patterns of barley and malt endosperm proteins. J Nat Inst Agric Bot 16:487–498

Zimmerman PA, Lang-Unnasch N, Cullis CA (1989) Polymorphic regions in plant genomes detected by an M13 probe. Genome 32:824–828

Analysis of Seed Proteins, Isozymes, and RFLPs for Genetic and Evolutionary Studies in *Phaseolus*

P. Gepts, V. Llaca, R. O. Nodari, and L. Panella

1 General Introduction

Common bean (*Phaseolus vulgaris* L.) has been used repeatedly as an experimental organism to derive important concepts in genetics. After discovering the rules of segregation and independent assortment governing the inheritance of heritable traits in pea (*Pisum sativum* L.), Mendel (1865) confirmed his observations on peas in crosses involving *P. nanus* (nowadays bush *P. vulgaris*) and *P. vulgaris* (nowadays climbing *P. vulgaris*). Johannsen (1909) was instrumental in introducing the distinction between phenotype and genotype following his investigations of seed size in pure lines of *P. vulgaris*. The first example to our knowledge of a linkage between a marker locus and a quantitative trait locus (QTL) was provided by Sax (1923) who established an association between a locus for seed pigmentation and a seed size factor in common bean. More recently, one of the first plant genes to be cloned was a gene for phaseolin seed protein (Sun et al. 1981). A comparison of the nucleotide sequence of cDNA and genomic clones of phaseolin established the presence of intervening sequences in higher plants in addition to other eukaryotic organisms. Our understanding of how plants defend themselves against pathogens has been considerably enhanced by the studies of Lamb and collaborators (reviewed in Dixon and Lamb 1990) on the relationship between common bean and *Colletotrichum lindemuthianum*, causal agent of anthracnose.

Common bean also constitutes an excellent model for evolutionary studies. The species consists of two major subdivisions, which are geographically separated (i.e., Mesoamerican vs. Andean) and show divergence for a wide range of characters, including morphological, agronomic, molecular, and adaptation traits. These two subdivisions are in the process of speciation as suggested by the simple genetic control of reproductive isolation between Mesoamerican and Andean genotypes (Shii et al. 1980; Gepts and Bliss 1985; Gepts 1988b, 1990a; Gepts and Debouck 1991). Hence, common bean can be used to investigate such speciation issues as the presence and configuration of hybrid zones and the molecular basis of reproductive isolation mechanisms. Furthermore, there is increasing evidence that domestication has induced a strong reduction in diversity at the molecular level in *Phaseolus* beans (e.g., Gepts et al. 1986; Schinkel and Gepts 1988). This reduction contrasts with the increase in diversity observed for morphological traits during and after domestication. Hence, common bean can also be used to study the apparent uncoupling between molecular and phenotypic evolution.

The study of several of these issues has been and will continue to be substantially enhanced through the use of molecular markers. We include among these molecular markers seed proteins, isozymes, and RFLPs. The qualitative analysis of variation for these markers by various electrophoretic techniques provides an assessment of the degree of genetic relatedness (i.e., common ancestry) and levels of genetic diversity free of environmental influence in most cases. In addition, electrophoretic techniques allow us to analyze the larger number of samples necessary in evolution and population studies.

In this review, we describe the procedures and protocols adopted or developed in our laboratory to analyze seed proteins, isozyme, and RFLP variation primarily in *Phaseolus*. We have worked primarily with *P. vulgaris* but our methodology has also been applied in other *Phaseolus* species and in cowpea (*Vigna unguiculata*). For each of these marker types, we briefly illustrate how they have been applied to investigate evolutionary concepts. We end our review by a comparison of the relative merits of the three types of markers as evolutionary markers.

2 Seed Protein Analysis

2.1 Introduction

Seed proteins of common bean contain predominantly globulins, which are salt-soluble (Osborne 1907). Among globulins, phaseolin is the major constituent followed by lectins or phytohemagglutinins, which account for 35% – 50% and 5% – 10% of total bean seed protein, respectively (Ma and Bliss 1978; Osborn et al. 1985). Phaseolin, also called globulin-1 (G-1: McLeester et al. 1973) or glycoprotein II (Pusztai and Watt 1970), is a 7S vicilin-like protein with homologies to vicilin of *Pisum sativum* and *Vicia faba*, β-conglycinin in *Glycine max*, and β-conglutin in *Lupinus angustifolius* (reviewed in Gepts 1990b). It undergoes a pH-dependent, reversible association between polypeptide subunits of approximately 44 kD (3S) at pH 12, protomers of 163 kD (7.1S) at pH 7, and tetramers of protomers of 653 kD (18.2S) at pH 3.6. The polypeptide subunits range in molecular weight between 43 and 54 kD and in isoelectric point between pH 5.6 and 5.8. Qualitative variation as assessed by electrophoresis is controlled by a single locus (*Phs*) exhibiting codominant expression in heterozygotes (Brown et al. 1981 a, b). The *Phs* locus is actually a complex locus constituted by six to eight tightly linked genes belonging to the α and β subfamilies of the phaseolin multigene family (Talbot et al. 1984; Slightom et al. 1985).

Seed lectin, also known as phytohemagglutinin (e.g., Staswick et al. 1986) or glycoprotein I (Pusztai 1966), has polypeptide subunits with molecular weights of 29 to 36 kD and isoelectric points between pH 4.9 and 7.9. In its native conformation, lectin is a tetramer with molecular weight ranging between 85 and 150 kD. Lectin qualitative electrophoretic variation is coded by a single locus (*Lec*) consisting of two tightly linked genes (Brown et al. 1981 a; Hoffman and Donaldson 1985; Osborn and Bliss 1985). The *Lec* locus is unlinked to the *Phs* locus (Brown et al. 1981 a).

Arcelin, a seed protein first described by Romero-Andreas et al. (1986), is found in both the globulin and albumin fractions (Osborn 1988). It has polypeptide subunits of molecular weight ranging between 35 and 42 kD and isoelectric points that are more basic than those of phaseolin and lectin. It exhibits 80% nucleotide sequence similarity with lectin (Osborn et al. 1988). The arcelin locus (*Arl*) is tightly linked to the lectin locus (*Lec:* Osborn et al. 1986). For further information on phaseolin, lectin, and arcelin the reader is referred to reviews by Bliss and Brown (1983) and Osborn (1988).

Numerous analytical methods have been used to characterize seed proteins. Of these, electrophoretic techniques have been used most frequently to characterize seed protein variation because of their rapidity, low, cost, and capacity to handle a large number of samples. Stegemann and Pietsch (1983) have outlined the major electrophoretic separation methods, including polyacrylamide gel electrophoresis in one and two dimensions. In the next paragraphs, we will describe these methods of polyacrylamide gel electrophoresis as they are applied specifically to bean seed proteins in our lab. A more general overview of gel electrophoresis of proteins can be found in Hames and Rickwood (1981) and manufacturers of gel apparatuses also provide protocols similar to the ones presented here but with quantities adjusted for the size of their equipment.

2.2 One-Dimensional Sodium Dodecyl Sulfate Polyacrylamide Gel Electrophoresis (SDS-PAGE)

Most electrophoretic analyses of common bean and cowpea seed proteins have used a discontinuous sodium dodecyl sulfate (SDS) polyacrylamide slab gel electrophoresis system as described by Laemmli (1970) because of its high resolving power and the possibility of measuring the molecular weight of individual polypeptides. Polyacrylamide is the preferred separation matrix because it offers fewer problems of reproducibility, is chemically inert and stable over a wide range of pH, temperature, and ionic strength, and it offers a wider range of pore sizes. The discontinuous system is a system in which proteins migrate first through a large-pore "stacking" gel and second through a small-pore "running" gel. When moving through the stacking gel (3% polyacrylamide; pH 6.8), proteins form thin bands (or "stacks") one above the other in order of decreasing mobility; when they reach the running gel (12%–15% polyacrylamide; pH 8.8), the proteins encounter a higher pH and much smaller pore sizes, hence, they will migrate according to their intrinsic charge and size. The discontinuous system results in much greater resolution of individual protein bands in the running gel and has, therefore, become the preferred system.

SDS is an ionic detergent which will bind to proteins in a constant ratio of 1.4 g SDS/g protein and confers a negative charge. As a consequence, migration will be proportional to polypeptide size and not charge density. The added advantage of SDS is that minute amounts of protein (of the order of a μg) are necessary. An alternative dissociating agent is urea at high concentrations (approx. 8 M); unlike SDS, however, it does not affect charge.

2.2.1 Protein Sample Preparation

A flour sample is prepared by removing a portion of the raphe end of the seed, which is the nonembryo end of the seed. If needed for further progeny studies, the seed can thus be saved and germinated. Removal of the raphe end can take place in several ways. It can either be scraped with the edge of a razor blade directly into a 1.5-ml microfuge tube or it can be cut with a razor blade and crushed between glassine paper with a pestle. In either case the presence of the seed coat does not interfere with electrophoretic separation. When seeds are very small, such as seeds of wild relatives, the entire seeds may have to be analyzed destructively by crushing them with a pestle between glassine paper.

The flour sample is suspended for at least 30 min, but usually overnight, in a mixture of equal volumes of 0.5 M NaCl pH 2.4 and cracking buffer (0.0625 M Tris-HCl, pH 6.8; 2 mM EDTA; 2% w/v SDS; 40% w/v sucrose; 1% v/v 2-mercaptoethanol, 0.01% w/v bromophenol blue; Brown et al. 1981 b), to dissolve the seed proteins. Approximately, 25 to 30 mg of flour are suspended in 150 μl of the NaCl − cracking buffer mixture. Protein samples, extracted with cracking buffer only, provide satisfactory analyses as well. The capped tube with the mixture is heated at 100 °C for 10 min and then centrifuged at 15 000 g for 5 min at room temperature. Five to 8 μl of the sample supernatant is loaded on the gel.

2.2.2 Gel Assembly

Discontinuous gel separation can only be done with a vertical electrophoresis apparatus, of which several designs exist. Usually slab gels are used because they can accommodate a larger number of samples. Besides commercially available slab gel apparatuses such as the Bio-Rad Protean II (Bio-Rad, Richmond, CA), a popular design has been the Studier (1973) model which can be easily and cheaply made in a laboratory workshop. For actual dimensions, the reader is referred to Hames (1981) or Schuler and Zielinski (1989).

Two glass plates of appropriate dimensions to fit the gel apparatus are used to mold the gel. One plate has a rectangular shape, whereas the other plate of the same overall dimensions has a 2.5-cm notch along one of the long sides leaving two 1-cm-wide "rabbit ears" at the edge of the plate (e.g., Hames 1981). Both plates are cleaned with 95% ethanol to remove any trace of protein and after drying are assembled to form a "sandwich". The plates are held at

Table 1. Composition of running and stacking gel solutions for one-dimensional SDS polyacrylamide gel electrophoresis

Running gel: 30 ml of solution for a 16 cm × 11 cm × 0.75 mm gel

Stock solution	13% w/v Acrylamide [ml]	15% Acrylamide [ml]
Deionized H_2O	13.44	11.44
Acrylamide : bisacrylamide, 30 : 0.15	13.00	15.00
Tris-HCl 3.5 M, pH 6.8	4.00	4.00
SDS 20%	0.20	0.20
EDTA 0.5 M	0.16	0.16
Mix well and add		
Ammonium persulfate 7.5% (*prepared freshly*)	0.32	0.32
N,N,N′,N′-tetraethylenediamine (TEMED)	0.04	0.04

Stacking gel

Stock solution	Aliquot
	[ml]
Deionized H_2O	4.40
Acrylamide : bisacrylamide, 30 : 1.5	0.50
Tris-HCl 1.25 M, pH 6.8	0.50
	[µl]
SDS 20%	25
EDTA 0.5 M	25
Ammonium persulfate 7.5%	80
N,N,N′,N′-tetraethylenediamine (TEMED)	5
Optional: trace of bromophenol blue	

a correct distance with spacers of uniform thickness (between 0.75 and 1.5 mm) and clamped together with binder clips on each side. The clamps are positioned in such a way that they will press on top of the spacers. To avoid any leakage of acrylamide during gel polymerization between the spacer and the glass plates, both sides of the spacers are coated with stopcock grease.

One ml of running gel solution (Table 1) is injected with a syringe into the bottom of the gel mold in order to establish an impermeable bottom border. An additional 4 µl of ammonium persulfate 7.5% is added along the spacers at the bottom of the gel to help polymerization of this bottom seal of the running gel. (Care should be exercised not to add too much stopcock grease on the spacers because the excess grease may interfere with polymerization.)

After polymerization of the bottom seal, the remainder of the running gel is poured on top of the bottom border, leaving a space of about 3.5 cm, half of which is filled with H_2O-saturated n-butanol to obtain a straight to gel edge. The gel is allowed to set for at least 45 min and may be left overnight.

Following polymerization of the running gel as shown by the presence of an interface, the butanol layer is removed and the space on top of the running gel is thoroughly rinsed with distilled water (three or four times) and dried with paper towels. Care should be exercised not to disturb the top edge of the running gel in this process as this will affect subsequent migration. The stacking gel (Table 1) is then poured on top of the running gel and a comb is inserted to form the wells. Air bubbles under the comb teeth should be removed. The upper corners of the comb are clamped with the uppermost glass plate. The stacking gel takes 45 to 60 min to polymerize.

The combs are carefully removed under running distilled water and the wells are rinsed to eliminate any excess acrylamide, after which excess water is also removed. Any well separations that have been disturbed can be repositioned using a needle or a spatula. The glass plates are placed in the gel apparatus and clamped with binder clips. To avoid leakage of the upper chamber buffer, the area around the apparatus notch is coated with stopcock grease. The position of the wells is marked with a felt-tip pen on the glass plates before adding the upper chamber buffer. Alternatively, a small amount of dye can be added to the stacking gel mixture before polymerization (Table 1). A small spatula can be used to clear any air bubbles in the wells.

2.2.3 Sample Loading and Running

Running buffer (0.025 M Tris; 0.192 glycine; 0.1% SDS; 0.002 M EDTA) is added in sufficient quantity to cover the wells in the top chamber by approximately 1 cm; at this time, one should check for any leaks between the apparatus and the gel plates. If none are found, running buffer is also added to cover the bottom of the gel in the lower chamber. The electrophoretic apparatus is then connected to a power supply, with the cathode connected to the top and anode to the bottom chamber. Initially, 7.5 mA per gel is used during gel loading. After all samples have been loaded, the current is increased to 12.5 mA per gel for 1 h or until the continuous blue dye line across the gel, formed by the loaded protein samples, reaches the running gel. For the remainder of the running time (approximately 4 h), a constant current of 30 mA per gel is used until the bromophenol blue dye line reaches the bottom of the running gel. If there is a provision for cooling the gel during operation, a current of up to 50 mA per gel may be used. The cooling of the gel will provide a better resolution of the bands.

2.2.4 Gel Staining and Drying

The glass plates are carefully pried apart and the gel is lifted from the plates and transferred to a glass dish containing Coomassie Brilliant Blue stain (0.15% w/v Coomassie Brilliant Blue R250; 45% v/v methanol; 9% v/v glacial acetic acid). The dish is placed on a shaker for overnight staining. The next day, the stain is replaced with destain solution (45% v/v methanol; 9% v/v glacial acetic acid) and after 5 to 8 h additional shaking (the destain solution can be changed if needed), the gel can be evaluated and photographed.

For permanent storage the gel can be dried according to two methods. In the first one, the gel is air dried by placing it between two sheets of cellophane and tightly clamping it in a plastic frame. Alternatively, the gel can be vacuum dried. In this case, both the gel and the cellophane are soaked overnight in a solution of 2% glycerol and 2% acetic acid on a shaker. If the soaking causes the gel to expand too much, it need not be soaked (but the cellophane must be). The next day, the gel is allowed to dry between the cellophane sheets during 2 h at 80 °C under vacuum. To avoid any acid damage to the pipes, the gel drier exhaust should be connected to a cold trap containing dry ice or liquid nitrogen.

Either before or after drying, molecular weights of the polypeptides can be determined according to the method of Weber and Osborn (1969). Molecular weight marker proteins can be purchased commercially and include, for example, phosphorylase B (94000 daltons), bovine serum albumin (68000 daltons), catalase (58000 daltons), fumarase (49000 daltons), aldolase (40000 daltons), malate dehydrogenase (34000 daltons), and soybean trypsin inhibitor (21000 daltons).

2.3 Two-Dimensional Polyacrylamide Gel Electrophoresis

Proteins can be separated either by mobility determined by size and net charge (as outlined in the previous section) or, alternatively, by equilibrium determined primarily by charge (electrofocusing). With both methods of separation, a given band can consist of different co-migrating proteins. It order to decrease the probability of this coccurrence, two-dimensional electrophoretic methods have been devised that will separate proteins at right angles according to different criteria. The two-dimensional method most used in bean seed protein analysis has been that of O'Farrell (1975) in which proteins are separated by isoelectric focusing in the first dimension and by SDS polyacrylamide gel electrophoresis (SDS/PAGE) in the second dimension. Separation in the first dimension provides an estimate of the isoelectric point (pI), whereas separation in the second dimension provides an estimate of the molecular weight. The specific methodology used to analyze common bean seed protein was first established by Ma and Bliss (1978) and Brown et al. (1981 b) and will be described in the next sections.

2.3.1 Protein Sample Preparation

A sample of flour from the raphe end of the seed is suspended in 0.5 M NaCl pH 2.4 for at least 30 min to dissolve seed proteins. After centrifugation at 14000 g for 5 min, the supernatant is mixed 1:1 with sample or lysis buffer (9.5 M urea, 2% w/v NP-40, 5% β-mercaptoethanol, 1.6% pH 5−7 ampholines, and 0.4% pH 3.5−10 ampholines.) Protein samples can be stored at −20 or −70 °C.

2.3.2 First Dimension

The first dimension is run in tube gels of 1–2 mm diameter or less. The tubes
are usually made of glass and should be extremely clean. We have used a 1:1
mixture of xylene and Micro detergent (Cole-Parmer), followed by rinses with
distilled H_2O and a 1:600 Photoflo solution. The bottom end of each tubes
is tightly wrapped in parafilm to avoid leakage of the gel solution and the tubes
are maintained in a vertical position during the entire polymerization procedure. The gel solution (Table 2) is injected into the tubes using a syringe to
which narrow plastic tubing has been attached. Care is exercised not to trap
air bubbles. The top surface of the gel solution in each tube is evened out to
obtain first dimension gels of similar length. Each tube gel is overlayed with
a few drops of distilled H_2O and left to polymerize for 45–60 min.

The distilled H_2O and the parafilm are discarded. The bottom end of each
tube is covered with dialysis tubing secured with rubber bands while care is tak-
en not to trap air bubbles. The tubes are placed in the first dimension gel apparatus with the bottom end of the tubes extending into the acidic buffer of the
lower chamber (0.01 M H_3PO_4). The top end of the gels is overlayed with
15 µl 8 M urea and the tubes are filled with the basic buffer of the upper chamber (0.02 M NaOH), after which the upper chamber is then filled with basic
buffer and checked for leaks. The pH gradient in the tube gels is created by
a pre-electrophoresis consisting of 30 min each at 100, 150, and 200 V (constant voltage).

After the pre-electrophoresis, the upper buffer chamber is discarded and the
top of the tubes is rinsed with distilled H_2O. The samples are loaded and
overlayed with 15 µl of 15% sucrose and 1% ampholines (7:3 v/v ratio of pH
5–7 and pH 3.5–10 ampholines). The tubes are filled with basic buffer solution, after which the upper chamber is also filled with basic buffer. Electrophoresis takes place – usually overnight – for 16 h at 200 V (constant voltage)
followed by 1 h at 400 V (constant voltage).

After electrophoresis, the gels can be stored in their tubes at −70°C until
further analysis in the second dimension or analysis can proceed immediately

Table 2. Composition of gel solution for the isoelectric
focusing dimension of two-dimensional IEF/SDS polyacrylamide gel electrophoresis

Urea	2.91 g
Acrylamide:bisacrylamide, 30:2.3	0.8 ml
NP-40 10%	1.2 ml
Deionized H_2O	0.8 ml
Ampholines pH 5–7	0.2 ml
Ampholines pH 3.5–10	0.1 ml
Ammonium persulfate 7.5%	8 µl
N,N,N′,N′-tetraethylenediamine (TEMED)	6 µl

as follows. Gels are removed from their tubes by injecting distilled H_2O with a needle between the glass and the gel and rotating the tube. Subsequently, gels are placed in equilibration solution (0.0625 M Tris-HCl pH 6.8, 2.3% w/v SDS, 5% v/v β-mercaptoethanol, 10% w/v glycerol) for at least 20 min.

2.3.3 Second Dimension

The second dimension is run in polyacrylamide slab gels, cast as described in Section 2.2, with two major exceptions. The first exception regards the glass plates used to cast the gel. While one plate is a normal, thin (2.5 mm) gel plate, the other gel plate is thicker and beveled at the top so that the first dimension gel can be positioned and sealed into place. The second exception concerns the stacking gel, which is cast to the top of the plates and without well comb in order to assure continuity with the positioned first-dimension gel. Before polymerization of the stacking gel, individual wells can be formed at the side of the stacking gel for reference to one-dimensional gels, in which case samples should be loaded when the blue front of the second dimension reaches the bottom of the wells.

After equilibration of the first-dimension gels, agarose solution (1% agarose, 0.125 M Tris-HCl pH 6.8, 0.01% SDS, trace of bromophenol blue) is poured to fill the gap between the top of the plate and the first-dimension gel. The tube gel is positioned in the bevel and a note is made of the position of the acid end of the gel. After all air bubbles are removed, the top side of the gel is coated with agarose solution. The second dimension is run similarly to the one-dimension gel (Sect. 2.2) and stained with Coomassie Blue. Gels can be dried and stored as described for one-dimensional gels.

2.4 Applications

Phaseolin electrophoretic type has been the first molecular marker used to identify patterns of genetic diversity and domestication in *Phaseolus vulgaris, P. lunatus* (lima bean), and *P. acutifolius* (tepary bean) (Brown et al. 1981 b, 1982 a; Manen and Otoul 1981; Gepts and Bliss 1986; Gepts et al. 1986; Gepts 1988 a; Schinkel and Gepts 1988, Debouck et al. 1989; Koenig et al. 1990; Lioi 1989 a, b). Phaseolin diversity data provided the initial evidence that domestication had induced a sometimes marked bottleneck in genetic diversity and, hence, that the current relative lack of diversity in crop plants traced back several thousands of years ago (Gepts et al. 1986; Schinkel and Gepts 1988; Koenig et al. 1990).

In *P. vulgaris,* phaseolin also suggested that the species consisted of two major gene pools, one from Mesoamerica and the other from the Andes. Figure 1 displays phaseolin diversity among some Mesoamerican and Andean cultivars as revealed by one-dimensional polyaycrylamide gel electrophoresis. Through multiple domestications in the two areas, two major groups of cultivars were

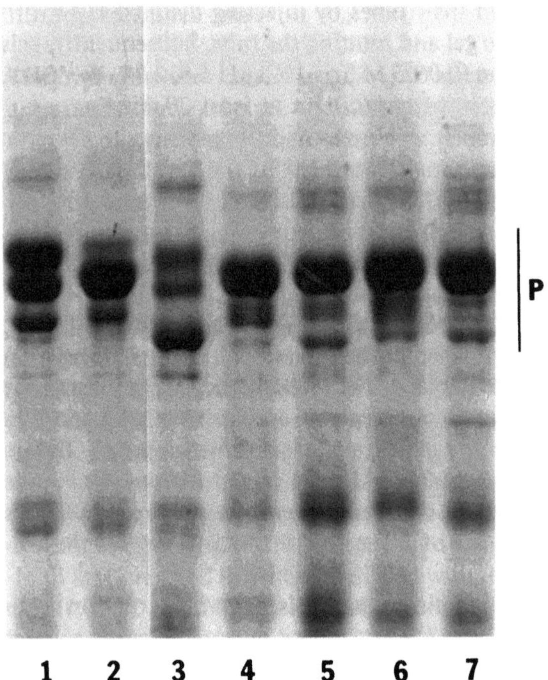

Fig. 1. Genetic diversity for phaseolin in common bean cultivars as determined by one-dimensional SDS polyacrylamide gel electrophoresis. *Lane 1*: "T" phaseolin (cultivar Jalo); *lane 2*: "H" (Cocacho); *lane 3*: "C" (Burrito); *lane 4*: "S" (BAT 93); *lane 5*: "B" (Orgulloso); *lane 6*: "Sb" (Black Turtle Soup); *lane 7*: "Sd" (Mexico 222). *Vertical bar (P)*: molecular weight range of phaseolin polypeptides. T, H, and C are Andean phaseolin types and S, B, Sb, and Sd are Mesoamerican phaseolin types as previously defined on the basis of one- and two-dimensional electrophoresis (Brown et al. 1981b; Gepts et al. 1986; Gepts and Bliss 1986; Koenig et al. 1990)

generated with distinctive agronomic traits (Gepts 1988b, 1990a) and some reproductive isolation (Singh and Guitérrez 1984; Gepts and Bliss 1985).

Lectin variation has also been analyzed by PAGE: 12 electrophoretic variants have been identified (Brown et al. 1981c, 1982b). In addition, a genetic association was identified between qualitative and quantitative lectin variation (Osborn et al. 1985). Arcelin was first identified by PAGE (Osborn et al. 1986; Romero-Andreas et al. 1986) and subsequently was found to possess insecticidal properties (Osborn et al. 1988).

3 Isozyme Analysis

3.1 Introduction

Isozymes have been defined as different variants of the same enzymes, having identical or similar functions, and present in the same individual (Markert and Moller 1959). Since their discovery by Hunter and Markert (1957), they have played an important role in numerous aspects of biological research (Stebbins 1989). In plants, in particular, isozymes have been used principally in population genetic and evolutionary studies (reviewed in Brown 1979; Gottlieb 1981; Tanksley and Orton 1983, Hamrick and Godt 1990). In recent years, several volumes have been published that include general descriptions of isozyme methodology and applications (Tanksley and Orton 1983; Pasteur et al. 1988; Soltis and Soltis 1989; Hillis and Moritz 1990). In this chapter, we will describe our isozyme methodology as applied to *Phaseolus*.

3.2 Preparation

3.2.1 Choice of Plant Tissue

When choosing the plant sample it is important that the enzyme of interest be present in sufficient quantity. Visualization of the enzyme is due to a successfully catalyzed reaction which results either in the deposition of a dye resulting in a colored band or the metabolization of a dye resulting in an achromatic area within the colored gel. In *Phaseolus,* young (2- to 4-week-old) seedlings provide tissue which is rapidly growing and has a large amount of enzymes present. Young tissues (before flowering) also contain less pigmentation than more mature plants and pigmentation can interfere with enzyme function. Only a small amount of tissue is needed (1 to 3 g) and the choice of tissue is generally a matter of trial and error for each species. Some enzymes are present in larger quantities in the roots, leaves, or cotyledons. In general, leaf tissue is easier to use, especially from field-grown plants. If the root tissue is needed, the plants, which may need to grow a little larger to provide enough tissue, can be grown in vermiculite or sand which is easily washed away without harming the plant. This is especially true of the small seeded wild taxa. The seed itself may also be used. It is imbibed overnight and the cotyledons are removed and used. This procedure necessitates the sacrifice of the seed of small seeded taxa but can be useful when screening homogeneous material, especially when time is an important consideration. Using root and leaf tissue allows us to assay a plant and still grow it to maturity. Table 3 lists the source of tissue that has given the best results for *P. vulgaris* under our conditions. Readers should be aware that under their specific conditions other tissues may also give satisfactory results (e.g., Sprecher and Vallejos 1989 for *P. vulgaris*). In addition, different species may require different sources of tissues or buffer systems (e.g., Schinkel and Gepts 1989; Gàrvin et al. 1989 in *P. acutifolius*).

Table 3. Isozyme systems, loci and alleles in *Phaseolus vulgaris* (Weeden 1984a; Sprecher 1988; Koenig and Gepts 1989a,b; Sprecher and Vallejos 1989)

Isozyme	Enzyme system			Genetic data		
	Buffer[a]	Tissue	Stain recipe	Loci	Alleles	Source
Aconitase (EC 4.2.1.3)	H	Leaves	Morden et al. (1987)	*Aco*	100, 102	R. Nodari (unpubl. data)
Aspartate aminotransferase (EC 2.6.1.1)	L	Root	Vallejos (1983)	*Aat-1*	100	Koenig and Gepts (1989b)
				Aat-2	100	
Diaphorase (EC 1.6.4.3)	H	Root	Vallejos (1983)	*Diap-1*	95, 96, 100, 102	Sprecher (1988), Koenig and Gepts (1989b)
				Diap-2	100, 105	
Fructokinase (EC 2.7.1.4)	H	Leaves	Koenig and Gepts (1989b)	*Fk*	100	Koenig and Gepts (1989b)
Glucose-6-phosphate dehydrogenase (EC 1.1.1.49)	H	Root	Shaw and Prasad (1970)	*G6pd*	100	Koenig and Gepts (1989b)
Glucose and phosphate isomerase (EC 5.3.1.9)	L	Leaves	Vallejos (1983)	*Gpi-c1*	100	Weeden and Liang (1985); Weeden (1986a,b)
				Gpi-c2	96, 100	
Leucine aminopeptidase (EC 3.4.11.1)	L	Leaves	Shaw and Prasad (1970)	*Lap-1*	100	Koenig and Gepts (1989a)
				Lap-2	100	
				Lap-3	100, 103	
Malate dehydrogenase (EC 1.1.1.37)	L	Leaves	Vallejos (1983)	*Mdh-1*	100, 103	Koenig and Gepts (1989a)
				Mdh-2	100, 102	Koenig and Gepts (1989b)
Malic enzyme (EC 1.1.1.40)	L	Roots	(Brewer (1970)	*Me*	98, 100, 102	Weeden (1984a); Koenig and Gepts (1989a)
Methylumbelliferyl esterase	H	Leaves	Pasteur et al. (1988)	*Mue*	100, 103	Garrido et al. (1991)
Peptidase (EC 3.4.-.-)	L	Leaves	Weeden (1984a)	*Pep*	100	Koenig and Gepts (1989b)
Peroxidase (EC 1.11.1.7)	L	Root	Vallejos (1983)	*Prx*	98, 100	Weeden (1986a,b)
6-Phosphogluconate dehydrogenase (EC 1.1.1.44)	H	Root	Vallejos (1983)	*6Pgd*	100	Koenig and Gepts (1989b)
Ribulose bisphosphate carboxylase (small subunit) (EC 4.1.1.39)	L	Leaves	Weeden (1984a)	*Rbcs*	98, 100	Weeden (1984a)
Shikimate dehydrogenase (EC 1.1.1.25)	L	Leaves	Vallejos (1983)	*Skdh*	100, 103	Weeden (1984a)
Triosephosphate isomerase (EC 5.3.1.1)	H	Leaves	Vallejos (1983)	*Tpi-1*	100	Koenig and Gepts (1989b)
				Tpi-2	100	

[a] L, lithium hydroxide tris borate pH 8.1 (Selander et al. 1971); H, histidine citrate pH 6.5 (Cardy et al. 1980) (see Table 4).

3.2.2 Gel Preparation

Adequate separation of enzyme variants will depend on the buffer system utilized. In order to streamline our analyses, we have attempted to limit the number of buffer systems to two: either lithium hydroxide tris borate pH 8.1 (Selander et al. 1971) or histidine citrate pH 6.5 (Cardy et al. 1980; Table 4). Other buffers have been suggested by Weeden (1984a) and Sprecher and Vallejos (1989). The gels are poured into a horizontal form similar to that described by Johnson and Shaffer (1974). The "legs" are immersed in the buffer reservoirs. The wicking action of the gel is utilized to draw the buffer from the reservoir into the gel.

Table 4. Recipes for isozyme buffers and gels

A. Buffers: *Lithium hydroxide − boric acid buffer*

Lithium hydroxide (0.03 M)	1.2 g
Boric acid (0.19 M)	11.9 g

Add to 800 ml dH$_2$O and stir until dissolved. Bring volume to 1 l and adjust pH to 8.1 with LiOH or boric acid.

Histidine buffer

L-histidine (0.065 M)	10.1 g

Add to 800 ml dH$_2$O and stir until dissolved. Bring volume to 1 l and adjust pH to 6.5 with citric acid.

Tris-citrate buffer

Tris	6.2 g

Add to 800 ml dH$_2$O and stir until dissolved. Bring volume to 1 l and adjust pH to 8.4 with citric acid monohydrate (about 1.2 g/l).

B. Grinding buffer

Glutathione (reduced)	1.5 g

Add to 30 ml dH$_2$O and stir until dissolved. Bring volume to 40 ml and adjust pH to 7.6 with 1 M unacidified Tris.

C. Gels: *LiOH-boric acid − Tris-citrate starch gel*

Tris-citrate buffer	320 ml
LiOH-boric acid	30 ml

Mix well and separate 150-ml aliquot. The remainder is brought to boiling in the microwave.

Hydrolyzed starch	35 g
Sucrose	10 g

The starch and sucrose are mixed in a 1-l Erlenmeyer flask with side arm. The 200-ml buffer remaining is added and the mixture gently swirled until the starch and sucrose are dissolved. Follow the protocol described in Section 3.2.2.

Histidine starch gel

Histidine buffer	87.5 ml
dH$_2$O	262.5 ml

Mix well and separate 150-ml aliquot. The remainder is brought to boiling in the microwave.

Hydrolyzed starch	35 g
Sucrose	10 g

The starch and sucrose are mixed in a 1-l Erlenmeyer flask with side arm. The 200-ml buffer remaining is added and the mixture gently swirled until the starch and sucrose are dissolved. Follow the protocol described in Section 3.2.2.

The gel forms are carefully washed and then rinsed with Photoflo [1:600 deionized water (dH$_2$O] dilution] to prevent the starch from sticking to the gel form. Masking tape is used to prevent the heated liquid starch from leaking out of the form before the starch has solidified.

The starch solution can be heated using a microwave. This allows enough starch solution for two gels to be prepared at once. The starch and sucrose (see Table 4 for recipes) are well mixed in an Erlenmeyer flask with side arm (to allow degassing). Half of the buffer (either lithium or histidine) is added. The solution is gently shaken to dissolve the starch and sucrose. The other half of the buffer is heated in the microwave until boiling. The boiling liquid is added to the dissolved starch solution and vigorously shaken. The mixture is then returned to the microwave and heated until it begins to boil. The mixture should be removed from the microwave at 60-s intervals and swirled to insure uniform heating and to prevent the starch from burning onto the bottom of the flask.

Upon boiling the flask is removed (use an insulated glove) and immediately degassed for 30 to 60 s, the point at which the rising air bubbles become larger. The degassed liquid starch is poured carefully into the gel form to avoid forming air bubbles. If a few bubbles form the may be quickly removed with a plastic disposable pipette. If the starch has been degassed too long or burned on the bottom of the flask, pieces of resolidified starch can be seen in the clear liquid. These pieces and air bubbles can interfere with the protein migration and must be avoided. This is the easiest place at which to start over again.

The starch should be allowed to cool on the bench until it begins to turn an opaque white (about 30–45 min). It should then be gently covered with plastic wrap and allowed to set up overnight.

3.3 Loading and Running the Starch Gels

3.3.1 Refrigerate the Gels and Buffer

The gels and running buffer should cool for at least 1 h before loading and running the gel. The gel, wrapped in the plastic wrap, is put into the refrigerator first thing in the morning. The running buffer needed should be measured out and put in a labeled graduated cylinder which is also set into the refrigerator to cool.

3.3.2 Preparing the Plant Tissue

A Plexiglass well board (with 2.5 cm diameter, numbered wells) is placed in a tub of ice at a slight angle and allowed to cool. One hundred fifty to 200 ml of grinding buffer (Table 4) is pipetted into each well to be used (grinding buffer should not be more than 2 weeks old). The plant tissue (leaf, root, or imbibed cotyledon) is put in the numbered wells and macerated in the grinding buffer using a plastic pestle. The plant tissue should be as fresh as possible. We bring the growing plants in flats into the lab to remove the tissue immediately before grinding it. Once all the samples have been well ground a small spatula is used to push the solid plant material to the uphill side of the well to allow the tissue extract with the freed plant protein to gather at the bottom of the well. Small filter paper wicks (1.5×0.2 cm) are carefully placed into the tissue extract and allowed to absorb the solubilized protein.

3.3.3 Inserting the Wicks in the Gel or Freezing Them

The gel is cut lengthwise 4 cm from the cathodal end (unless an cathodal system is to be run, then it is cut 5 cm from the cathodal end) and wiped dry on either side of the cut (to make it easier to spread the gel at the cut). The wicks are removed from the wells with forceps, dabbed on paper towels to remove adhering plant tissue and excess tissue extract, and placed in the cut through the gel about 2–3 mm apart. The wicks containing the protein can be frozen at this time to be used at a latter date or as a backup in case of problems with the gels. We have kept wicks frozen in microtiter well plates wrapped in cellophane for up to 2 weeks at $-20\,^\circ$C.

Wicks soaked in bromophenol blue dye are placed on the outside of the samples. They are used to mark both where the samples are (when the gel is cut to be removed from the form) and how far the protein has migrated. A plastic spacer is placed between the frame and the short (4 cm) piece to hold the wicks firmly in the gel.

3.3.4 Running the Gel

The two buffer reservoirs are filled with the cooled running buffer. The gels run from the negative (black) to the positive (red) and the forms are placed on the buffer reservoirs with the legs in the buffer. Histidine-citrate gels are run at 10 mA until the proteins (i.e., the dye) has migrated about 1 cm from the wick (30 min). Lithium hydroxyide Tris borate gels are started at 20 mA and run for 40 min. Once the protein has migrated for that time, the wicks are carfully removed using forceps and the gels are covered with plastic wrap. It is always best to turn the power off when handling the gels. Histidine citrate gels are then run at 20 mA and the lithium hydroxide Tris borate gels at 40 mA. The gels run 5 to 6 h; for good reproducibility all gels should be run at the same amperage and for the same length of time each run.

3.3.5 Preparing and Staining Gel Slices

The gel is removed from the reservoirs and cut with a spatula. The front and back area which were in contact with the reservoir are removed. A small piece is cut from the side to make it easier to remove the gel from the form. The gel is placed on the "cutting board" which consists of two pieces of Plexiglass (1 cm wide×0.15 cm thick, which are about 10 cm longer than the gel) glued to a glass plate a little wider apart than the width of the gel which is laid between them. Four more equally sized spacers are laid on the glued spacers on either side of the gel to serve as guides. The 4-cm anodal piece of the gel is kept and used to press against the portion of the gel containing the migrated protein. This helps assure an even cut. A "cheese cutter" which has been made using a hack saw frame and a fine guitar string (metal) is used to cut the gel horizontally in 0.15-cm slices. It is slowly and continuously pulled through the gel guided by the spacers. One spacer from each side is removed and the gel is sliced again. This is repeated until there are five cuts, leaving a top and bottom piece (which will be discarded) and four 0.15-cm thick slices. The gel slices are stained in plastic boxes (19×11×3.5 cm). The stain boxes are lined with plastic wrap and the gel slices are carefully placed in them. The stain is mixed and placed over the gels (or vice versa) which are then set on the shaker to incubate (shake gently or they may break apart). The bands appear in anywhere from 0.5 h to overnight, depending on the enzyme-staining system, with the exception of methylumbelliferyl esterase (MUE), which has to be read on a UV transilluminator within 5–10 min after addition of the stain.

3.4 Stains

3.4.1 Staining Preparation

While the gels are running, the stain recipes should be prepared. It is best to start 1 or 2 h before the gel is done, depending on the number of different systems to be stained. Always measure the chemicals and solutions kept at room temperature first, followed by the refrigerated chemicals, and those stored in the freezer last. Whenever possible use stock solutions because it is both faster and more accurate to pipet small amounts. We have found that NAD and NADP solutions (10 mg/1 ml) can be aliquoted and frozen at $-70\,^{\circ}$C, PMS at $-20\,^{\circ}$C, and MTT refrigerated, with no loss of activity. When needed, they can be thawed and kept on ice until mixed into the stain.

It is best to measure the reagents beforehand, keeping the temperature-sensitive ones at the proper temperature until the stain solution is mixed. We line the boxes with plastic wrap because it makes the gel slices easier to handle after they have been stained. The staining solution should be mixed and added immediately to the gel slice in the box. It is then incubated on the shaker the requisite time needed for the staining reaction to occur. It is important to check each gel at regular intervals because some will reach an optimum point rather quickly (within 0.5 h) and after that point the background increases, making scoring more difficult.

3.4.2 Stain Recipes

For the sake of conciseness, stain recipes will not be repeated here. They have been detailed by several authors among which Shaw and Prasad (1970), Vallejos (1983), Pasteur et al. (1988), Wendel and Weeden (1989), and Murphy et al. (1990). Table 3 lists the particular recipe we have used in our *Phaseolus* analyses.

3.5 Interpretation of Gels

3.5.1 Naming Loci and Alleles

Nomenclature for isozyme loci and alleles in *Phaseolus* follows the proposal of Myers and Weeden (1988). Loci are designated using the appropriate two to four italicized letter abbreviation of the biochemical name of the enzyme, the first letter of the abbreviation being capitalized (e.g. *Me* for malic enzyme, *Skdh* for shikimate dehydrogenase). The accepted names for most enzymes are given in the International Union of Biochemistry (1984) and appropriate abbreviations have been established for most enzymes (Wendel and Weeden 1989). When more than one isozyme locus is identified, they are distinguished by a numeric suffix, the most anodal isozyme being given the suffix 1 (e.g., *Diap-1, Diap-2*).

The most common allele at each locus is designated 100 and all other alleles are designated by their migration distance measured in millimeters from the standard (e.g., slower migrating: 98; faster migrating: 102) (Koenig and Gepts 1989a, b). Figure 2 shows polymorphisms for various enzymes analyzed as described in Table 3. The two *P. vulgaris* reference cultivars included in all our gels have the following genotype:

ICA-Pijao: $Rbcs^{100}$, $Skdh^{103}$, Prx^{100}, Me^{100}, $Mdh\text{-}1^{100}$, $Mdh\text{-}2^{100}$, $Diap\text{-}1^{95}$, $Diap\text{-}2^{105}$, Lap^{100}, $Aco\text{-}2^{100}$, Mue^{100}
California Dark Red Kidney: $Rbcs^{98}$, $Skdh^{100}$, Prx^{98}, Me^{98}, $Mdh\text{-}1^{103}$, $Mdh\text{-}2^{100}$, $Diap\text{-}1^{100}$, $Diap\text{-}2^{100}$, $Lap\text{-}3^{103}$, $Aco\text{-}2^{102}$, Mue^{103}

3.5.2 Genetic Control

Table 3 lists the number of loci and their respective alleles as observed in our analyses of *P. vulgaris* genetic diversity. Additional polymorphic loci have been identified by Weeden (1984a, b; 1986a, b) and Weeden and Liang (1985). They include adenylate kinase (*Adk*; two alleles), esterase (*Est-1* and *Est-2*; each two alleles), glucose phosphate isomerase (*Gpi-c1*; two alleles), and N-acetyl glucoseaminidase (*Nag*). Most loci are unlinked with the possible exception of *Rbcs* and *Me* (Weeden 1984b; Koenig and Gepts 1989a), which in certain populations form a linkage group with the lectin-arcelin locus: *Rbcs-[Lec-Arl]-Me*.

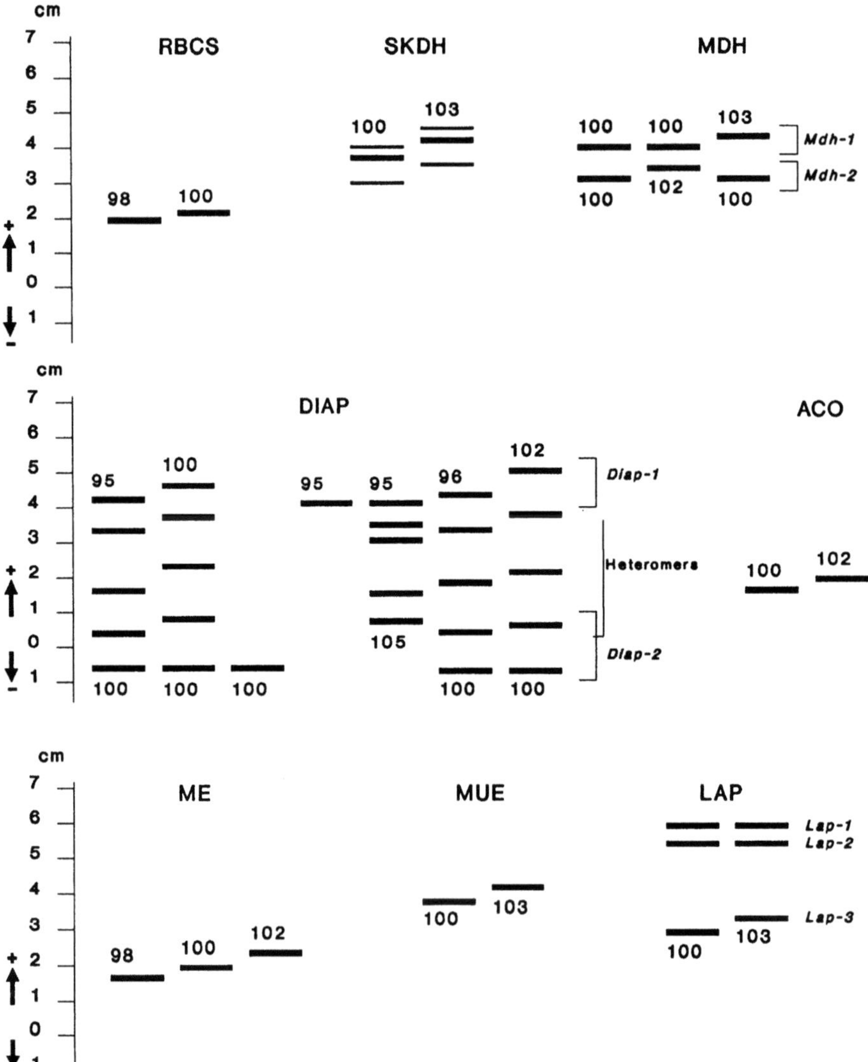

Fig. 2. Allozyme diversity for eight enzyme systems in wild and cultivated common bean. *RBCS* ribulose bisphosphate carboxylase (small subunit); *SKDH* shikimate dehydrogenase; *MDH* malate dehydrogenase; *DIAP* diaphorase; *ACO* aconitase; *ME* malic enzyme; *MUE* methylumbelliferyl esterase; *LAP* leucine aminopeptidase. For locus and allele designations, see Section 3.5.1

This general absence of linkage indicates that the polymorphic isozyme loci characterize different regions of the genome, an important feature in evolutionary studies.

3.6 Preservation of Gels

It is important that gels be preserved in order to compare results across experiments. Gels can be placed overnight in 50% ethanol and then stored indefinitely at 4 °C. It is important that the gels be scored before this is done since some resolution is lost through this processing. Photographs can also be taken at this time. A 35-mm camera with 100 ASA Ektachrome film is adequate to photograph the gel on a light box. These slides provide a more easily stored copy of the gel and are especially useful when a centimeter scale is included in the photograph. The lab has also used a Bio-Rad gel dryer at 70 °C for 1 h. The gels are dried between cellophane in the same manner as polyacrylamide gels (and on the same drying cycle), however, the dyes stain the backing and cover of the gel dryer quite severely.

3.7 Applications

The earliest isozyme analyses in *Phaseolus* were performed by Wall and Wall (1975) and Bassiri and Adams (1978a, b). More recently, several more detailed studies have been conducted on the patterns of genetic diversity in *P. vulgaris* and *P. acutifolius* (Weeden 1984a; Sprecher 1988, Koenig and Gepts 1989b; Singh et al. 1991b).

In *P. vulgaris*, these studies have revealed or confirmed that: (1) the species consists of two diverged geographical subdivisions – Mesoamerican vs. Andean – which, based on independent data, appear to represent two subspecies (Sprecher 1988; Koenig and Gepts 1989b; Singh et al. 1991b); (2) the species is characterized by extensive multilocus associations even among unlinked loci (Gepts, unpubl. observ.); (3) even in presumed secondary centers of diversity where both Mesoamerican and Andean genotypes are grown together, there is a dearth of recombinants between the two groups for reasons unknown so far (Sprecher 1988); (4) snap beans, which as a group are very homogeneous morphologically, appear to be relatively more diverse than other commercial bean classes (Weeden 1984a); and (5) certain allozyme variants correlate with morphological and geographical variation and have allowed us to identify races of related cultivars in both Andean and Mesoamerican cultivars (Singh et al. 1989, 1991a).

In *P. acutifolius*, isozyme analyses have confirmed the reduction of diversity during domestication revealed by phaseolin (Schinkel and Gepts 1989). They also identified geographic patterns of variation, such as the divergence between populations west and east of the Sierra Madre Occidental in Mexico.

4 RFLP Analysis

4.1 Introduction

The methodology for restriction fragment length polymorphism (RFLP) analysis is based on the properties of Type II restriction enzymes. Restriction enzymes are sequence-specific bacterial endonucleases. Of the three types of restriction enzymes that have been described (Yuan 1981), the Type II enzymes comprise ATP-independent endonucleases cleaving DNA at specific sites within or very close to short (usually 4 or 6 bp) recognition sequences. Hence, this cleavage activity will produce equimolar amounts of the same discrete fragments for a given DNA molecule. The fragments obtained by restriction digestion can then be separated by size using agarose gel electrophoresis and visualized by ethidium bromide staining. Defined fragments can be easily detected from the bulk of DNA fragments by hybridization procedures using radiolabeled fragment-specific sequences as probes and exposure to X-ray films.

Differences in the array of restriction fragments between two related sequences is indicative of modification in the DNA primary structure. These modifications can be due to gain/loss of recognition sites, rearrangements, or insertion/deletion of sequences between sites. RFLPs are thus molecular markers providing data about quantitative and qualitative differences between two sequences. Currently, they constitute one of the most efficient methods for detection of variation at the DNA level, being surpassed only by direct sequencing. Although RFLP analysis is less informative than direct sequencing, it is faster and less expensive. Moreover, in cases where many different samples must be analyzed, it is the only feasible procedure. As molecular markers, plant unclear RFLPs have already been broadly used in applied and basic genetics, systematics, and evolution (reviewed in Helentjaris and Burr 1989).

Variation at the DNA sequence level as determined by RFLP analysis has been examined for various seed proteins of common bean. These include phaseolin (e.g., Talbot et al. 1984), lectin (e.g. Staswick et al. 1986), and arcelin (Osborn et al. 1988). For example, phaseolin types can be recognized by specific restriction fragment patterns (Fig. 3). Each of these patterns is produced by the hybridization of a radiolabeled phaseolin gene fragment to the $7-8$ genes in the phaseolin gene family. Nevertheless, because of tight linkage at the *Phs* locus, all the bands corresponding to the various phaseolin genes tend to segregate as a single Mendelian unit and represent a simple co-dominant marker (Brown et al. 1981a).

4.2 Methodology

Our methodology for RFLP analysis of bean DNA sequences, and, in particular, bean seed protein genes can be divided as follows:

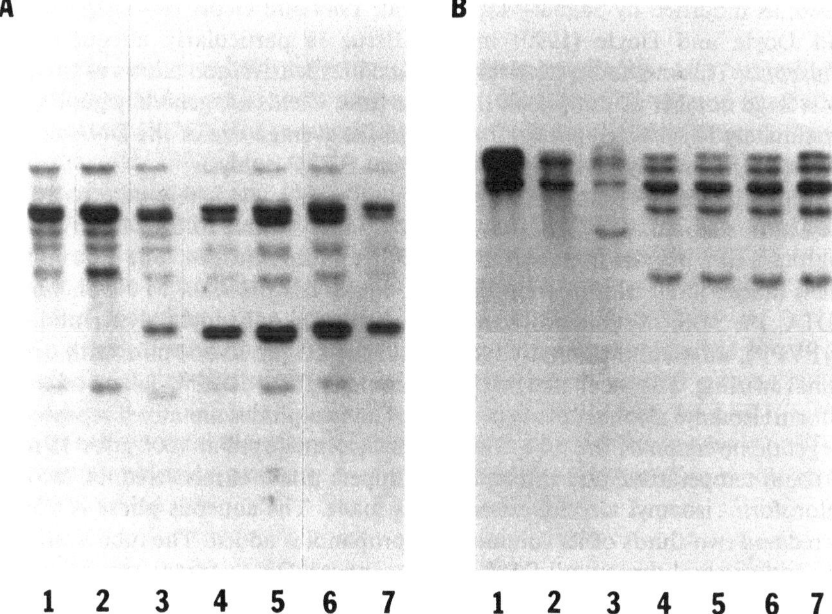

Fig. 3. Genetic diversity for phaseolin in common bean cultivars as determined by RFLP analysis. Genomic DNA of lines with various phaseolin types was digested with *Eco*RI (**A**) or *Hind*III (**B**), electrophoresed in a 0.8% agarose gel, Southern blotted, and hybridized to phaseolin probe MC31 (provided by JL Slightom). *Lane 1*: "T" phaseolin (cultivar Jalo); *lane 2*: "H" (Cocacho); *lane 3*: "C" (Burrito); *lane 4*: "S" (BAT 93); *lane 5*: "B" (Orgulloso); *lane 6*: "Sb" (Black Turtle Soup); *lane 7*: "Sd" (Mexico 222). T, H, and C are Andean phaseolin types and S, B, Sb, and Sd are Mesoamerican phaseolin types as previously defined on the basis of one- and two-dimensional electrophoresis (Brown et al. 1981b; Gepts and Bliss 1986; Gepts et al. 1986; Koenig et al. 1990)

1) Total or nuclear DNA is obtained from leaf tissue.
2) The DNA is digested with restriction enzymes.
3) The restriction fragments are separated by agarose gel electrophoresis.
4) Fragments in the gel are denatured and transferred to a nylon or nitrocellulose membrane.
5) Probes (genomic or cDNA) are radioactively labeled.
6) Hybridization of DNA on the membranes with the labeled probes.
7) The membrane is exposed to an X-ray film in order to visualize the restriction patterns for the specific DNA region.

4.2.1 DNA Extraction

Currently, a number of protocols for genomic DNA extraction from plants have been developed (Bendich et al. 1980; Murray and Thompson 1980; Saghai-Maroof et al. 1984; Rogers and Bendich 1988; Doyle and Doyle 1990). We have found that a modification of the method by Murray and Thompson

(1980; as modified by Saghai-Maroof et al. 1984 and Gepts and Clegg 1989) and Doyle and Doyle (1990) in leaf tissue is particularly adequate for *Phaseolus*. This method is relatively fast and inexpensive, and allows us to handle a large number of samples at the same time. Yields are generally good (approximately 50–100 µg/g fresh tissue) and the average size of the DNA molecules (approximately 40 kb) is sufficient for RFLP analysis.

Young leaves (4–6 cm) are harvested and used immediately (alternatively, they can be stored at −70 °C for several months). Three to 5 g of leaf tissue is ground to a powder in a mortar containing liquid nitrogen. The ground tissue is placed in 10 ml extraction buffer (50 mM Tris pH 9, 0.7 M NaCl, 1 mM EDTA, 1% SDS, 1% β-mercaptoethanol, 50 mg/ml polyvinyl polypyrrolidone or PVPP), mixed and promptly incubated at 60 °C for 30–60 min, with occasional swirling. The incubated mixture is cooled off for 10 min. Ten ml of chloroform:isoamyl alcohol (24:1) is added. The two phases are mixed repeatedly by gentle inversion of the tube. The sample is centrifuged at 1600 g for 15 min at room temperature, and the aqueous (upper) phase is recovered. A second chloroform:isoamyl alcohol extraction is made. The aqueous phase is transferred and two-thirds of its volume in isopropanol is added. The tube is inverted gently several times until DNA strands appear. DNA strands are recovered by hooking them out with a bent Pasteur pipet (if the strands are too sheared to be hooked out, DNA is pelleted by centrifuging at 1600 g for a few seconds and the supernatant is discarded). DNA is removed to a sterile tube containing washing solution (76% ethanol, 10 mM ammonium acetate). At this stage, the DNA strands may have a greenish or brownish tint, which is usually eliminated after incubating 2–12 h at room temperature in washing solution with occasional swirling. If needed, the washing solution can be replaced with fresh solution. After washing, the solution is discarded, the DNA is air-dried and resuspended in 300–500 µl TE (10 mM Tris·HCl pH 8, 1 mM EDTA). After the DNA is dissolved the solution is centrifuged at 3500 g for 10 min and the supernatant is saved.

In some cases, the DNA appears brownish even after a long washing period, probably due to the oxidation of polyphenolics (Rogers and Bendich 1988). In this case, we suggest a digestibility test for the specific sample prior to further work. Extracted DNA may be further purified by a cesium chloride gradient (Sambrook et al. 1989). This procedure yields highly purified DNA, eliminating RNA, polysaccharides, and other contaminants, but it is time-consuming and expensive. In practice, we have seen that most of the samples can be efficiently digested without cesium chloride gradients.

The DNA concentration of samples purified by cesium chloride gradient can be estimated by UV spectrophotometry at 260 nm (1 unit, equivalent to 50 µg/ml DNA; Sambrook et al. 1989). However, if the sample is going to be used without purification, UV estimates are not reliable. In that case, DNA can be more precisely quantified using a fluorometer (Brunk et al. 1979; Cesarone et al. 1979). The final concentration for the extracted DNA in the TE solution should be around 0.4 to 1 µg/µl. For short periods, DNA solutions can be stored at 4 °C; long-term storage takes place at −70 °C.

4.2.2 Digestion

Restriction enzymes have different optimal assay conditions. All require Mg^{2+} as cofactor, and most of them have a similar pH range (7.4−8) and incubation temperature (37 °C). The most critical difference lies in the ionic strength (Fuchs and Blakesley 1983). Most of the manufacturers provide a buffer of the appropriate concentration and instructions to achieve optimal activity for each enzyme. Alternatively, a standard buffer (50 mM Tris·HCl pH 8, 10 mM $MgCl_2$, 1 mM dithiothreitol, 100 µg/ml bovine serum albumin) can be used, in which low, medium or high concentrations (0, 50, and 100 mM, respectively) of NaCl are used according to the enzyme (Fuchs and Blakesley 1983).

In a typical assay, 5−10 µg DNA is digested in a total volume of 30−50 µl. Generally, the restriction enzyme is added in five- to six-fold excess (25−30 units for a 5 µg digestion), and the reaction mix is incubated for 4−7 h, in order to achieve total digestion. If the sample to be used has not been purified by cesium chloride gradient, 1−4 µg DNAse-free RNAse (Sambrook et al. 1989) is added.

4.2.3 Gel Electrophoresis

Electrophoresis is performed according to Sambrook et al. (1989). Agarose at a concentration of 0.8% is dissolved in TAE (40 mM Tris-Acetate pH 7.4, 1 mM EDTA) to which 0.5 µg/ml ethidium bromide is added. For a large sample number, we use large horizontal gels (25×20 cm, available from several companies (e.g., Bethesda Research Laboratories, Gaithersburg, MD), allowing us to run up to 60 samples at the same time. The digested samples are mixed (6:1) with a solution containing glycerol (50%) and bromophenol blue (0.5%), and loaded into the wells. One µg lambda phage DNA digested with *Hind*III is used as size marker (Sambrook et al. 1989). The optimal voltage to be used is 1 V/cm.

After the overnight electrophoresis run, DNA is visualized by ethidium bromide fluorescence using a UV transilluminator to check for digestion. Completely digested DNA should be seen as a smear along the lane; some bands (e.g., cpDNA, highly repetitive DNA) may be brighter in relation to the smeared background. Incompletely digested or undigested DNA samples will show a strong band above the 23 000 base pairs molecular weight marker band. The positions of the lambda *Hind*III fragments should be marked by puncturing the gel with an 18-gauge needle containing India ink in order to establish the correct position of the restriction fragments in the next steps.

4.2.4 Transfer

Restriction fragments can be transferred to either nylon- or nitrocellulose-based membranes. Nylon membranes are more desirable than nitrocellulose paper because they have a much higher mechanical strength than the former and, hence, can be reused several times. In our hands, Zetabind membranes (AMF-CUNO, Meriden, CT) can be reused 10 to 15 times. Protocols can change slightly according to the properties of the membranes and it is therefore recommended to follow the protocol of the manufacturers. We describe here the Zetabind protocol.

The gel is washed for 30 min at 40–60 rpm in a tray with 0.2 N NaOH, 0.6 M NaCl to denature the DNA. The denaturing solution is replaced then by 0.5 M Tris pH 7.5, 1.5 M NaCl and again shaken gently (40–60 rpm) for 30 min, to neutralize the gel. Capillary transfer is made overnight according to Southern (1975; cited in Sambrook et al. 1989).

Membranes onto which DNA has been transferred are prehybridized in sealed plastic bags for 2–24 h at 42 °C in 5×SSC (20×SSC = 3 M NaCl, 0.3 M sodium citrate, pH 7), 10×Denhardt's solution (100×Denhardt's = 2% Ficoll, 2% polyvinyl pyrrolidone, 2% bovine serum albumin), 50 mM NaPO$_4$ pH 6.7, 5% dextran sulfate, 50% formamide, and 500 µg/ml sonicated or sheared salmon sperm DNA. Sonicated or sheared salmon sperm should be heated in a water bath at 100 °C for 10 min and immediately cooled off on ice for 10 min prior to addition to the prehybridization solution in order to maintain denaturation. Five ml of prehybridization solution is used for each 100 cm^2 of membrane. Several membranes can be place in the same bag. If not used immediately, they can be stored in their plastic bag at 4 °C.

4.2.5 Probes

Probes include either genomic or cDNA cloned fragments. They can be either anonymous sequences or sequences of specific genes, some of which are expressed in seeds such as phaseolin (Slightom et al. 1985), lectin (Hoffman and Donaldson 1985), or arcelin (Osborn et al. 1988). Plasmids containing clone sequences are amplified in *E. coli* liquid cultures and isolated by the quick-boiling minipreparation (Holmes and Quigley 1981). The vector is then cut with the appropriate restriction enzymes. Once digested, the fragment is isolated by low melting point agarose gel electrophoresis or by DEAE-NA45 paper affinity (Selden and Chory 1987).

The cloned sequence probe is radioactively labeled by the random primer method (Feinberg and Vogelstein 1984); kits for random priming are available from several manufacturers, who also give instructions for use. Alternatively, the component solutions such as the nucleotide solutions or the reaction buffer can be prepared in the lab. After labeling, unincorporated nucleotides are removed with a Sephadex G50 spun column as described by Sambrook et al. (1989). Specific activity for the probe is determined by measuring a small aliquot (1–2 µl) in a scintillation counter and should be at least $1\times10^8 - 10^9$ dpm/µg of probe to minimize background and exposure time. A total of $0.6-1\times10^7$ dpm of probe is used for each 100 cm^2 of membrane to be hybridized.

4.2.6 Hybridization

The prehybridization solution is totally removed and the following hybridization solution is added: $5 \times$ SSC, $1 \times$ Denhardt's, 20 mM NaPO$_4$ pH 6.7, 10% dextran sulfate ml, 50% formamide, and 100 mg/ml boiled sonicated or sheared salmon sperm DNA. This solution also contains the probe, previously denatured by boiling and cooled on ice.

Hybridization is carried on in a temperature-controlled orbital shaker (42 °C, 150 rpm) for 20 h according to Sambrook et al. (1989). After hybridization, the membrane is removed from the bag and briefly washed at low stringency in $2 \times$ SSC, 0.1% SDS at room temperature. Membranes are then washed twice in $0.1 \times$ SSC, 0.1% SDS at 60 °C. Membranes are lightly blotted to remove excess washing solution, wrapped in plastic wrap, and exposed to a Kodak X-Omat X-ray film. Exposure times can vary from 1 to several days, depending on the amount of DNA sample used and/or probe-specific activity.

4.2.7 Reuse of Solutions and Membranes

Nylon membranes can be reused by washing for 30 min in 0.4 M NaOH, neutralizing with $0.1 \times$ SSC, 0.5% SDS, 0.2 M Tris pH 7.5 for 30 min, and prehybridizing as described.

Prehybridization and hybridization solutions can be recovered after use and stored at -20 °C. Before reusing, the solutions are heated at $75-80$ °C for 15 min in a water bath, and immediately added to the incubation bag. Prehybridization solution can be reused $3-4$ times and stored for weeks. Hybridization solution should not be stored for more than 1 week.

4.3. Applications

This methodology has been applied successfully to several areas of study in our program that are still under way: (1) RFLP levels in wild and cultivated *Phaseolus vulgaris*. Figure 3 shows restriction fragment length polymorphisms detected with a phaseolin probe (R. Nodari, E. Koinange, V. Becerra, J. Kelly, and P. Gepts, in prep.); (2) chloroplast DNA evolution in the *P. vulgaris* − *P. coccineus* complex (V. Llaca and P. Gepts, in prep.); and (3) mapping of the common bean genome (Gepts et al. 1990).

5. Conclusion: Comparison of the Three Classes of Evolutionary Markers in Evolutionary Studies

Each of the three types of markers discussed in this chapter possess advantages and disadvantages as evolutionary markers. In general, an ideal marker should exhibit a sufficiently high level of polymorphism to identify genetic diversity among the entries of the studies; yet, this level of polymorphism should not be so high as to include repeated or parallel mutations that would obscure patterns of common ancestry. The specific marker(s) most suitable for a given study will depend on the inherent variability of the marker and the genetic distance among entries of the study. For example, chloroplast DNA is generally highly conserved and is therefore most useful at higher taxonomic levels (Palmer 1987); on the other hand, hypervariable markers such M13 or human minisatellite sequences are usually sufficiently variable to detect differences among individuals of the same species, whether animal (Jeffries et al. 1985) or plant species (Dallas 1988; Rogstad et al. 1988). On the other hand, rare variants may be useful in detecting introgression and gene flow, for example between wild ancestral beans and cultivated descendant beans (e.g., the *Mdh-2*[102] allele in *P. vulgaris;* Singh et al. 1991 b). An empirical preliminary study will have to be conducted to identify the most suitable markers.

The second desirable attribute of evolutionary markers is environmental stability. Ideally, the expression of evolutionary markers should be independent of any confounding environmental influence. If this is the case, differences can be attributed exclusively to genetic causes. The third attribute is the number of loci detected by any category of markers. Categories that detect larger numbers of markers, preferably unlinked, will afford a more complete coverage of the genome; this will permit the detection of recombinants, translocations, inversions, linkage drag, etc.

A fourth attribute is the molecular basis of the variation. The more complex the molecular basis, the less likely that a particular variant will have appeared more than one. Hence, individuals that share a variant with a complex molecular basis, will most probably have a common ancestor (Gepts 1990a, b). Finally, preference should be given whenever possible to markers that are easy and cheap to analyze. In order to increase confidence levels, evolutionary studies require fairly large sample numbers that can only be achieved with markers that can be analyzed in a routine fashion.

How do these three categories of markers – phaseolin, isozmyes, and RFLPs – match these five attributes (Table 5)? As argued by Gepts (1990b), the advantages of phaseolin as a marker include a high level of polymorphism, a high degree of environmental stability, a complex molecular basis of the electrophoretic banding pattern, along with the simplicity and low cost of the

Table 5. Comparison of molecular electrophoretic markers in evolutionary studies

	Seed protein (phaseolin)	Isozymes	RFLPs
Polymorphism	High	Low	High
Environmental stability	High	Moderate	High
Number of loci	Single locus	Low (15 – 20 loci)	High
Molecular basis	Complex	Simple	Intermediate
Practicality	Quick, cheap	Quick, cheap	Slow, expensive

analysis. Its major disadvantage is the simple genetic control since all variation in the banding pattern is controlled by a single locus (*Phs*; Brown et al. 1981a; Bassett 1989).

The advantages of isozymes are the larger number (15–20) of (unlinked) loci and the simplicity and low cost of analysis. Their disadvantages include the low levels of polymorphism, their environmental sensitivity (which can be alleviated by working under standardized conditions), and their simple molecular basis (which can be alleviated by considering several isozmye loci simultaneously).

The advantages of RFLPs are their high levels of polymorphism, not only between the Mesoamerican and Andean subspecies, but also within these subspecies (Nodari, Koinange, Kelly, and Gepts, in prep.), their high level of environmental stability, and the high (potentially infinite) number of loci. Their main disadvantages is the high cost and cumbersomeness of the analysis and to a lesser extent the simple molecular basis (in many cases, single nucleotide substitutions), which can also be overcome by considering several RFLPs simultaneously.

Because each of these markers have advantages and disadvantages, none represents a straightforward choice when initiating a study. The particular marker(s) chosen will depend principally on the resources available (personnell, equipment, funds) and prior informtaion about molecular variation in the plant materials to be studied. Our own approach has been to start with the easiest and cheapest marker (phaseolin), continue with isozymes on a subset of the entries analyzed for phaseolin, and, finally to analyze an even smaller subset for RFLPs. Both isozymes and RFLPs have provided new information (e.g., strong multilocus associations shown by isozyme data) or confirmed previous findings (e.g., reduction of diversity during domestication by RFLPs). Hence, our repeated analyses with different markers have not been futile exercises but are providing a more in-depth and dependable picture of the evolutionary history of *Phaseolus* beans.

Acknowledgments. Our research on diversity at the molecular level in *Phaseolus* beans is funded by the U.S. Agency for International Development, the International Board for Plant Genetic resources, and the Lindbergh Fund. V. Llaca holds a fellowship of DGAPA, UNAM, Mexico. R.O. Nodari held, while at Davis, a fellowship of CNPq, Brazil.

References

Bassett MJ (1989) List of genes. Annu Rep Bean Improv Coop 32:1–15
Bassiri A, Adams MW (1978a) An electrophoretic survey of seedling isozymes in several *Phaseolus* species. Euphytica 27:447–459
Bassiri A, Adams MW (1978b) Evaluation of bean cultivar relationships by means of isozyme electrophoretic patterns. Euphytica 27:707–720
Bendich AJ, Anderson RS, Ward BL (1980) Plant DNA: long, pure, and simple. In: Leaver CJ (ed) Genome organization and expression. Plenum, New York, pp 31–33

Bliss FA, Brown JWS (1983) Breeding common bean for improved quantity and quality of seed protein. Plant Breed Rev 1:59–102

Brown AHD (1979) Enzyme polymorphism in plant populations. Theor Popul Biol 15:1–42

Brown JWS, Bliss FA, Hall TC (1981 a) Linkage relationships between genes controlling seed proteins in French beans. Theor Appl Genet 60:251–259

Brown JWS, Ma Y, Bliss FA, Hall TC (1981 b) Genetic variation in the subunits of globulin-1 storage protein of French bean. Theor Appl Genet 59:83–88

Brown JWS, McFerson JR, Bliss FA, Hall TC (1982a) Genetic divergence among commercial classes of *Phaseolus vulgaris* in relation to phaseolin patterns. HortScience 17:752–754

Brown JWS, Osborn TC, Bliss FA, Hall TC (1982b) Bean lectin 2: relationship between qualitative lectin variation in *Phaseolus vulgaris* L. and previous observations on purified bean lectins. Theor Appl Genet 62:361–367

Brown JWS, Osborn TC, Bliss FA, Hall TC (1981 c) Genetic variation in the subunits of globulin-2 and albumin seed proteins of French bean. Theor Appl Genet 60:245–250

Brunk C, Jones K, James T (1979) Assay for nanogram quantities of DNA in cellular homogenates. Anal Biochem 92:497–500

Cardy BJ, Stuber CW, Goodman MM (1980) Techniques for starch gel electrophoresis of enzymes from maize (*Zea mays* L.). Department of Statistics Mimeo Series No. 1317, North Carolina State University, Raleigh

Cesarone C, Bolognesi C, Santi L (1979) Improved fluorometric DNA determination in biological material using 3358 Hoechst. Anal Biochem 100:188–197

Dallas JF (1988) Detection of DNA "fingerprints" of cultivated rice by hybridization with a human minisatellite DNA probe. Proc Natl Acad Sci USA 85:6831–6835

Debouck DG, Maquet A, Posso CE (1989) Biochemical evidence for two different gene pools in lima beans. Annu Rep Bean Improv Coop 32:58–59

Dixon RA, Lamb CJ (1990) Molecular communication in interactions between plants and microbial pathogens. Annu Rev Plant Phys Plant Mol Biol 41:339–367

Doyle JJ, Doyle JL (1990) Isolation of plant DNA from fresh tissue. Focus 12:13–15

Feinberg AP, Vogelstein B (1984) A technique for radiolabelling DNA restriction endonuclease fragments to high specific activity. Anal Biochem 137:266–267

Fuchs R, Blakesley R (1983) Guide to the use of type II restriction endonucleases. Methods Enzymol 100:3–38

Garrido B, Nodari R, Debouck DG, Gepts P (1991) *Uni-2*, a dominant mutation affecting leaf development in *Phaseolus vulgaris*. J Hered 82:181–183

Garvin DF, Roose ML, Waines JG (1989) Isozyme genetics and linkage in tepary bean, *Phaseolus acutifolius* A. Gray. J Hered 80:373–376

Gepts P (1988a) Phaseolin as an evolutionary marker. In: Gepts P (ed) Genetic resources of *Phaseolus* beans. Kluwer, Dordrecht, pp 215–241

Gepts P (1988b) A Middle American and an Andean gene pool. In: Gepts P (ed) Genetic resources of *Phaseolus* beans. Kluwer, Dordrecht, pp 375–390

Gepts P (1990a) Biochemical evidence bearing on the domestication of *Phaseolus* beans. Econ Bot 44(3):S28–S38

Gepts P (1990b) Genetic diversity of seed storage proteins in plants. In: Brown AHD, Clegg MT, Kahler AL, Weir BS (eds) Plant population genetics, breeding, and genetic resources. Sinauer, Sunderland, MA, pp 64–82

Gepts P, Bliss FA (1985) F_1 hybrid weakness in the common bean: differential geographic origin suggests two gene pools in cultivated bean germplasm. J Hered 76:447–450

Gepts P, Bliss FA (1986) Phaseolin variability among wild and cultivated common beans (*Phaseolus vulgaris*) from Colombia. Econ Bot 40:469–478

Gepts P, Clegg MT (1989) Genetic diversity in pearl millet (*Pennisetum glaucum* [L.] R. Br.) at the DNA sequence level. J Hered 80:203–208

Gepts P, Debouck DG (1991) Origin, domestication, and evolution of the common bean, *Phaseolus vulgaris*. In: Voysest O, Van Schoonhoven A (eds) Common beans: research for crop improvement. CIAT, Cali, Colombia, pp 7–53

Gepts P, Osborn TC, Rashka K, Bliss FA (1986) Phaseolin-protein variability in wild forms and landraces of the common bean (*Phaseolus vulgaris*): evidence for multiple centers of domestication. Econ Bot 40:451–468

Gepts P, Singh S, Nodari R, Garrido B, Koinange E (1990) Towards an integrated linkage map of common bean (*Phaseolus vulgaris* L.). J Cell Biochem (Suppl) 14E:281

Gottlieb LD (1981) Electrophoretic evidence and plant populations. Progr Phytochem 7:1–46

Hames BD (1981) An introduction to polyacrylamide gel electrophoresis. In: Hames BD, Rickwood D (eds) Gel electrophoresis of proteins. IRL, Oxford, pp 1–91

Hames BD, Rickwood D (1981) Gel electrophoresis of proteins: a practical approach. IRL, Oxford

Hamrick JL, Godt MJW (1990) Allozyme diversity in plant species. In: Brown AHD, Clegg MT, Kahler AL, Weir BS (eds) Plant population genetics, breeding, and genetic resources. Sinauer, Sunderland, MA, pp 43–63

Helentjaris T, Burr B (1989) Development and application of molecular markers to problems in plant genetics. Cold Spring Harbor Laboratory, Cold Spring Harbor, NY

Hillis DM, Moritz C (1990) Molecular systematics. Sinauer, Sunderland, MA

Hoffman LM, Donaldson DD (1985) Characterization of two *Phaseolus vulgaris* phytohemagglutinin genes closely linked on the chromosome. EMBO J 4:883–889

Holmes DS, Quigley M (1981) A rapid boiling method for the preparation of bacterial plasmids. Anal Biochem 114:193–197

Hunter RL, Markert CL (1957) Histochemical demonstration of enzymes separated by zone electrophoresis in starch gels. Science 125:1294–1295

International Union of Biochemistry (1984) Enzyme nomenclature. Acad Press, New York

Jeffreys AJ, Wilson V, Thein SL (1985) Hpyervariable 'minisatellite' regions in human DNA. Nature 314:67–73

Johannsen W (1909) Elemente der exakten Erblichkeitslehre. Fischer, Jena, Germany

Johnson FM, Schaffer HE (1974) An inexpensive apparatus for horizontal gel electrophoresis. Isozyme Bull 7:4–6

Koenig R, Gepts P (1989a) Segregation and linkage of genes for seed proteins, isozymes, and morphological traits in common bean (*Phaseolus vulgaris*). J Hered 80:455–459

Koenig R, Gepts P (1989b) Allozyme diversity in wild *Phaseolus vulgaris:* further evidence for two major centers of diversity. Theor Appl Genet 78:809–817

Koenig R, Singh SP, Gepts P (1990) Novel phaseolin types in wild and cultivated common bean (*Phaseolus vulgaris,* Fabaceae). Econ Bot 44:50–60

Laemmli UK (1970) Cleavage of structural proteins during the assembly of the head of bacteriophage T4. Nature 227:680–685

Lioi L (1989a) Geographical variation of phaseolin patterns in an old world collection of *Phaseolus vulgaris*. Seed Sci Technol 17:317–324

Lioi L (1989b) Variation of the storage protein phaseolin in common bean (*Phaseolin vulgaris* L.) from the Mediterranean area. Euphytica 44:151–155

Ma Y, Bliss FA (1978) Seed proteins of common bean. Crop Sci 17:431–437

Manen JF, Otoul E (1981) Etudes électrophorétiques et détermination des fractions protéiques principales chez quelques cultivars élites de *Phaseolus lunatus* L. et de *Phaseolus vulgaris* L. Bull Rech Agron Gembloux 16:309–326

Markert CL, Moller F (1959) Multiple forms of enzymes: tissue, ontogenetic and species specific patterns. Proc Natl Acad Sci USA 45:753–763

McLeester RC, Hall TC, Sun SM, Bliss FA (1973) Comparison of globulin proteins from *Phaseolus vulgaris* with those of *Vicia faba*. Phytochemistry 2:85–93

Mendel G (1865) Experiments on plant hybrids. In: Stern C, Sherwood ER (eds) The origin of genetics. Freeman, San Francisco, pp 1–48

Murphy RW, Sites JW Jr, Buth DG, Haufler CH (1990) Proteins I: isozyme electrophoresis. In: Hillis DM, Moritz C (eds) Molecular systematics. Sinauer, Sunderland, MA, pp 45–126

Murray MG, Thompson WF (1980) Rapid isolation of high molecular weight plant DNA. Nucl Acid Res 8:4322–4325

Myers JR, Weeden NF (1988) A proposed revision of guidelines for genetic analysis in *Phaseolus vulgaris* L. Annu Rep Bean Improv Coop 31:16–19

O'Farrell PH (1975) High resolution two-dimensional electrophoresis of proteins. J Biol Chem 250:4007–4021

Osborn TC (1988) Genetic control of bean seed protein. CRC Crit Rev Plant Sci 7:93–116

Osborn TC, Bliss FA (1985) Effects of genetically removing lectin seed protein on horticultural and seed characteristics of common bean. J Am Soc Hortic Sci 110:484–488

Osborn TC, Brown JWS, Bliss FA (1985) Bean lectin 5: quantitative genetic variation in seed lectins of *Phaseolus vulgaris* and its relationship to qualitative genetic variation. Theor Appl Genet 70:22–31

Osborn TC, Blake T, Gepts P, Bliss FA (1986) Bean arcelin. 2. Genetic variation, inheritance and linkage relationships of a novel seed protein of *Phaseolus vulgaris*. Theor Appl Genet 71:847–855

Osborn TC, Alexander DC, Sun SSM, Cardona C, Bliss FA (1988) Insecticidal activity and lectin homology of arcelin seed protein. Science 240:207–210

Osborne TB (1907) The vegetable proteins. Longmans and Green, London

Palmer JD (1987) Chloroplast DNA evolution and biosystematic uses of chloroplast DNA variation. Am Nat 130:S6–S29

Pasteur N, Pasteur G, Bonhomme F, Catalan J, Britton-Davidian J (1988) Practical isozyme genetics. Ellis Horwood, Chichester

Pusztai A (1966) The isolation of two proteins, glycoprotein I and a trypsin inhibitor, from seed of kidney bean (*Phaseolus vulgaris*). Biochem J 101:379–384

Pusztai A, Watt WB (1970) Glycoprotein. II. The isolation and characterization of a major antigenic and non-hemagglutinating glycoprotein from *Phaseolus vulgaris*. Biochim Biophys Acta 207:413–431

Rogers SO, Bendich AJ (1988) Extraction of DNA from plant tissues. In: Gelvin SB, Schilperoort RA (eds) Plant molecular biology manual. Kluwer, Dordrecht, pp A6:1–10

Rogstad SH, Patton JC II, Schaal BA (1988) M13 repeat probe detects DNA minisatellite-like sequences in gymnosperms and angiosperms. Proc Natl Acad Sci USA 85:9176–9178

Romero-Andreas J, Yandell BS, Bliss FA (1986) Bean arcelin. 1. Inheritance of a novel seed protein of *Phaseolus vulgaris* L. and its effect on seed composition. Theor Appl Genet 72:123–128

Saghai-Maroof MA, Soliman KM, Jorgensen RA, Allard RW (1984) Ribosomal DNA spacer-length polymorphisms in barley: Mendelian inheritance, chromosomal location, and population dynamics. Proc Natl Acad Sci USA 81:8014–8018

Sambrook J, Fritsch E, Maniatis T (1989) Molecular cloning. Cold Spring Harbor Laboratory Press, Cold Spring Harbor, New York

Sax K (1923) The association of size differences with seed coat pattern and pigmentation in *Phaseolus vulgaris*. Genetics 8:552–560

Schinkel C, Gepts P (1988) Phaseolin diversity in the tepary bean, *Phaseolus acutifolius* A. Gray. Plant Breed 101:292–301

Schinkel C, Gepts P (1989) Allozyme variability in the tepary bean, *Phaseolus acutifolius* A. Gray. Plant Breed 102:182–195

Schuler MA, Zielinski RE (1989) Methods in plant molecular biology. Acad Press, San Diego

Selander RK, Smith MH, Yang SY, Johnson WE, Gentry JR (1971) Biochemical polymorphism and systematics in the genus *Peromyscus*. I. Variation in the old-field mouse (*Peromyscus polionotus*). Stud Genet VI Univ Texas Publ 7103:49–90

Selden RF, Chory J (1987) Isolation and purification of large DNA restriction fragments from agarose gels. In: Ausubel FM, Brent R, Kingston R, Moore DD, Seidman JG, Smith JA, Struhl K (eds) Current protocols in molecular biology. Wiley, New York, pp 2.6.1–2.6.8

Shaw CR, Prasad R (1970) Starch gel electrophoresis: a compilation of recipes. Biochem Genet 4:297–320

Shii CT, Mok MC, Temple SR, Mok DWS (1980) Expression of developmental abnormalities in hybrids of *Phaseolus vulgaris* L. J Hered 71:218–222

Singh SP, Gutiérrez AJ (1984) Geographical distribution of the DL_1 and DL_2 genes causing hybrid dwarfism in *Phaseolus vulgaris* L., their association with seed size, and their significance to breeding. Euphytica 33:337–345

Singh SP, Debouck DG, Gepts P (1989) Races of common bean, *Phaseolus vulgaris* L. In: Beebe S (ed) Current topics in breeding of common bean, Proc Int Bean Breed Workshop. CIAT, Cali, Colombia, pp 75–89

Singh SP, Gutiérrez JA, Molina A, Urrea C, Gepts P (1991 a) Genetic diversity in cultivated common bean II. Marker-based analysis of morphological and agronomic traits. Crop Sci 31:23–29

Singh SP, Nodari R, Gepts P (1991 b) Genetic diversity in cultivated common bean. I. Allozymes. Crop Sci 31:19–23

Slightom JL, Drong RF, Klassy RC, Hoffman LM (1985) Nucleotide sequences from phaseolin cDNA clones: the major storage proteins from *Phaseolus vulgaris* are encoded by two unique gene families. Nucl Acid Res 13:6483–6498

Soltis DE, Soltis PS (1989) Isozymes in plant biology. Dioscorides, Portland, OR

Sprecher SL (1988) Allozyme differentiation between gene pools in common bean (*Phaseolus vulgaris* L.), with special reference to Malawian germplasm. PhD Thesis, Michigan State University, East Lansing, MI

Sprecher SL, Vallejos CE (1989) System/tissue/enzyme combinations for starch gel electrophoresis of bean. Annu Rep Bean Improv Coop 32:32–33

Staswick P, Chapell J, Voelker T, Vitale A, Chrispeels M (1986) Molecular biology of seed storage proteins and lectins. In: Shannon LM, Chrispeels M (eds) Proc 9th Annu Symp Plant Phys Univ of California, Riverside, 9–11 January 1986. Am Soc Plant Phys, Rockville, MD, pp 107–115

Stebbins GL (1989) Introduction. In: Soltis DE, Soltis PS (eds) Isozymes in plant biology. Dioscorides, Portland, OR, pp 1–14

Stegemann H, Pietsch G (1983) Methods for quantitative and qualitative characterization of seed proteins of cereals and legumes. In: Gottschalk W, Müller HP (eds) Seed proteins: biochemistry, genetics, nutritive value. Martinus Nijhoff/Dr. W. Junk, The Hague, pp 45–75

Studier FW (1973) Analysis of bacteriophage T7 early RNAs and proteins on slab gels. J Mol Biol 79:237–248

Sun SM, Slightom JL, Hall TC (1981) Intervening sequences in a plant gene – comparison of the partial sequence of cDNA and genomic DNA of French bean phaseolin. Nature 289:37–41

Talbot DR, Adang MJ, Slightom JL, Hall TC (1984) Size and organization of a multigene family encoding phaseolin, the major seed storage protein of *Phaseolus vulgaris* L. Mol Gen Genet 198:42–49

Tanksley SD, Orton TJ (1983) Isozymes in plant genetics and breeding. Elsevier, Amsterdam

Vallejos CE (1983) Enzyme activity staining. In: Tanksley SD, Orton TJ (eds) Isozymes in plant genetics and breeding. Elsevier, Amsterdam, pp 469–516

Wall JR, Wall SW (1975) Isozyme polymorphisms in the study of evolution in the *Phaseolus vulgaris-P. coccineus* complex of Mexico. In: Markert CL (ed) Isozymes IV. Acad Press, New York, pp 287–305

Weber K, Osborn M (1969) The reliability of molecular weight determinations by sodium dodecyl sulphate-polyacrylamide gel electrophoresis. J Biol Chem 244:4406–4412

Weeden NF (1984 a) Distinguishing among white-seeded bean cultivars by means of allozyme genotypes. Euphytica 33:199–208

Weeden NF (1984 b) Linkage between the gene coding the small subunit of ribulose biphosphate carboxylase and the gene coding malic enzyme in *Phaseolus vulgaris*. Annu Rep Bean Improv Coop 27:123–124

Weeden NF (1986 a) Enzyme loci defined in *Phaseolus vulgaris*. Annu Rep Bean Improv Coop 29–53

Weeden NF (1986 b) Genetic confirmation that the variation in the zymograms of 3 enzyme systems is produced by allelic polymorphism. Annu Rep Bean Improv Coop 29:117–118

Weeden NF, Liang CY (1985) Detection of a linkage between white flower color and *Est-2* in common bean. Annu Rep Bean Improv Coop 28:87–88

Wendel JF, Weeden NF (1989) Visualization and interpretation of plant isozymes. In: Soltis DE, Soltis PS (eds) Isozymes in plant biology. Dioscorides, Portland, OR, pp 5–45

Yuan R (1981) Structure and mechanisms of multifunctional restriction endonucleases. Annu Rev Biochem 50:285–316

Determination of the Nitrogen-to-Protein Conversion Factor in Cereals

M. YAMAGUCHI

1 Introduction

The nitrogen-to-protein (N-to-P) conversion factor has been popularly used for protein determination in both the basic and applied fields of food and nutrition. This factor originated from the work of Mulder (1839), in which he prepared several proteins in a highly purified state from natural substances and presented a common elemental composition of $C_{40}H_{62}N_{10}O_{12}$ for protein. Since the nitrogen content of the formula is 16%, its reciprocal, 6.25, was used as a N-to-P factor at that time.

In 1931 Jones examined the nitrogen content of 121 proteins of major occurrence in foods and biological materials, and proposed that a specific N-to-P factor other than 6.25 should be used for some food proteins. The FAO (1970) factor, i.e., one of the most representative N-to-P factors in the world, is derived mainly from the Jones' values. However, the examined foods or materials in the Jones' report were limited, for the most part, to cereals, nuts, and seeds. Moreover, his work depended on the following two assumptions. One was that the purity of the prepared proteins should be close to 100% and the other was that the prepared proteins used for calculation should be representable in weight percentage for the whole protein of the foods. However, these two assumptions are usually difficult to fulfill.

Among the N-to-P conversion factors of foods, much attention has been paid to those of cereals and oilseeds because of their food or feed value. Heathcote (1950) ant Tkachuk (1966) tried independently to calculate the N-to-P conversion factors of oat and wheat, respectively, from their amino acid compositions. In these studies, however, amide nitrogen of the samples was not determined, but estimated from ammonium nitrogen in the course of amino acid analysis. Yamaguchi (1979) applied this method to calculate the conversion factors for various kinds of foods, thus newly determining their amide nitrogen content.

Recently, reports on the N-to-P factor of foods determined by their amino acid compositions increased and a symposium titled "N-to-P factors for Food and Feeds Components" was held at the AACC (American Association of Cereal Chemists) 74th Annual Meeting in 1989. Although several problems regarding the present N-to-P factors of foods or feeds have been pointed out in these studies, it is difficult to solve these problems due to the historical background of their usage.

Here, the analytical method of determining the N-to-P factor of foods will be described and the obtained values, not only for cereals but also for other foods, will be presented. Thus, the significance of the determined values for cereals can be elucidated.

2 Analytical Method

In the present study, the N-to-P conversion factors of foods were determined by dividing the weight of amino acid residues of the foods by the weight of their nitrogen, including the amide form. The amino acid residues and their nitrogen were calculated from the amino acid composition tables of FAO (1970) and Japan (1966), and the amide nitrogen was newly determined by the method of Bailey (1937). About 50 Japanese foods were used for the samples. Accordingly, nitrogenous components, except protein and the free amino acids of the samples, were excluded from the calculation of N-to-P factors, which will be discussed in later sections.

2.1 Determination of Amide Nitrogen

The protein fraction was separated as residue after extracting the freeze-dried sample with a chloroform and methanol mixture (2:1) as shown in Fig. 1. First, about 3 g of the sample was weighed accurately and transferred to a 50-ml stoppered centrifuge test tube. After the sample was extracted with 30 ml of the above solvent, using a Politron homogenizer, and centrifuged at 3000 rpm for 10 min, the supernatant was removed by suction and set aside for *joining*. This extraction was repeated three times and the supernatant was *join-ed* and filled to 100 ml with the above solvent for futher fractionation.

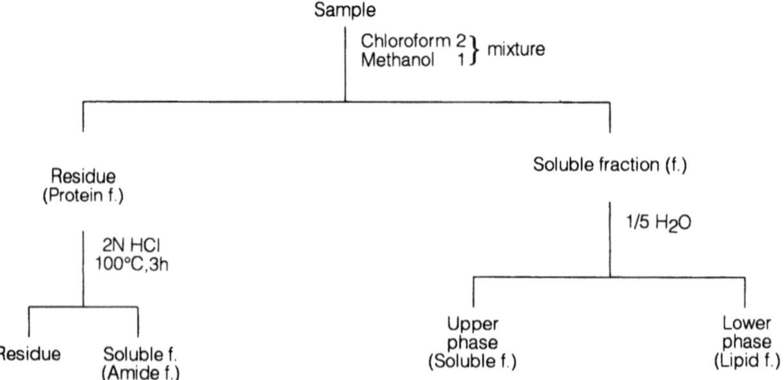

Fig. 1. Fractionation for determination of the amide nitrogen content of foods

Table 1. Distribution of amino acids into the three fractions (%)

	Fraction		
	Protein	Soluble	Lipid
Asparagine	97.4	2.0	0.7
Glutamine	97.9	1.5	0.6
AA mix[a]	79.2	20.8	0.0

[a] Amino acid mixture of whole egg pattern.

The residue was dried over a water bath and used for the protein fraction. About 100 mg of the protein fraction was weighed accurately in a 25 ml stoppered test tube and 5 ml of 2 N hydrochloric acid was added. The protein fraction was then heated in boiling water for 3 h and filtered through filter paper after cooling. An aliquot of the filtrate was analyzed for amide nitrogen by steam distillation according to the method of Bailey (1937).

On the other hand, one-fifth of the water was added to the supernatant, which was left to stand after shaking to obtain the upper and lower layers, i.e., the soluble and lipid fractions, respectively. After centrifugation (3000 rpm, 10 min), the upper layer was sucked with vacuum pump to isolate it into a flask. These fractions were used to examine the distribution of amide nitrogen and other nitrogenous compounds concerned.

Table 1 shows that nearly whole parts of asparagine (Asn) and glutamine (Gln) and a greater part of the amino acid mixture, which was simulated to the pattern of whole egg protein, were present in the protein fraction. Since in ordinary food materials the amount of free Asn and Gln is low compared to that in protein, it is reasonable to determine the amide nitrogen content only for the protein fraction. In addition, it was acertained that the proportion of amino acids, including those of the amide form, per gram of total nitrogen of foods was nearly equal to the proportion of nitrogen recovered in the protein fraction of the samples.

2.2 Calculation of the N-to-P Conversion Factor

An example of the calculation of the N-to-P conversion factor for highly milled rice is shown in Table 2. The first two columns (a and b) indicate the proportion (%) of amino acid residue (AAR) to amino acid (AA) and the nitrogen content (%) of the amino acid residue (AAR), respectively. These values are constant and specific for each amino acid and were used to calculate the conversion factor of any sample. Column c gives the amino acid compositions of the sample which are derived from the amino acid composition table. Column d, i.e., the amino acid residue (AAR, mg) per nitrogen (N, g) of the sample, and column e, i.e., the amino acid residue nitrogen (AAR-N, mg) per nitrogen (N, g) of the sample, are calculated from the values of columns a, b, and c as

Table 2. Calculation of conversion factor for highly milled rice from its amino acid composition and amide nitrogen content

	(a) AAR[a] AA (%)	(b) AAR – N AAR (%)	(c) AA gN (mg)	(d) AAR gN (mg) a·c/100	(e) AAR – N gN (mg) b·d/100
His	88.4	30.6	146	129	39.5
Ile	86.3	12.37	262	226	28.0
Leu	86.3	12.37	514	444	55.0
Lys	87.7	21.8	226	198	43.2
Met	87.9	10.67	133	117	12.5
Cys	92.5	12.60	96	89	11.2
Phe	89.1	9.51	303	270	25.7
Tyr	90.1	8.58	200	180	15.4
Thr	84.9	13.85	207	176	24.4
Val	84.6	14.11	361	305	43.0
Trp	91.2	15.04	84	77	11.6
Arg	89.7	35.9	473	424	152.2
Asp	86.5	12.16	596	516	62.7
Ser	82.9	16.07	291	241	38.7
Glu	87.8	10.84	1195	1049	113.7
Pro	84.4	14.42	293	247	35.6
Gly	76.0	24.5	272	207	50.7
Ala	79.8	19.69	355	283	55.7
Total				5178	819

Amide-N measured (mg/gN) = 64

$$\frac{5178}{819 + 64} = 5.86$$

[a] AA = amino acid; AAR = amino acid residue.

indicated in Table 2. Amide nitrogen (mg) per nitrogen (N, g) of the sample is measured as described above.

Then, the N-to-P factor of the sample is calculated by dividing the total amino acid residues (AAR) by the total amino acid residue nitrogen (AAR-N) plus amide nitrogen as shown at the bottom of the table. Since the difference in molecular weight between the dicarboxylic acids (Asp and Glu) and their amide form (Asn and Gln) is only one, total amino acid residues can be calculated from the amino acid composition table within the limits of errors without determining Asn and Gln. In this way, the conversion factor for highly milled rice was determined to be 5.86.

The validity of this method was verified earlier using pancreatic ribonuclease, whose amino acid sequence is completely known, thus: calculated 5.83, determined 5.87.

3 Determined Values for Cereals

3.1 Rice, Wheat, Barley, and Buckwheat

Specific conversion factors, other than 6.25, are currently applied to these cereals, except the FAO factor of buckwheat. They are indicated in the right-hand column of Table 3. Comparatively little difference was found between the determined factors and current ones for rice and barley, but a slightly larger difference existed for wheat and buckwheat.

3.2 Varieties of Rice

The determined factors for six varieties of rice in Japan are shown in Table 4. They are not significantly different from each other, but slightly smaller than the current factor of 5.95.

3.3 Different Sorts or Grades of Wheat Flour

Table 5 shows the results of hard, medium, and soft types and four different milling grades of wheat flour. Although there was only a little variance between their values, the value for the fourth grade flour was clearly higher than others.

4 Determined Values for Other Foods

4.1 Foods with Current Conversion Factors Other than 6.25

In addition to cereals, for several food groups such as legumes, nuts, seeds, and milk products it is usually recommended to use the specific conversion factors other than 6.25. Their determined values, together with the current ones, are shown in Table 6, where the current values for red bean and kidney bean (6.25) are an exception. On the whole, the determined values were close to their current values.

1) Legumes
   The determined values for the three beans, i.e., soybean, red bean, and kidney bean, were close to 5.6 or 5.7, but that for processed soybean, i.e., soy paste, was a little over 5.8.
2) Nuts and seeds
   The determined values ranged from 5.2 to 5.7. Many of the values were very close to the respective current ones.

Table 3. Determined conversion factors for cereals

Cereals	Determined factor	Current factor (FAO)
Rice, well-milled	5.86	5.95
Wheat, flour (1st, and 2nd grades)	6.03	5.70
Barley, milled, pressed	5.94	5.83
Buckwheat, flour	5.48	6.25
		5.75 (Jones')
		6.31 (Japan's)
	$\bar{m}$ 5.83	$\bar{m}$ 5.82

Table 4. Determined conversion factors for rice varieties

Varieties of rice (current factor, 5.95)	Determined factor
Katsura wase	5.81
Honen wase	5.79
Reimei	5.65
Koshihikari	5.82
Nakada no. 63	5.71
Fujihikari	5.88
	$\bar{m}$ 5.78
	$\pm$ 0.08

3) Milk products

The determined values for skim milk and yogurt were very close to their current value, 6.38, but cheese a processed milk product, showed a lower value.

4.2 Foods with the Current Conversion Factor of 6.25

Table 7 shows the results on other foods to which the current factor of 6.25 has been tentatively applied. Their determined values, especially those of potatoes, vegetables, and fruits, were clearly much lower than 6.25.

1) Potatos and vegetables

The determined values for potatoes, especially that for sweet potato, were extremely low, and those for vegetables were around 5.5 except turnip.

2) Fruits

The determined values for fruits were slightly lower than those for vegetables, showing a mean value of 5.16.

Table 5. Determined conversion factors for different sorts or grades of wheat flour

Different sorts or grades of wheat flour (current factor, 5.70)	Determined factor
Hard flour	6.05 ± 0.05
Medium flour	6.07 ± 0.01
Soft flour	5.98 ± 0.09
First grade	6.03 ± 0.09
Second grade	6.04 ± 0.03
Third grade	6.08 ± 0.08
Fourth grade	6.23 ± 0.06

Table 6. Determined conversion factors for other foods having current factors other than 6.25

	Determined factor	Current factor (FAO)
Legumes:		
Soybean	5.61	5.71
Soy paste (Miso)	5.85	5.71
Red bean (Azuki)	5.69	(6.25)[b]
Kidney bean (Ingenmame)	5.67	(6.25)[b]
Nuts and seeds:		
Almond	5.20	5.18
Cashew nut	5.55[a]	5.30
Ginkgo nut	5.74[a]	5.30
Peanut	5.39	5.46
Pine nut	5.28[a]	5.30
Walnut	5.39[a]	5.30
Sesame seed	5.55	5.30
	$\bar{m}$ 5.44	$\bar{m}$ 5.31
Milk products:		
Skim milk	6.36	6.38
Yogurt	6.40	6.38
Cheese (processed)	6.13	6.38

[a] Based on the Amino Acid Composition Table for Use in East Aisa (FAO 1972).
[b] Exception.

3) Meats and liver

The determined values for beef, pork, and chicken were around 5.8 and those for ham and sausage were 5.6. The values for liver were close to these values.

4) Fish and shellfish

The values for the majority of these foods were around 5.8, but processed foods like fish cake gave slightly different values.

Table 7. Determined conversion factors for other foods having the current factor of 6.25

	Determined factor		Determined factor
Potatoes and vegetables:		Fish and shellfish:	
Potato	5.01	Cod and pollack	5.86
Sweet potato	4.33	Flatfish	5.90
Carrot	5.27	Horse mackerel	5.84
Turnip	4.68	Mackerel	5.67
Cabbage	5.44	Salmon	5.76
Cucumber	5.48	Sardine	5.95
Tomato	5.70	Sea-bream	5.93
Asparagus	5.57	Skipjack	5.84
		Squid and cuttlefish	5.76
	m̄ 5.19	Tuna	5.89
Fruits:		Short-neck clam	5.86
Apple	5.18		
Banana	5.19	Kamaboko (fish cake)	6.15
Grape	5.11	Hanpen (fish cake)	5.56
Orange	4.76		
Pear	5.37		m̄ 5.84
Strawberry	5.34		
		Egg and others:	
	m̄ 5.16	Egg	5.85
Meats and liver:		Shiitake (Lentinus edodes)	5.49
Beaf	5.82	Susabinori (Laver)	5.90
Pork	5.72	Tangle	6.23
Chicken	5.86	Yeast	5.88
	m̄ 5.80		
Ham	5.63		
Sausage	5.60		
Cattle liver	5.64		
Swine liver	5.60		
Chicken liver	6.06		
	m̄ 5.77		

5) Egg and others

The value for egg was also close to 5.8. The values for other miscellaneous foods varied between about 5.5 and 6.2.

5 Further Application to Determine the Value for Different Food Groups Using Their Average Amide Contents

The left-hand column of Table 8 shows the proportion (%) of amide nitrogen to total nitrogen of the protein fraction of samples and the right-hand column shows the molar percentage of total Asn and Gln to total Asn, Asp, Gln, and Glu, which corresponds to the total Asp and Glu taken from an animo acid

Table 8. Contents of amide nitrogen in different food groups

Food groups	Amide-N / Protein-N (%)	Asn + Gln / Asn + Asp + Gln + Glu (%)
Cereals	11.2 ± 3.8[a]	50 ± 15
Pulses	9.3 ± 1.2	49 ± 5
Vegetables	19.1 ± 6.0	69 ± 14
Fruits	17.4 ± 4.1	79 ± 10
Fish and shellfish	5.5 ± 1.2	32 ± 7
Meats and liver	6.7 ± 1.3	41 ± 10
Milk and milk products	10.0 ± 2.6	49 ± 10
Egg	7.0	47
Others	7.2 ± 2.8	44 ± 15

[a] Mean $\pm$ SD.

Table 9. Conversion factors in different food groups

Food groups	Calculated factors	Estimated values by use of amide-N	Those including other food items
Cereals	5.8 (4)[a]	5.8 (4)	5.9 (13)
Pulses	5.7 (4)	5.7 (4)	5.6 (9)
Vegetables	5.2 (8)	5.3 (8)	5.3 (12)
Fruits	5.2 (6)	5.2 (6)	5.3 (9)
Fish and shellfish	5.8 (6)	5.8 (6)	5.9 (11)
Meats and liver	5.8 (4)	5.7 (4)	5.7 (6)
Milk and milk products	6.3 (3)	6.4 (3)	6.3 (5)
Egg	6.0 (1)	6.0 (1)	6.0 (4)

[a] Numbers of samples.

composition table. These dicarboxylic acids contain Asn and Gln deaminated in the course of acid hydrolysis.

It is clear that the amide nitrogen contents gave different characteristic values for the different food groups, where vegetables and fruits showed a particularly higher percentage. Since the molar percentages of amidation of Asp and Glu, as seen in the right-hand column, are rather specific to each food group with comparatively little variance, it seems possible that these values can be used to calculate the N-to-P factors of other foods in each food group.

Table 9 shows the factors for food groups, including the foods in which amide nitrogen was not determined. The values in the first column are the mean of values for the foods indicated in the previous tables (Tables 3, 6 and 7), but they are expressed as the mean to one significant place of decimals. The values in the second column are those calculated for the same samples using the assumed percentage of amidation for each food group as indicated in Table 8. Similar values to those in the first column were obtained. The values in the third column are those applied to the extended varieties of foods in addition to the examined ones. The values were also similar to those in the first and second columns.

Thus, the determined factors of the food groups were in fairly good agreement with the current ones in the cases of specific factors other than 6.25. On the other hand, those of the 6.25 groups, especially potatoes, vegetables, and fruits, showed values much lower than 6.25. The factors for the main protein sources, such as cereals, meats, egg, fish, and shellfish, were found to be around 5.8. The weight average of factors for the ingested proteins, as examined by the Japanese Nutrition Survey, was calculated to be 5.82.

6 Discussion

There have been several reports on the N-to-P conversion factors of foods which were calculated from their amino acid composition and the determined or estimated amide nitrogen content. However, reappraisals of the determination or definition of the factor, especially that of 6.25, have been suggested (Tkachuk 1966, 1969; Yamaguchi et al. 1976–1979; Boisen et al. 1987; Mossé 1990; Sosulski and Imafidon 1990). Moreover, since the methods of determination or calculation of the factor and the species, sorts, or grades of the samples examined were not uniform among the investigators, a definite concept concerning a reappraisal of the conversion factor has not yet been obtained. Since the problems pertaining to the factor for cereals are essencially the same as for other foods, they will be solved more easily by discussing all the data.

Yamaguchi (1976–1979) first introduced the chemically determined amide nitrogen, mostly Apn and Gln, into the calculation of the N-to-P factor of foods by their amino acid composition. Since there are some nitrogenous compounds other than Asn and Gln, which produce ammonium in the course of amide analysis according to Bailey's method, only the protein fraction of the sample was analyzed for amide nitrogen. The protein fraction contained the majority of Apn and Gln of the free form in addition to those of the protein and peptide forms. Nonprotein nitrogen in the soluble and lipid fractions amounted to about 10% to 15% of total nitrogen, possibly reducing an overestimation of amide nitrogen.

The N-to-P factor thus determined refers to the conversion of nitrogen to true protein, but not to crude protein. As stated earlier, the current FAO factors are mostly derived from the Jones's factors which were determined on the prepared proteins of biological materials. Therefore, it can be said that the newly determined factors are characterized to have the same meaning as those of FAO and Jones, which represent the metabolically available nitrogenous compounds.

The nonprotein nitrogenous compounds other than free amino acids and oligo-peptides are represented by ammonium, nitrate, nucleotide, porphyrin, carnosine, anserine, creatine, and others. Their conversion factors as expressed by molecular weight/nitrogen are as follows, for example: nucleotides (yeast) 6.21, porphyrin 5.54, creatine 3.12, carnosine 4.04, and nitrate 4.43 (potasium nitrate 7.72). Since these values are not so different from 5.8 of the average of conversion factors for usual foods, the factor multiplied by total nitrogen of the sample roughly corresponds to the total weight of nitrogenous compounds in the sample.

The values of the N-to-P factors for cereals from the literature, which were calculated from their amino acid composition and the determined or estimated amide nitrogen, were as follows: 5.61 for wheat flour and 5.7 for wheat (Tkachuk 1966); 5.66 for wheat, 5.59 for wheat flour, 5.67 for barley, 5.50 for hulled oats, 5.64 for rye flour, 5.53 for buckwheat and 5.60 for millet (Tkachuk 1969); 5.74 for barley (Boisen et al. 1987); 5.4 for oat bran and 5.5 for barley bran (Hidvegi and Tkachuk 1989); 5.56 for wheat, 5.61 for triticale, 5.58 for rye, 5.72 for barley, 5.53 for pearl millet and 5.52 for oats (Mossé 1990); 5.75 for wheat, 5.61 for rice, 5.72 for corn, and 5.93 for sorghum (Sosulski and Imafidon 1990). These values ranged from about 5.5 to 5.8 and the conversion factors for cereals determined in the present study were slightly higher than these values except for buckwheat. The difference in value for the same food item can arise mainly from the differences in species, sorts, or grades of the samples and in the amino acid composition and amide contents determined or estimated.

However, an overall view of the data, including other food groups, will show that a fairly good agreement in the value of each food group can be found among different investigators. Thus, the validity of this method of calculating the N-to-P factor is apparent. In this report, the factors for foods providing the main protein sources, such as cereals, meats, egg, fish, and shellfish, were found to be around 5.8 and the weight average of the factors for ingested proteins as determined by the Japanese Nutrition Survey was 5.82. Therefore, the value of 5.8 is recommended as the common N-to-P conversion factor for foods instead of 6.25. The value of 5.8 was also suggested in another approach to this problem by Gnaiger and Bitterlich (1984), however, the value of 5.7 was suggested at the same time by Dintzis et al. (1988) and Sosulski et al. (1990).

In the marketing of agricultural products, it is common to use the factor 6.25 for standardization, but it has created difficulties occasionally. In the field of nutrition, on the other hand, protein requirements are estimated as the required nitrogen multiplied by 6.25, but protein intake as examined in the nutrition survey is normally calculated according to the food composition table in which specific factors are used. As a result, a 7% to 8% difference can arise between protein requirements and protein intake based on the same amount of nitrogen. In animal nutrition, a factor of 6.25 has been indiscriminately used for feedstuffs, resulting in some discrepancies between human and animal nutrition.

In the studies of the N-to-P factor, the problems regarding nonprotein nitrogenous compounds other than amino acids in foods have often been discussed. Although it is possible to estimate the factors, including these compounds, this problem rather depends on whether or not the protein content of foods should be defined as the amino acid residues from a nutritional point of view. In any case, further discussion will be required to avoid confusion in both basic and applied fields of food and nutrition.

7 Summary

The nitrogen-to-protein conversion factors for cereals, together with other foods, were calculated by dividing the weight of amino acid residues of the samples by the weight of their nitrogen including the amide form. The amide nitrogen of the samples was newly determined according to the method of Bailey (1937).

The determined factors for cereals were as follows: fully milled rice 5.86 instead of 5.95 (FAO factor), wheat flour 6.03 instead of 5.70, barley 5.94 instead of 5.83 and buckwheat flour 5.48 instead of 6.25. The factors for rice varieties (six samples) were 5.78 and those for different sorts or grades of wheat flour (seven samples) ranged from 5.98 to 6.23, both showing slight variance.

The factors of the main protein sources such as cereals, meat, egg, fish, and shellfish were around 5.8 and the weight average of factors for the ingested proteins as examined by the Japanese Nutrition Survey was 5.82.

Although the determined factors of foods, including cereals, were in fairly good agreement with the current ones in the cases of specific factors other than 6.25, further studies and discussions will be needed to obtain more accurate values of individual factors and to determine a common factor for foods and feedstuffs in order to apply these values in practice.

References

AACC (1989) Abstracts of the AACC (American Association of Cereal Chemists) 74th Annu Meeting, Washington, DC, pp 211–214

Bailey K (1937) Composition of the myosins and myogen of skeletal muscle. Biochem J 31:1406–1413

Boisen S, Bech-Anderson S, Eggum BO (1987) A critical view on the conversion factor 6.25 from total nitrogen to protein. Acta Agric Scand 37:299–304

Dintzis FR, Cavins JF, Graf F, Stahly T (1988) Nitrogen-to-protein conversion factors in animal feed and fecal samples. J Anim Sci 66:5–11

Gnaiger E, Bitterlich G (1984) Proximate biochemical composition and caloric content calculated from elemental carbon hydrogen nitrogen analysis – a stoichiometric concept. Oecologia 62:289–298

FAO (1970) Amino acid content of foods and biological data on proteins. FAO Nutritional Studies No 24, Rome

FAO (1972) Amino acid composition tables for use in east asia, Rome

Heathcote JG (1950) The protein quality of oats. Br J Nutr 4:145–154

Hidvegi M, Tkachuk R (1989) N-to-P factors and amino acid composition of oat and barley bran. In: AACC (1989), p 213

Japan (1966) The amino acid composition of foods in Japan. Resources Bureau, Science and Technology Agency, Tokyo

Jones DB (1931) Factors for converting percentage of nitrogen in foods and feeds into percentage of proteins. US Dept Agric Circ, No 183, Washington, DC, pp 1–21

Mossé J (1990) Nitrogen to protein conversion factor for ten cereals and six legumes or oilseeds. A reappraisal of its definition and determination. Variation according to species and to seed protein. J Agric Food Chem 38:18–24

Mulder GJ (1839) Über die Zusammensetzung einiger tierischer Substanzen. J Prak Chem 16:129–152

Sosulski FW, Imafidon GI (1990) Amino acid composition and nitrogen-to-protein conversion factors for animal and plant foods. J Agric Food Chem 38:1351–1356

Tkachuk R (1966) Note on the nitrogen-to-protein conversion factor for wheat flour. Cereal Chem 43:223–255

Tkachuk R (1969) Nitrogen-to-protein conversion factors for cereals and oilseed meals. Cereal Chem 46:419–423

Yamaguchi M (1979) Studies on determination of amide nitrogen compounds in dairy products. In: Report of the studies on establishment of analytical method for special components of foods (in Japanese), Resources Bureau, Science and Technology Agency, Japan

Yamaguchi M, Matsuno N, Miyazaki M (1976) Studies on determination of amide nitrogen compounds in dairy products. In: Report of the studies on establishment of analytical method for special components of foods (in Japanese), Resources Bureau, Science and Technology Agency, Japan, pp 168–177

Yamaguchi M, Matsuno N, Miyazaki M (1977), pp 139–144

Yamaguchi M, Matsuno N, Miyazaki M (1978), pp 55–62

Yamaguchi M, Matsuno N, Ohno I, Miyazaki M (1978) Nitrogen-to-protein conversion factors with special regard to food proteins. Abstr 5th Int Congr Food Science and Technology, p 244

Protein Analysis of Wheat
by Monoclonal Antibodies
and Nuclear Magnetic Resonance

J.-C. Autran

1 Introduction

Wheat ranks first among our cultivated plants. Production in 1990 approached 600 million metric tons. Despite increasing industrial end uses, most wheat is used for food. In addition to providing a range of nutrients, it also possesses remarkable technological properties which allow the production of a variety of different processed foodstuffs such as bread, biscuit, and pasta. In addition, upon removal of the water-soluble components of a flour, *wheat proteins* have the unique property to form, with a few percent lipids, an insoluble and viscoelastic proteinaceous mass termed *gluten*, which forms the basis of the rheological properties of dough. Dry gluten, also, is increasingly used as improver or additive in flours and in various foods.

This great nutritional and functional importance of wheat proteins has stimulated investigation of the genetics and biosynthesis of wheat storage proteins, while the traditional interests of physical chemists in understanding the contributions of wheat protein components to milling, dough-forming, and baking properties of wheat have continued undiminished (Kasarda et al. 1976). The protein content of wheat grain is one of the basic measurements of its quality in marketing, while protein composition is primarily responsible for quality differences among different varieties. For instance, the presence or ratio of some allelic variants of gliadin or glutenin fractions are valuable indicators of the bread-making potential in wheat breeding programs.

There is some evidence that the conformations adopted by some specific protein components (e.g. HMW or LMW subunits of glutenin) play an important role in dictating the functional properties of wheat gluten. However, these proteins are highly heterogeneous and largely insoluble: their functionality appears in a weakly hydrated dough medium in which hundreds of constituents interact to determine cohesive, extensible, and elastic characteristics. The understanding of their functional properties cannot be derived from studies of protein solutions; and the detailed investigations of their basic components cannot be carried out while respecting the integrity of their native structure. These are the reasons why many conventional methods based on solubility, electrophoretic or chromatographic fractionations, and aimed at exploring gluten structure or detailed composition, have been only partially successful.

Despite many years of study, therefore, we do not yet have a detailed understanding at the molecular level of the basis for the unique properties of doughs and the way in which gliadins, glutenins, and other constituents contribute to the functional properties of different wheat flours.

Several recent advances provide the potential to make a significant step forward in a more complete understanding of the fundamental bases of quality as well as and in the development of improved wheat varieties and wheat products or dietary foods that come within legal requirements.

This chapter reviews the potential of two especially powerful techniques: *monoclonal antibodies* and *nuclear magnetic resonance*, which offer alternative means to study conformational aspects of cereal storage proteins, to yield information on the functionally important sites on proteins while approaching the quantification, composition, structure, and function of some important wheat protein fractions.

An understanding of the nomenclature and basic physical chemistry of wheat proteins is assumed; the reader is referred to the published reviews on cereal proteins (Shewry and Miflin 1984; Feillet 1988, Wrigley and Bietz 1988; Bushuk and MacRitchie 1989) and on physical (Tatham et al. 1990) and immunochemical (Skerritt 1988) aspects.

2 Monoclonal Antibodies

2.1 Immunochemistry in Wheat Proteins

The immunochemical studies of wheat proteins have been pursued by two main groups of researchers. Physical chemists have used immunochemical methods to study protein structures. Simultaneously, clinical immunochemists have investigated the serum antibody response to wheat and other cereal proteins in gluten intolerances such as celiac disease.

Approaches to immunochemical studies of wheat proteins and enzymes are numerous, and each of them can incompass various techniques. They include immunoprecipitation in solution or in gel, immunoabsorption, immunoaffinity chromatography, enzyme-linked immunosorbent assay (ELISA), radioimmunoassay (RIA), dot-binding and immunoblotting assays, and immunohistological techniques.

Early immunological studies of wheat relied on less informative techniques requiring precipitin lines to indicate antibody-antigene reaction and polyclonal antisera comprising a multiplicity of antibodies. Much more specific information can now be obtained with monoclonal antibodies, and with radio- or enzyme-linked assays together with direct testing of zones separated by gel electrophoresis. Fuller reviews of the principles of these techniques and their particular use in the study of cereal proteins and enzymes have been recently provided by Daussant and Bureau (1988) and Skerritt (1988).

2.2 General Principles of Monoclonal Antibody Production and Utilization

Immunochemistry makes use of special seric proteins (immunoglobulins), called antibodies, which appear in vertebrates in response to the injection of foreign constituents called antigens. During immunization, antigens induce the proliferation of specific clones of B-lymphocytes and the production of antibodies specific for various discrete parts of the antigen molecules. These parts consist of a small number of amino acids and are called antigenic determinants, or epitopes (Daussant and Bureau 1988). Because proteins generally have several structurally different epitopes, several B-cells with different receptors can be selected. They multiply and give birth to several clones of cells. All cells from the same clone synthesize antibody with the same specificity. But, as several clones are involved in this production, the antibodies of an immune serum are called *polyclonal antibodies*. As the population of antibody-producing cells is not constant, the antibody sets vary. Consequently (even if a single antigen is used for immunization), immune serums sampled at different times, or from different animals immunized with the same antigen, generally contain an antibody specific for the same epitope and others specific for different epitopes.

In contrast, antibodies specific for a single epitope can be proliferated in tissue culture by the selection and multiplication of a single-immune competent cell. Such antibodies produced by cells of the same clone and possessing therefore exactly the same specificity are called *monoclonal antibodies*. Presently, a clone can be obtained from a single cell by fusion of myeloma cells with B-lymphocytes taken from the spleen of an immunized mouse. The resulting hybridoma retains the properties of both mother cells: it secretes antibodies with the same specificity and multiplies indefinitely. Hybridomas are selected by culturing the cells in a medium in which the growth of either parental line is not possible. The hybridomas are then separated, and the antibodies produced by the resulting clones are tested for the desired specificity by ELISA and immunoblotting. In contrast to polyclonal antibodies, which constitute a collection of immunoglobulins with different specificities, monoclonal antibodies are constant in their specificity; they can be produced in unlimited quantities; their secreting cells can be stored in liquid nitrogen to be later expanded at will (Daussant and Bureau 1988).

However, monoclonal antibodies from one clone can only precipitate if this antigene possesses only two copies of the particular epitope that is recognized by the monoclonal antibodies. If only one copy is found per antigen molecule, only small, nonprecipitating complexes consisting of one antibody molecule combining with identical epitopes on two separate antigen molecules can be generally formed (Daussant and Bureau 1988). In contrast to the polyclonal antibodies, therefore, monoclonal antibodies from one clone alone cannot be used in immunochemical techniques based on the formation of large antigen-antibody complexes. They are, however, well suited for other analytical techniques, where detection of antibody-antigen recognition is based on other means. Such methods include the immunofluorescence method and the ELISA (Vaag and Munck 1987; Gallant et al. 1989).

2.3 Methodological Problems in the Immunochemical Study of Wheat Proteins and Use of Monoclonal Antibodies

One factor that has complicated the use of immunochemical techniques in cereal protein analysis was the difficulty to obtain good serum antibody responses to certain protein fractions such as prolamins. For instance, the titer of the hordein antiserum obtained by Asano et al. (1983) was 1/100 of that of an albumin-globulin antiserum, and, to produce rabbit antisera to gliadin, Ciclitira et al. (1985) performed over a dozen immunizations.

According to Skerritt (1988), the *poor imune response* to prolamins was due to immune tolerance or suppression resulting from antigen feeding rather than to poor immunogenicity of prolamins as such.

A second major difficulty resulted from the *insolubility* of storage proteins. Initial attempts based on diffusion-in-gel methods appear best suited for water- or salt-soluble extracts. Since they required long periods for diffusion and consume large amounts of antisera, they were unsuited to prolamins or glutelins. Fortunately, the enzyme-immunoassay methods developed over the last years with monoclonal antibodies proved adaptable to water-insoluble proteins, they are rapid (a few minutes), amenable to automation, sensitive, and suitable for analysis of large numbers of samples (Skerritt and Henry 1988).

However, because many of the common solvents for wheat storage proteins (urea, alcohols, acids, detergents) are protein denaturants, antigenic reactivity may still be lost or reduced. It has been often necessary, therefore, to optimize conventional procedures in order: (1) to retain native epitope structures; (2) to produce an acceptable signal; and (3) to have a good retention on the solid phase, making the monoclonal antibodies able to function with convenient materials (microwell plastic trays). In addition, the screening assays had to be chosen with greatest care, for instance, to select monoclonal antibodies able to recognize the spatial (as opposed to sequential) organization of residues and, in any case, to select antibodies designed to specific purposes (EIA, immunoblotting, immunocytochemistry, etc.).

The effects of various solvents of wheat gliadins and glutenins, and the nature of the solid phase used for antigen immobilization were thoroughly studied by Skerritt and Martinuzzi (1986). Antigen solvents affected both protein retention on the solid phase and the sensitivity of detection: alcohol- and urea-based extractants were far superior to SDS solutions. A panel of 12 monoclonal antibodies also showed marked differences in their abilities to bind to antigen immobilized on nitrocellulose and on plastic. Despite better convenience of microwell plastic trays for screening large numbers of samples, the lower binding capacity of gluten proteins (some clones did not produce a signal with gliadin bound to microwell EIA plates, even at high concentrations) and the possible alteration of epitope structure made the use of nitrocellulose disks (prepared from a nitrocellulose solid phase soaked in food extracts) a more suitable solid phase for use with trace antigens in protein mixtures or with certain monoclonal antibodies (Skerritt and Martinuzzi 1986). Recent developments from Skerritt and Hill (1990a), however, including optimization of coating and washing polystyrene microwells and choice of solvent and extracting conditions, allowed the development of novel high-affinity monoclonal antibodies able to function for the first time in microwell sandwich EIA and to accurately quantitate gluten in all types of foods.

Mills et al. (1989a) confirmed that direct absorption of proteins onto the surface of microtitration plates may cause considerable loss of native structures. To retain absorbed gliadin without disrupting conformational epitopes, they developed another two-site, enzyme-linked immunosorbent assay for wheat gliadins in which chicken antibodies against gliadin are absorbed to the microtitration plate as capture antibodies.

2.4 Main Applications of Monoclonal Antibodies to the Study of Wheat Proteins

Interest in the application of monoclonal antibodies to wheat proteins has come from two main directions: investigation of structural relationships between protein fractions of wheat and other cereals by physical chemists and study on clinical sensitivity to cereals. Alternatively, other applications of these methods have given insights in various aspects of cereal knowledge:

- Genetic studies including gene expression, genome relationships, and varietal identification.
- Studies of grain development and localization of specific protein components.
- Aid in molecular biology studies.
- End-use quality studies.
- Immunological detection of gluten in foods.

The following sections evaluate the present impact of monoclonal antibody methods, pointing out the various applications, limitations, and trends.

2.4.1 Structural Homologies Between Wheat Proteins

Immunochemical techniques have great potential for studying antigenic homologies between proteins and thus structural homologies of regions on the molecular surface. Monoclonal antibodies raised to gluten proteins have been used to examine the extent of structural homology of gluten proteins within or between genotypes of bread wheat, and also between similar proteins in related cereal species.

In earliest studies, Skerritt et al. (1984) prepared monoclonal antibodies to a gliadin protein extract of *Triticum aestivum* and specific antibody-cereal protein interactions were detected using horseradish peroxidase-coupled second antibodies after transfer of proteins to nitrocellulose following electrophoresis. Many clones showed broad specificity, while several clones secreted antibodies selective for smaller families of gliadins such as ω-gliadins and bound neither with HMW subunits of glutenin, albumins, globulins, nor a variety of other proteins.

However, certain related species such as durum wheat, barley, and rye contained endosperm proteins recognized by these monoclonal antibodies. This observation confirmed that *sequence homologies exist between prolamins from wheat and related cereal species* as well as between certain gliadins from hexaploid wheat (du Cros et al. 1984; Skerritt et al. 1984).

Skerritt and Underwood (1986) have used immunoblotting methods to classify the specificities of a library of monoclonal antibodies. While most anti-gliadin monoclonal antibodies bound to all gliadin bands separated by PAGE, several antibodies binding to smaller gliadins were identified. At higher concentrations, however, these specific antibodies bound to an increasing number of α-, β-, and γ-gliadins, indicating very high sequences homologies between groups of gliadins.

In recent investigations, Skerritt and Lew (1990) prepared a library of monoclonal antibodies to wheat gluten proteins and studied quantitatively the interactions with extracts of total seed storage proteins from related cereals using immunoblotting and EIA techniques. They found antibodies giving four cross-reaction types that were generally in agreement with structural homologies determined from DNA sequencing: (1) selective binding to wheat proteins; (2) similar specificity with gliadins with binding to prolamins from other species (rye, barley, oats); (3) specific binding to certain γ- and ω-gliadins and HMW subunits of glutenin, with strong binding to rye and barley proteins; and (4) anomalous cross-reactivities such as binding to wheat or maize proteins but no binding to rye or barley proteins. In these studies, however, the interpretation of quantitative data was limited by a number of constraints such as differences in binding according to the cereal variety within a species, solvent, assay format (indirect, competition, or sandwich ELISA), solid phase (plastic, nylon, or nitrocellulose membrane).

The effects on apparent antibody specificity and cross-reaction have been investigated in detail by Skerritt and Hill (1990b). Because specificity could be manipulated by variation of one of these parameters, these authors strongly recommended *to define cross-reaction of antibodies with respect to the assay format used* when reporting any immunological homology. Moreover, because cross-reactivity of monoclonal antibodies to cereal proteins showed considerable variation with either broad or narrow specificities, Freedman et al. (1988) stressed the point that, in any attempt to determine homologies or classification of cereal proteins, *a very careful selection of antibodies* (and of proteins or peptides used for their characterization) is needed.

Although *glutenins* have been the subject of fewer studies than gliadins, monoclonal antibodies to glutenins were also produced, with specificity for all major groups of glutenins, although at low concentration some bound selectively to a single subunit (Skerritt and Underwood 1986). Other anti-glutenin antibodies bound to a variety of proteins including γ-gliadins or HMW subunits of glutenin, suggesting the presence of identical or at least similar epitopes in these different groups (Mills et al. 1990) and weakening the conclusion of Ewart (1977) that such homologies could largely result from mutual contamination. Moreover, Skerritt and Robson (1990) compared the immunological homologies of low molecular weight (LMW) subunits of glutenin with the other major gluten polypeptides (HMW and gliadins) by one-step, or two-step SDS-PAGE in concert with immunoblotting and EIA methods. Many antibodies raised to gliadins and HMW bound to LMW and antibodies with specificities for similar groups of gliadins bound to similar groups of glutenins (Table 1), supporting conclusions from gene sequencing studies that LMW may be responsible for many biochemical properties and quality effects previously attributed to gliadins. On the other hand, some antibodies bound to each of the major gliadins, LMW and HMW, but not to other grain proteins, suggesting the existence of "common gluten" amino acid sequences or conformations (Skerritt and Robson 1990).

Another class of wheat endosperm nongluten protein, which is increasingly considered, is *surface starch granule protein (SGP)*, a fraction (0.2% − 0.3% by

Table 1. Cross-reaction of monoclonal antibodies with gliadins and glutenins − immunoblotting results (Skerritt and Robson 1990)

Gliadin specificity	Glutenins[a]		
	HMW (A)	LMW	
		B	C
$\alpha\beta\gamma$ (high-mobility)-Gliadin binding			
221/23 $(\alpha>)$[b]	−	−	+ +
230/9 $(\alpha\beta)$	−	+	+
403/8 $(\alpha\beta>\gamma)$	−	−	+
227/22 (β)	−	−	−
404/6 $(\beta>\gamma)$	−	−	±
222/5 $(\alpha\beta\gamma)$	−	−	+ +
$\gamma\omega$-Gliadin binding			
218/17	±	+ +	+
236/9	+	+	±
237/24	+	+	±
246/21	+	+	±
401/16	+	+ +	+ +
ω-Gliadin binding			
122/24 $(\omega$ slow)[b]	+	+	+
304/13 $(\omega$ fast)	+ +	−	−
401/21 $(\omega$ slow)	+	+	+
405/7 $(\omega$ slow)	+ +	+ +	+ +
Broad specificity			

[a] No reaction (−), rather weak reaction (±), weak-moderate reaction (+), very strong reaction (+ +). HMW and LMW, high and low molecular weight, respectively.
[b] Specificity determined by acidic-buffer PAGE and immunoblotting.

weight) that is retained by washed wheat starch. Monoclonal antibodies have been raised to SGP by Ariss (1986), giving surprisingly cross-reactions with HMW subunits of glutenin and ω-gliadins. Recently, Skerritt et al. (1990a), using immunoblotting, ELISA, and immunocytochemical techniques, observed that antibodies with similar gliadin and glutenin specificities had similar SGP specificities. For instance, some antibodies to α-, β-, or γ-gliadins labelled both protein bodies and the periphery of starch granules in sections of immature grains. In addition, antibodies binding broadly to all major gluten protein classes also bound most SGPs, although the Mr 15000 fraction, which has been associated with endosperm softness, appeared immunologically distinct.

2.4.2 Genetic Studies: Genome or Variety Relationships and Gene Expression

Storage proteins show considerable polymorphism and prove useful as genetic markers. In connection with various immunochemical studies applied to taxonomy and genome-species relationships, monoclonal antibodies (especially through immunoblotting methods and quantitative immunoassays) have recently permitted further understanding of genetic linkages between gliadins genes on specific chromosomes (Skerritt 1988). Interestingly, removal of one chromosome pair and substitution of a homoeologous pair did not alter the binding of certain antibodies, so that genes on several chromosomes seemed to be necessary for full binding of (for instance) one β-gliadin-specific monoclonal antibody. This finding suggests that the corresponding epitopes were present on proteins coded by any of a large number of genes on different chromosomes and supports the theory that the large number of gliadin polypeptides in each wheat variety results from duplication and divergence of a few ancestral genes (Kasarda 1984). This possibility and additional information on the possible genome-species origin of various epitopes were reinforced by the study of the primitive wheats such as *Triticum monococcum* (A-genome donor) or *Triticum speltoides* (putative B-genome donor) (Skerritt 1988).

Several important applications of this breakthrough in genome-species origin of wheat protein epitopes may be expected: measurement of the level of specific protein products of quality-related genes; determination of the gene copy number by immunological measurement of the level of a gene product; new insights in the regulation of gliadin gene expression by genes on distant chromosomes, whose study should be of high priority with regard to better control the expression of quality-related genes in wheat (Skerritt 1988).

Despite the availability of methods (immunoblotting, quantitative EIA, and RIA) for determining antibody specificities, fewer authors have been successful in preparing variety-specific antibodies to gluten proteins. Fritschy et al. (1985), using a "sandwich" EIA with antigen extracts of several *Triticum aestivum* and *Triticum durum* cultivars found large varietal differences in the binding of an antiserum to aggregable A-gliadin, but the varietal differences of an antiserum to α-, β-, and γ-gliadins were lower. In connection with similar studies on various cereal varieties (Skerritt et al. 1986; Wrigley et al. 1987a), and supplementing electrophoresis, HPLC, or turbidity techniques, the binding of various antibodies (including an ω-gliadin-specific monoclonal antibody) was examined by du Cros et al. (1984) and Skerritt et al. (1988) among Australian cultivars, showing that a specific binding of this antibody to a pair of slow-moving ω-gliadins occurred in all varieties. Quantitative antigen-competition immunoassays, developed by Skerritt et al. (1987) to measure the effect of sulfur deficiency in wheats, were also reported by Skerritt (1988) with several monoclonal antibodies to assess varietal differences in binding.

Recently, with respect to discriminating cultivars and overcoming the high degree of homologous amino acid sequences between gliadins which result in a high degree of cross-reactivity when polyclonal or even monoclonal antibodies are prepared against whole gliadin, Dawood et al. (1989) prepared monoclonal antibodies against two purified gliadin components (γ-45 and α-74). While protein blotting of total gliadin separated by SDS-PAGE showed that monoclonal antibodies prepared against α-74 bound to a large region corresponding to α- and β-gliadins, those prepared against γ-45 bound to one discrete region corresponding to the location of the starting antigene, indicating that γ-45 had an unique epitope. According to Dawood et al. (1989), the higher specificity of the latter monoclonal antibody compared with the results of Skerritt et al. (1984) was presumably due to the use of a pure antigen, which gave the immunized animal a chance to develop antibodies against a unique epitope that may be present at only a low concentration in a total gliadin extract. Thus, such clones may have the potential to establish an immune-based test to differentiate wheat cultivars using a battery of monoclonal antibodies prepared against several purified gliadins, as well as to determine the proportions of wheat, rye, and other cereal grains in various food and feed products.

2.4.3 Identification of Translation Products of mRNAs and Characterization of cDNA Clones Expressing Specific Wheat Endosperm Proteins

The identification of in vitro translated proteins needs highly sensitive and specific techniques. When classical alcohol extraction and SDS-PAGE methodology are used to characterize gliadins, other proteins of similar molecular mass such as LMW-glutenins are also solubilized. On the other hand, highly aggregative and insoluble fractions such as HMW glutenins cannot be studied and prematurely terminated translation products or precursors cannot be recognized either.

Due to their high specificity, immunoprecipitation and immunoblotting are therefore increasingly used in concert with molecular biological techniques to identify the protein products of mRNAs, to study the relationship between mRNA levels and protein product level, to identify precursors, to detect post-translational processes, or to establish whether the synthesis of particular

Table 2. Specificities of monoclonal antibodies with blots from one-dimensional SDS-PAGE blots, double one-dimensional SDS-PAGE blots, and with starch granule proteins (Donovan et al. 1989)

Clone	Subclone	Isotype	SDS-blots (one-dimensional)	SDS blots (LMW glutenins double one-dimensional)	Starch granule proteins
218/17	7E8	IgG1	γ/Fast ω + HMW glutenin	+ +	+, Surface
221/23	8D11C8	IgG1	α-gliadin	+	+, Surface
222/5	9F9	IgG1	Gliadin, mainly α, β, γ	+ +	+, Surface and intrinsic
227/22	12H12	IgG1	β-Gliadin	–	+, Extrinsic
228/20		IgG1	α- and β-gliadin	+ +	–
236/9	13C6	IgM	γ, ω + HMW glutenin	+	+ +, Intrinsic
237/24	4H11	IgM	γ, ω + HMW glutenin	–	+, Intrinsic
246/21	18F2	IgM	γ, ω + HMW glutenin	+	+ +, Intrinsic
304/13	B2	IgG1	Fast ω, weak HMW glutenin	–	+, Intrinsic

endosperm proteins is under transcriptional or translational control (Reeves et al. 1986; Skerritt 1988).

In addition, molecular biology has provided potential tools for separating and characterizing both cDNA and genomic clones encoding wheat storage proteins and has allowed both the primary sequences of the proteins and the putative controlling regions of the genes to be deduced. The extensive homologies of gliadin or glutenin components, however, have resulted in a high frequency of cDNA cross-hybridization making the screening and character-ization of the clones laborious and tedious. Following various studies on zein and other maize proteins, monoclonal antibodies have proved invaluable for the screening of *expression vector* libraries (such as in lambda gt 11), i.e., in identifying hybrid proteins after their synthesis by bacterial colonies bearing recombinant DNA coding for wheat proteins (Donovan et al. 1989; Table 2).

2.4.4 Studies on Grain Development and Localization of Specific Protein Components by Immunocytochemical Methods

Immunological probes have been used to follow the developmental accumula-tion pattern of wheat storage proteins. The use of immunoblotting methods in concert with several monoclonal antibodies of different specificities allowed the recording of temporal differences in the deposition of various protein groups and showed that maturity patterns were altered by environmental stresses such as sulfur deficiency (Skerritt et al. 1988a).

Another application has been the localization of specific proteins on sections of grain, dough, or baked goods by immunocytochemical methods. These include tissue fixation, incubation with antisera labeled with either a fluorescent dye, an enzyme (peroxidase), or gold particles and examination by light or electron microscopy (Gallant et al. 1989). The use of monoclonal antibodies coupled directly or indirectly to colloidal gold has therefore enormous potential for locating individual proteins at the ultrastructural level and offers an alternative approach to conventional physical techniques for investigating the changes taking place during dough development and bread-making (Parker et al. 1990). In an early study, Ariss (1986) prepared monoclonal antibodies to various wheat proteins; some monoclonal antibodies to gluten proteins were protein body-specific, while some labeled both glutenins and starch granule proteins; certain antibodies labeled protein bodies unevenly, suggesting the presence of internal substructures. In addition, antibody staining showed concentric rings of integral starch granule proteins in A-type granules, but not in B-type granules.

On the other hand, because of the very well-defined structural specificity of monoclonal antibodies, structural changes could also be demonstrated using antibody cytochemical studies of doughs (Ariss 1986). In baked dough, this study confirmed the existence of a gluten network in which starch granules are embedded. In contrast, changes from a protein matrix in which starch granules are deposited to a starch matrix surrounding aggregates of proteins were observed in extruded products. In addition, a loss of antigenicity of gluten proteins (but not of starch granule proteins) at high extrusion temperatures was observed.

In a recent report, Parker et al. (1990) developed an indirect, two-step, immunolabeling procedure to locate gluten proteins in wheat and bread. In developing grain, storage protein bodies were immunolabeled and there was no cross-reactivity with aleurone proteins or nonstorage proteins associated with starch. In addition, some epitopes were not destroyed by the baking process: a specific antibody was bound strongly to the cut surface of bread crumbs but did not recognize the gluten-gas cell interface.

2.4.5 Study of Protein Structure and Interactions

Much of the challenge of the 1990s remains in obtaining a more accurate picture of the relationship between the presence or amount of specific proteins in wheat and end-use quality (baking strength, grain hardness etc.). Complementary to electrophoretic or HPLC techniques, it is possible to exploit the potential of immunochemical methods, especially those based on monoclonal antibodies, to recognize protein conformation, to yield information on the functionally important sites, or to quantitative specific endosperm polypeptides.

The development of monoclonal antibodies has an important impact on the study of protein structure and interactions because *immunochemical analysis relates to the surface structure of proteins which are involved directly in inter-*

actions. Since they bind to single specific points of the antigen, monoclonal antibodies are especially valuable and may constitute surface markers of very high specificity. According to Feillet and Popineau (1990), raising monoclonal antibodies against various conformation states of a protein will yield *a series of different structural probes whose binding behavior can be related to the conformational state of the antigen.* Moreover, specific monoclonal antibodies of particular sequence domains can be prepared by raising them against peptides, natural or synthetic. Because the binding of monoclonal antibodies can be modified by conformational changes or protein-protein interactions that cause modifications in the access to antigenic sites, monoclonal antibodies (especially those raised against specific peptides) will allow a better identification of the interacting areas of wheat and dough proteins.

Because dough strength and baking properties are largely determined by both the subunit composition of glutenin polypeptides and the amounts of particular subunits, the effects of varying glutenin-subunit compositions on flour properties can be studied using genetic variants. With this objective, Skerritt and MacRitchie (1990) quantified certain glutenin subunits using a monoclonal antibody-based ELISA technique. With a number of sets of wheat varieties, both methods yielded positive correlations between the amount of glutenin aggregate or of subunits, and measures of dough strength. Studies, using lines lacking specific high and low molecular weight subunits, have shown that antibodies binding to D-genome HMW glutenin subunits (i.e., allelic types 5-10/2-12) yield the best correlations. In addition to providing valuable information on the structure of the gluten complex, these techniques can be applied to quality assessment in wheat breeding and by milling and bakery laboratories (Skerritt and MacRitchie 1990).

2.4.6 Breeding for Quality

As monoclonal antibodies can be utilized in very rapid (ca. 10 min) tests, which are easy to perform and allow efficient screening of large numbers of new lines for desirable characteristics, antibodies specific for proteins involved in quality may aid progeny selection in plant breeding programs. In addition, as single seeds could be used for source material in screening of breeders' samples for quality, the lengthy process of multiplication at present required for such characterization would be eliminated (Mills et al. 1989b).

Earlier attempts from Skerritt et al. (1987) were centered around the measurement of the detrimental effect of sulfur deficiency, which is associated with a change in relative proportions of gliadins, resulting in a decrease in dough extensibility. In this study, an antibody (clone 227/22) specific for a sulfur-rich fraction (β-gliadin) was identified whose binding in a competition enzyme-immunoassay showed significant correlation with flour sulfur and dough extensibility (Fig. 1).

In another study (Skerritt 1989) the screening of 71 wheat varieties from 26 countries with several gliadin- or HMW-glutenin-binding antibodies was de-

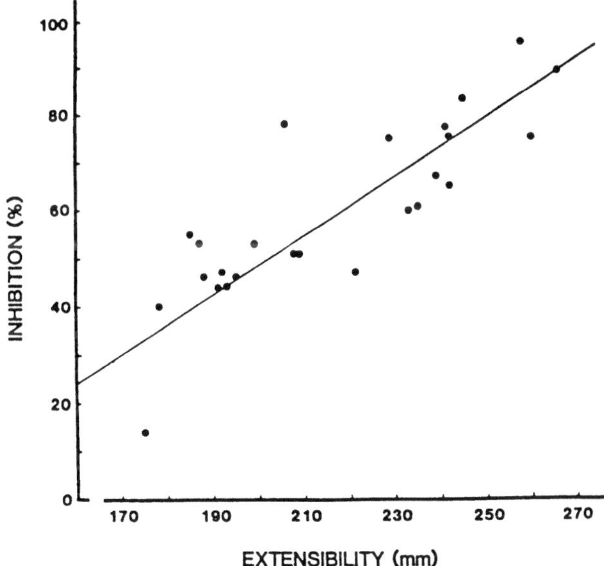

Fig. 1. Relationship between dough extensibility and inhibition of binding to antibody 227/22 (measured by competition immunoassay) for 25 samples of flour milled from field-grown Olympic grain (1980 trial). (Skerritt et al. 1987)

scribed, showing three types of antibody-binding patterns: (1) no varietal difference in binding other than that due to differences in total protein content; (2) varietal differences in binding noted, but no correlations found between antibody binding and quality parameters; and (3) varietal differences in binding, correlated with quality parameters. This final group of antibodies may be of use in quality testing. Of seven gliadin-specific antibodies tested by competition enzyme-immunoassay using flour extracts made with 1 M urea, two bound in a manner which correlated significantly ($P < 0.02$) with dough resistance and dough-mixing work input. These were antibodies 230/9 and 404/6, with specificity to α/β- and β/γ-gliadins, respectively.

Another indication of the feasibility of using monoclonal antibody techniques to quantify specific proteins associated with quality traits was recently reported by Chan et al. (1990). Monoclonal antibodies were prepared to detect HMW-glutenin sequences that are assumed to indicate desired wheat characteristics such as the presence of a cystein residue at position 97 or of the amino acid sequence Thr-Cys-Pro, a characteristic of the HMW subunit No. 5. The use of monoclonal antibodies would therefore allow a rapid prediction of good baking quality and would be of benefit to millers wishing to control the quality of the flour which they buy, and also to cereal breeders who could readily determine the bread-making quality of a new wheat lines and thereby accelerate their cross-breeding programs.

2.4.7 Immunological Detection of Gluten in Foods –
Adulteration of Wheat Products

Wheat gluten is used as an additive in foods, especially meat products or soups and desserts for economic reasons and for its functional properties, e.g.: viscoelasticity, thickening or binding capacity, and stabilization of oil-water emulsions. Consequently, there is a need to determine the amount of added gluten, which is strengthened by the existence of some individuals with gluten intolerances, including celiac disease (see Sect. 2.4.8).

Despite limitations due to the great variability of physical and chemical nature of foods and to the various mechanical or heat treatments that often result in a loss of solubility and antigenicity of proteins, immunochemical techniques have a great potential for gluten detection owing to their specificity and sensitivity.

Especially as conventional antisera used in food testing may fail to recognize antigens after cooking and because immunizing animals with heat-stable proteins may give rise to a large number of molecules of different specificity, Skerritt and Smith (1985) stressed the need to develop monoclonal antibodies, in particular to slow-moving ω-gliadins (which lack cysteine), the most heat-stable gluten fraction. Wheat and rye prolamins could thus be detected from a wide range of foods including baked goods and processed meats by simple transfer from PAGE patterns to nitrocellulose membranes, followed by treatment with enzyme-conjugated monoclonal antibody (EIA). Upon addition of the appropriate enzyme substrate, gluten-containing foods yield purple spots, while other prolamins from barley, oats, maize, or rice are weakly or not detected, and a wide range of nongluten common food proteins do not react (Skerritt and Smith 1985). This immunoassay was then improved to give quantitative results by soaking small disks of nitrocellulose in food extracts, incubating with an antibody and a horseradish-peroxidase substrate yielding a soluble product, and estimating the gluten content by photometric measurements using standard curves for gliadin (Skerritt 1985a, b;, Fig. 2). Further results and discussions on the factors governing the choice of suitable food and grain extractants, the methods for detecting a bound antibody, and the choice of primary or secondary antibodies were also reported by Skerritt et al. (1985), suggesting that artifact-free detection could be obtained using the peroxidase-antiperoxidase technique or by direct conjugation of horseradish peroxidase to the monoclonal antibodies. Because the assay based on handling of nitrocellulose disks proved tedious and lengthier than the use of microwells (which could not be used for quantitation of gluten in foods because of low affinities), novel types of high-affinity ω-gliadin-binding monoclonal antibodies that can be used in microwell sandwich EIA with an extremely high sensitivity have now been developed (Bony 1990; Skerritt and Hill 1990a).

Some other applications of monoclonal antibodies to detect adulteration of wheat products, including the detection of nonbaking varieties in baking wheats or of bread wheat in pasta, are of greatest relevance in countries with specific legislation prohibiting such additions (Skerritt 1988). Special attention

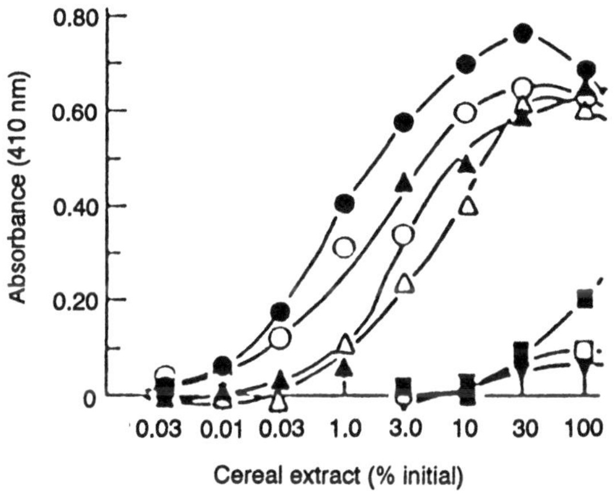

Fig. 2. Assessment of "gliadin-like immunoreactivity" in various cereals. Cereal extracts are indicated as follows: ● bread wheat, ○ durum wheat, ▲ rye, △ barley, ■ oats, □ maize, ▼ rice. The cereals were all assayed together. (Skerritt 1985 a)

must be paid to the *determination of bread wheat in durum wheat pasta submitted to high temperature drying or precooking processes.* In such products, conventional electrophoretic methods (developed from low-temperature dried pasta and involving heat-sensitive albumins or enzymes) are useless. The difficulty may be overcome, however, by using antibodies against slow-moving ω-gliadins whose advantage is to bind to those proteins that are both highly resistant to heat treatments and encoded by specific genes on the D-genome of bread wheats. Alternatively, it could be considered to develop monoclonal antibodies against other − even heat-sensitive − proteins, which would recognize epitopes common to native and denatured structures. Should such antibodies be selected, accurate bread wheat determination will be possible in all types of pasta whether raw or processed.

2.4.8 Immunological Responses to Gluten in Humans: Celiac Disease

Celiac disease (gluten-sensitive enteropathy) is characterized by malabsorption of nutrients as a consequence of small intestinal mucosal injury. The incidence of celiac disease is thought to range between 1 in 300 and 1 in 1000 according to age group and country. Although further studies are still needed to understand the basic mechanisms of toxicity and to identify the toxic molecular structures, it is well accepted that these abnormalities are induced by *specific amino acid sequences contained in the gliadin fraction of wheat gluten* and, to a lesser extent, in the prolamin fraction of rye, triticale, and barley. Since as little as 100 mg of gliadin appeared to cause changes in the absorptive epithelial cells, the disease is treated by strict exclusion of dietary gluten; this treatment is lifelong. In order *to ensure the gluten-free nature of food* prescribed for such patients, or to monitor the legal limit of gluten that may be added in foods for economic or technological reasons, a rapid, specific, and reliable method for gluten detection is required (Ayob et al. 1988).

Because techniques such as electrophoresis, HPLC, or microscopy are slow or not really quantitative, and often unreliable with cooked or processed foods, immunochemical techniques are potentially the best to pick up traces of specific proteins among a large excess of others. Following several studies using polyclonal antisera against α-gliadin or total gliadin that failed to yield quantitative results in processed foods, rapid and more reliable immunoassays using monoclonal antibodies in two-site immunoassays have been developed by Skerritt (1985a, b) and Mills et al. (1989a). Recently, Skerritt and Hill (1990a) have obtained quantitative results by using a 40% ethanol extraction over a wide range of gluten contents (0.015% – 10%) and two simple monoclonal antibody-based test kits are now available for the quantitation of gluten. The first one is able to quantify gluten in all types of uncooked, cooked, and processed foods and is intended for laboratory use by the industry or government regulatory bodies, while the second provides rapid or semiquantitative results and is suitable for either home use or in-process quality control by manufacturers (Skerritt et al. 1990b). For these tests, antibodies to heat-stable ω-gliadins have been selected, so that binding is not influenced by food processing or cooking. In addition, these antibodies have been selected for binding to certain wheat, rye, triticale, or barley prolamins, with little cultivar variation, but not to proteins from maize or rice.

Panels of monoclonal antibodies have been developed to recognize the putative epitope that exacerbates celiac disease in individual cereal proteins. Through studies on the differential binding of the various fractions to specific antibodies, several research groups have found that α- and β-gliadins were more active than γ- or ω-gliadins in exacerbating the disease (Freedman et al. 1988). Moreover, by raising antibodies to synthetic peptides, some active sequences (e.g. PQPFPSQQPYLD) have been recently mapped to particular domains within the α-gliadin sequence (Ellis et al. 1989; Skerritt et al. 1990b). However, the enigma stands because closely homologous sequences were not found in barley or rye samples that were still celiac-toxic.

Only a few methods for accurate measurement of gliadin in food matrices have been proposed, however. RIAs using a polyclonal gliadin antibody showed some reactivity against barley, rye, and oat prolamins (Ciclitira et al. 1985). Monoclonal gliadin antibodies raised by Freedman et al. (1987a, b) also displayed reactivity against glutenins, indicating that the antigenic epitope may be present in several proteins but that it may not represent the polypeptide structure that is toxic to celiacs. Moreover, celiac disease may not be associated with a single characteristic pattern or amino acid sequence and distinct epitopes exacerbating the disease may exist in different regions. According to Friis (1988), the use of monoclonal antibodies to test the suitability of food products for the celiac diet may be insufficient, since they may be unable to measure all toxic polypeptides, so that a polyclonal prolamin-antibody reacting with various gliadin-like proteins might be preferred.

3 Nuclear Magnetic Resonance

3.1 What is NMR? Principles and Techniques

3.1.1 Why "Nuclear" and "Magnetic"?

As indicated by the qualifier "Nuclear", NMR concerns the atomic nucleus. A nucleus comprises uncharged neutrons and positively charged protons. Nuclei of certain isotopes with odd charge and mass numbers (^{1}H, ^{13}C, ^{31}P, etc.) possess a nonzero magnetic momentum μ, so that they behave as micromagnets. They are characterized by a spin quantum number and they have the capacity to interact with a magnetic field. See also Linskens and Jackson (1986).

If, for instance, hydrogen nuclei are placed in an external and uniform magnetic field, B_0, it is found that protons have only two allowable situations in the magnetic field (e.g. $+1/2$ and $-1/2$). All the nuclei will therefore distribute into these energy levels, which will result in the magnetization, M_0. *NMR is based on the population difference between these energy states.* It must be noted, however, that this difference is always very weak: a few units per millions of nuclei.

On the other hand, because the magnetic momentum results from the nucleus spinning, the momenta of the nuclei placed in B_0 will be in line with the field direction and, as in a top or a gyroscope motion, when pulled aside from its rotational axis, it slowly draws out a precessional orbit around B_0 with a frequency v_0 called the Larmor frequency.

3.1.2 How to Attain the Resonance Condition

When a second oscillatory magnetic field B_1 (with a radiofrequency v) is applied over a short time at $90°$ from the first one, an energy $E = h.v$ may be transferred to the system. By sweeping the frequency v_0, the nuclear magnetic resonance condition can be attained and the energy transferred to the system can induce transitions between the energy states. Thus, for instance, many protons having the lower energy level ($+1/2$) will reach the higher energy level ($-1/2$) due to this net absorption of energy.

On stop of the radiofrequency field B_1, nuclei from the upper spin state will progressively return to the lower state according to various types of transitions called *relaxation processes*. These transitions will involve a release of energy as an emission wave that will be exchanged between close spins or between spins and the lattice or molecular nuclei. This emission wave may be detected in a perpendicular coil and it will correspond to the NMR signal.

3.1.3 What Information Can Be Deduced from the NMR Signal?

According to Guillou-Charpin et al. (1988), the versatility of the NMR signal is the result of its three main characteristic parameters: (1) initial intensity;

(2) spin-lattice relaxation, associated with time T1 (longitudinal spin relaxation time); and (3) spin-spin relaxation, associated with time T2 (transveral spin relaxation time). The first parameter depends on the density of a given type of nuclei in the field-excited region of the sample, and it can be exploited for quantitative determinations. On the other hand, *relaxation parameters are intimately related to molecular motions in the environment of the nuclei* (viscosity delays the return to equilibrium), and on *chemical structure and conformation of the microscopic surroundings of nuclei and molecules* (energy exchanges between precessing nuclei in close proximity shorten the lifetime of the spins in the higher energy state).

For instance, chemical shifts (usually reported in parts per million, ppm) of the resonance frequency of the nucleus may be observed. These arise because a given nucleus is shielded from the externally applied magnetic field to a different extent in different molecules (Wüthrich 1976).

The utility of the NMR method results therefore from the sensitive variation of the resonance frequency with chemical, microdynamic, and functional properties of polymers and their environments. In addition, a reconstitution of patterns, indicating the density and energy decay of a given type of nucleus, can be regarded as a photograph of the cell biochemistry.

NMR has recently gained a new dimension with the advent of various improvements such as computer-assisted processing of the signals, introduction of the Fourier transform method (FT-NMR) that turns the time-domain signal into the more familiar spectrum, use of color pictures of signals coming from several directions that allow one to draw two- or three-dimensional pictures, and the development of high-resolution methods using much higher frequencies (several tens or hundreds of MHz). Because a broadening of NMR lines was often observed, various technical improvements have been proposed. For instance, interactions with neighboring protons can be eliminated by irradiating the protons with high-power radio frequency (*dipolar decoupling*). On the other hand, chemical shift anisotropy can be eliminated by the process of *"magic angle spinning"* in which the sample is packed into a rotor and rotated at an angle of 54° 44' in the direction of the magnetic field, yielding a narrower line provided the spinning rate is fast enough (Tatham et al. 1990).

Although nuclei with zero spin such as ^{12}C, ^{14}N, or ^{16}O are not accessible to NMR analysis, nonzero spin nuclei are omnipresent in biological materials. However, because the most sensitive signal is obtained with protons and small nuclei (resonance phenomena being more difficult to obtain with heavy nuclei), typical nuclei of interest in biology or agriculture and food industry are: ^{1}H, ^{13}C, ^{15}N, or ^{31}P. These have been used, for instance, to determine the amount of moisture, to investigate water movements and free/bound water, to control the origin and quality of food products by quantitation of the natural isotopes, and to carry out nondestructive studies on various biochemical mechanisms, or on the structure and dynamics of proteins.

3.2 General Interest of NMR in the Study of Wheat Proteins

In contrast to the considerable information available on genetics and biochemistry of wheat proteins, much less is known about their physical properties such as conformation and structural dynamics. This lack of information essentially results from the fact that most wheat proteins are soluble only under relatively severe conditions that affect their native structure. According to Belton et al. (1987a), wheat gluten represents a very intractable material from the spectroscopic point of view. The great complexity of wheat proteins has accounted for the delay in the characterization of their physical nature, and the residual heterogeneity of purified protein fractions may have also hampered progress in their crystallization and characterization in the solid state by X-ray diffraction techniques.

Physical studies of wheat proteins in solution have been undertaken using circular dichroism and optical rotatory dispersion, however, whether these proteins exhibit the same properties in solution as in the solid state is a question that has not yet been answered (Schofield and Baianu 1982). The use of other physical techniques which can provide additional information about the molecular properties of wheat proteins in relation with functionality are therefore needed. *NMR is one of the few techniques which can be used to investigate the whole gluten viscoelastic mass because it has the potential to provide information on both molecular dynamics and chemical environment and may determine the chemical configurations associated with certain dynamic properties* (Belton et al. 1987a). In addition, the resolved resonances can be used as probes for local environmental conditions since the parameters extracted from NMR spectra depend greatly upon the conformational and dynamic characteristics of the compound, as well as upon its interaction with the surroundings (Lecomte et al. 1982).

Two main methods have been used: ^{13}C high-resolution NMR and ^{1}H relaxation time measurements. ^{13}C solid-state NMR can provide direct evidence for the chemical identity and mobility of the components (Belton et al. 1989). In addition, two alternative schemes have been experienced: the single-pulse excitation (SPE) that discriminates in favor of mobile components of the sample, while cross-polarization magic-angle spinning (CP-MAS) deals with signals arising from solid-like regions. On the other hand, ^{1}H relaxation time measurements allow the quantitation of the number and degree of mobility of protons in each environment.

3.3 Main Applications of NMR to the Study of Wheat Proteins

3.3.1 Physical Characterization of Gluten Proteins

3.3.1.1 Assignment of NMR Signals

In the first NMR studies on wheat proteins, the assignments of the proton peaks were difficult due to the sample heterogeneity and the superimposition

of proton lines in a narrow range of chemical shifts, so that only partial assignments were proposed. Further studies with respect to physically characterizing
gluten and glutenin- or gliadin-enriched subfractions of gluten were carried
out by Schofield and Baianu (1982) using solid-state CP-MAS ^{13}C NMR. It
was found that the glutenin-enriched fraction gave sharp resonances corresponding to aliphatic and aromatic amino acids, whereas the gliadin-enriched
fraction gave broad peaks in these regions. This feature suggested that a much
greater level of hydrophobic interaction existed in the gliadin-enriched fraction
than in the glutenin-enriched fraction, indicating that these two populations
had different degrees of mobility and different levels of interaction.

Under similar CP-NMR conditions, however, Moonen et al. (1985) could
not observe sharp resonances. This was confirmed by Belton et al. (1985), who
could only observe sharp resonances when SPE was used. It was suggested,
therefore, that these sharp resonances (very unlikely to exist in the solid state)
might have arisen rather from small organic contaminants or from mobile side
chains. The exact contribution of lipids to gluten NMR signals has been prob-
lematic for several years. In the beginning, signals from lipids were discounted

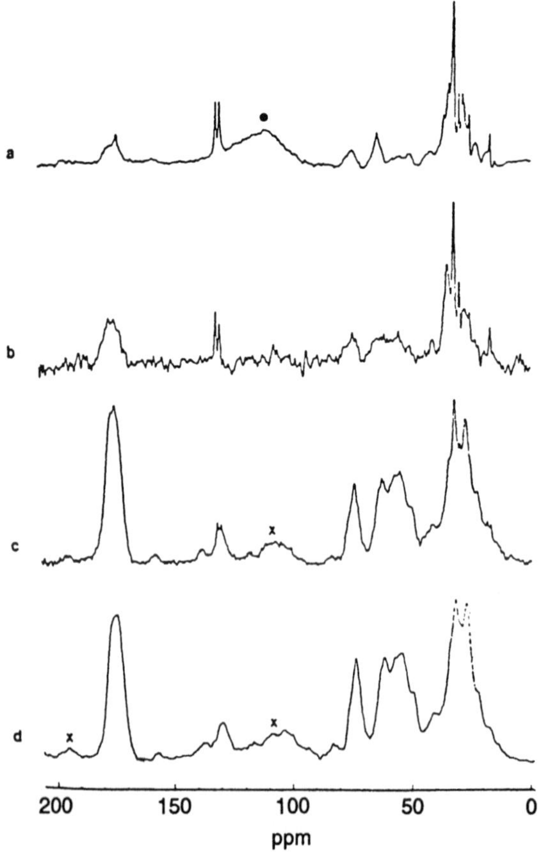

Fig. 3. Solid-state ^{13}C single-pulse excitation spectrum
of gluten (**a**) and cross-polarization spectra with contact
times of 20 (**b**), 5 (**c**), and 1
(**d**) μs. The peak marked with
a *black dot* is a background
signal; those marked with an *x*
are spinning side bands.
(Belton et al. 1988b)

because chloroform-methanol fractions gave broader bands than those observed by Schofield and Baianu (1982) from gluten. However, in another study (Belton et al. 1988a), the contribution of lipids was confirmed since sharp signals were completely lost upon total lipid extraction.

Belton et al. (1988b) and Tatham et al. (1990) interpreted the different results in terms of the *relative immobility in the proteins and mobility in the lipids.* They reported that the two sets of signals observed consist of a protein CP spectrum at short contact times and of a mobile lipid SPE spectrum or CP spectrum at long contact times (Fig. 3). Thus, whether gluten is studied as solution-state or solid-state, a signal from lipids may or not become visible.

3.3.1.2 Functional Properties of Dough and Gluten

In pioneer studies, Baianu (1981) and Schofield and Baianu (1982) reported high-field and high-resolution proton and ^{13}C NMR of wheat proteins (gliadins) in solution. Because a weighted summation of the spectra of the two fractions could not give a spectrum equivalent to that for the whole gluten, the authors suggested that hydrophobic interactions between gliadins and glutenins may be important in the gluten complex. In another study, Baianu et al. (1982) investigated changes in the NMR spectra of two cultivars and observed marked changes in gliadin spectra when temperature or concentration was increased. This suggested the possibility that ^{13}C (or ^{15}N) NMR spectra could provide a direct approach to *wheat variety identification*, *gluten microrheology*, and *nondestructive analysis of wheat proteins.*

However, ^{13}C solid-state methods could not give quantitative estimates of either mobility or the amounts of material in the different environments, making it necessary to use ^{1}H NMR relaxation methods and to interpret both transverse relaxation and spin-lattice relaxation. For instance, Le Grys et al. (1981) used ^{1}H-NMR and recognized the importance of exchange mechanisms in gluten and the possible contribution of mobile proteins to the relaxation process. More recently, Belton et al. (1987a, b, 1988a, b) could identify at least three motional regimes in transverse relaxation of dry gluten samples:

1) Component I (75% – 95% of the protons present) is characteristic of protons in very slow motions and originates from a rigid lattice condition.

2) Component III (3% – 23% of the protons present) indicates a mobile environment within the gluten and is assumed to arise from mobile protein.

3) Component IV, which disappears upon lipid extraction, is likely to represent the lipid content.

The transverse relaxation rate component III is of primary importance in protein dynamics because it may represent some part of the protein which is plasticized by interaction with the lipids (Tatham et al. 1990): lipids would interact with the hydrophobic side chains of the proteins, and act as a solubilizing medium, which would result in an increased mobility. In addition, the subset of protons that give rise to component III depends on the origin of the gluten. As shown in Fig. 4, component III significantly differs according to the

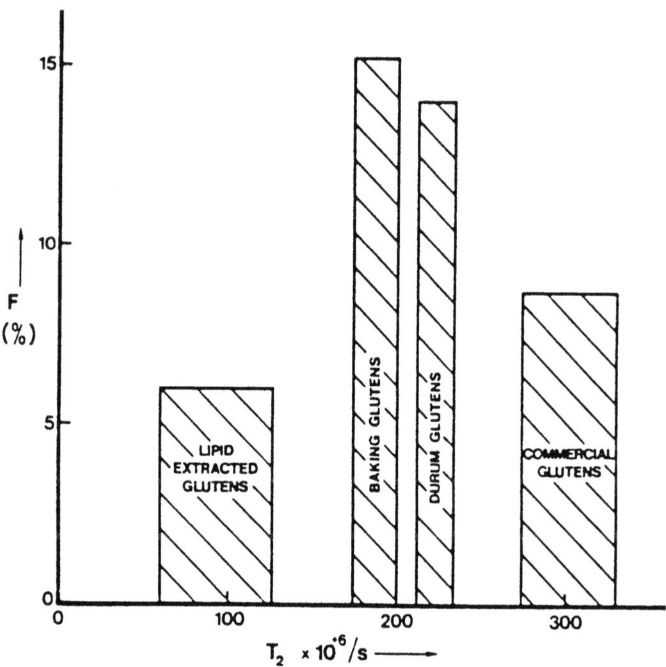

Fig. 4. The transverse relaxation rate component III observed in different glutens. F is the normalized population of the component with relaxation time T_2. (Belton et al. 1988a)

type of gluten (commercial, durum, baking, or lipid-extracted). Whether this difference represents the effects of heat or mechanical work or whether it can be used to assess or predict industrial gluten quality is not known, however.

On the other hand, spin-lattice relaxation has been investigated by Belton et al. (1988a, b) through both the rotating and laboratory frame. Whereas in the rotating frame, the slow- and fast-decaying signals have been assigned to lipids and proteins, respectively, the laboratory frame showed components insensitive to lipid content and gluten origin, in a domain of about 10 nm (in the order of the size of a small protein molecule).

Although it is still difficult to make general statements on the dynamics of the gluten system as a whole, a discussion of these phenomena has been reported by Tatham et al. (1990). These authors stress that *comments on mobility must be viewed in the context of the motional regime to which measurement responds.* They confirm the anisotropy of molecular motions in the gluten system, as motions of the gluten polymer chain are very slow in some directions but quite rapid in other directions. Whereas the *lipids affect the slow motions* of the polymers, leaving the faster motions unchanged, *disulfide bonds leave the slower motions unchanged but restrict the faster motions.*

Another approach has been attempted by Ablett et al. (1988) in the area of polymer dynamics based on relaxation data from ^{1}H-NMR with gluten hy-

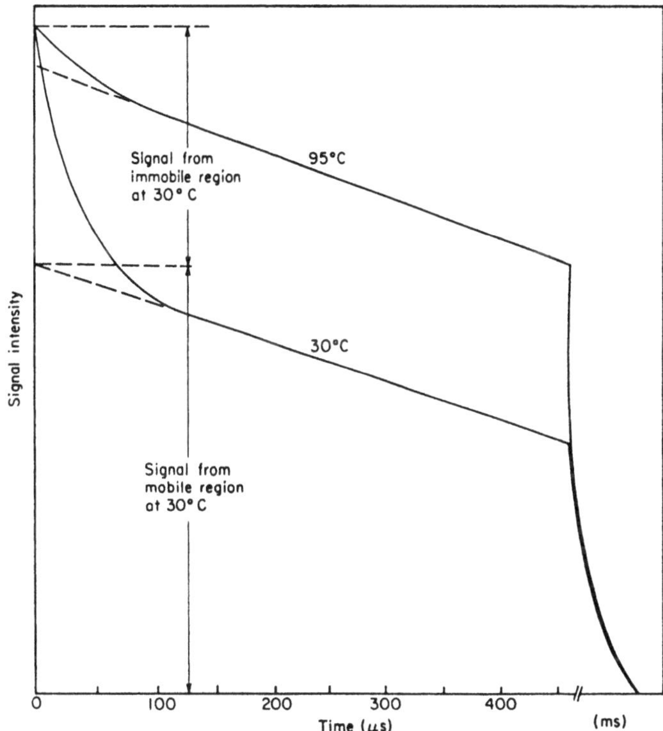

Fig. 5. Temperature dependence of the proton NMR-free induction decay signal for 40% gluten in D_2O. (Ablett et al. 1988)

drated with D_2O. (D_2O was used so that the NMR signal arises solely from the nonexchangeable protons of the protein, with no interfering water resonance). From the overall decay pattern, as shown in Fig. 5, they could separate an "immobile" component (fast-decaying) and a "mobile" component (slower-decaying). Furthermore, an approach to the changes that occur upon heating was possible since the mobile region covered about 95% of the protons at 90 °C instead of 65% at 30 °C. On this basis, Ablett et al. (1988) assumed that the changes in NMR spectra that occur upon heating are consistent with rheological changes that would be caused by the replacement of labile, non-covalent interactions by permanent disulfide bonds. Furthermore, both ^{1}H- and ^{13}C-NMR results would indicate that wheat proteins have open structures with a high degree of mobility, both at ambient and after heat setting, even though the formation of new disulfide bonds dramatically changes the rheological properties.

According to Wrigley and Bietz (1988), the potential of NMR in the study of hydrated dough systems must be pointed out because it can provide results likely to relate more directly to the conditions of actual dough mixing and baking. However, it still proved extremely difficult to show any consistent difference between molecular structures related directly to baking quality. For instance, in contrast to certain studies based on electron-spin resonance (ESR), Moonen et al. (1985) could not demonstrate any difference between glutenin subunits, as cultivars belonging to extremely different allelic acid types, such as Arminda $(2-7-12)$ and Sicco $(1-5-7-9-10)$, had identical CP-MAS spectra.

NMR spectroscopy is suitable in many other respects since it is theoretically possible, in the spectra of proteins, to determine that many signals arise from given amino types which can often be assigned to specific residues in the polypeptide chain. For instance, in an attempt to relate amino acid sequence and specificity of action in toxicity to phytopathogenic bacteria and insect larvae, the analysis of the NMR spectra of wheat and barley thionins enabled Lecomte et al. (1982) to identify all of the 84 methyl resonances by amino acid type and to assign 51 to specific sites.

3.3.2 Lipid-Protein Interactions in Wheat Gluten

It has been suggested that gluten is not composed of discrete molecules but of water-insoluble aggregates, including both proteins and lipids, so that a lipid-protein complex is formed during gluten preparation and dough formation. Lipid-protein interactions have been considered to play important roles in the physical and rheological properties of doughs since Grosskreutz (1961) postulated a gluten model involving the Danielli-type membrane to account for its extensibility. To give new insights in the dynamic properties of protein-lipid interactions in wheat gluten, nonintrusive physical techniques such as phosphorus NMR are the most desirable since the sensitive nucleus of phosphorus is naturally 100% abundant and no extrinsic probes are needed (Marion et al. 1987).

By applying ^{13}P-NMR spectroscopy in association with freeze-fracture electron microscopy, Marion et al. (1987) demonstrated that:

1) In aqueous gluten, extracted by hand-washing, phosphorus accounted for $160-190$ mg/100 g dry matter and that more than 90% of the phosphorus NMR signal in gluten was due to phospholipids.

2) One of the features of ^{31}P-NMR spectra was the symmetrical shape of the signal, reflecting an isotropic motional averaging of phospholipids in gluten, which was retained during a heating-cooling cycle.

3) The line shape and broadening of the NMR signal was quite different when wheat flour lipids were extracted before gluten washing and the missing intensity in defatted samples was postulated to be due to phospholipids immobilized in the gluten protein framework as in biological membranes.

Furthermore, the interpretation of the line shape of NMR spectra, in concert with freeze-fracture electronic microscopy observations, gave further information on the macromolecular organization of gluten constituents. For instance, Le Roux (1987) and Akoka et al. (1988) suggested (1) that phospholipids in dough and gluten were organized in bilayer vesicles, whose radii ranged from 30 to 150 nm; (2) that these vesicles were formed during flour hydration from hexagonal phospholipid structures (membrane remnants) of the starting flour; and (3) that the hexagonal phases themselves derived from a reversed-phase transition that occurred during the drying phase of grain development.

Moreover, the suppression of dipolar interactions and/or the decrease in the residual chemical anisotropy, which are indicated by the resonance broadening in ^{31}P-NMR, may be accounted for by the expulsion of lipid vesicles into the protein matrix upon gluten heating (Marion et al. 1987, 1989). Because heating also affects wheat glutenins (by promoting sulfhydryl-disulfide interchange reactions), the phospholipid rearrangements that are suggested by the increase in the resonance line from 25 °C to 70 °C may be concomitant to changes in gluten protein aggregates. Interestingly, the fact that vesicles would influence gluten protein aggregation may not necessarily involve interactions between classical gliadins or glutenins and lipids. In contrast to previous models, that of Marion et al. (1989) discounts the role (other than mechanical trapping) of Osborne's classical (1907) protein fractions in the lipoprotein interactions, and emphasizes rather the involvement of smaller (16.5 kD) and highly interactive molecules that showed homology with interface-stabilizer membrane proteins.

3.3.3 Nondestructive Protein Content Determinations in Wheat Seeds

There are various methods to measure the cereal protein content and several NMR techniques have been described. When the proteins can be solubilized, classical NMR methods measure the relaxation of water protons submitted to two (uniform and oscillatory) magnetic fields. Because proteins can complex metals such as Cu or Fe, resulting in different relaxation kinetics, it is possible to correlate NMR signals with the protein concentration. However, because wheat samples are sometimes needed for further analyses, the determination of the protein content may have to be nondestructive and possible from single kernels, which is far more difficult.

In contrast to lipids or water, which constitute a highly mobile phase, the more rigid constituents of the seed, such as proteins or starch, cannot contribute quite as well to the spectrum obtained by liquid-state techniques because of the long ^{13}C spin-lattice relaxation times and the broadening influences of ^{1}H-^{13}C dipole-dipole interactions (O'Donnell et al. 1981).

^{13}C-NMR spectra of rigid bioorganic materials have been obtained, however, which contain only signals from carbons in structures of low mobility owing to suppression of the intense liquid signals. Such spectra consist of two groups of resonances centered at ~75 and ~25 ppm, representing respectively, carbohydrates and proteins, so that the protein content may be estimated by comparing these peak intensities. Further improvement in the quantitative analysis was obtained by using proton-enhanced MAS, making the resonance lines narrower and resolving the spectrum better (Rutar 1982). Under these conditions, ^{13}C-NMR is of great potential value for the nondestructive determination of the protein content in viable seeds. It may promote, therefore, the rapid breeding of new varieties with higher protein content.

Similar studies based on the amplitude of signal relaxation were also reported in an attempt to determine the gluten content in individual seeds and hybrids (Bebyakin and Lutsishina 1987 a, b), indicating a possible checking of the gluten content (and perhaps gluten quality) in breeding programs.

Serious limitations and technical problems, however, are still to be faced in the development of CP-MAS as a routine technique. Results might be influenced by the size and shape of the seed, making the use of an outstanding instrument and high spinning speeds (3000 rotations/s) necessary. Another limitation would be the length of the analysis: a minimum of 15 – 30 min/seed seemed to be required.

4 Conclusion and Future Research

The practical importance of wheat proteins is well recognized. In contrast to the rapid technological development and diversification in the application of wheat in various foods, the precise molecular basis of the functional properties of wheat proteins is not clearly understood. Several recent advances, however, provide the possibility to explore in more detail the wheat protein structure, thus permitting a more effective quality control of wheat products.

4.1 Monoclonal Antibodies

With the absolute reproducibility of specificity or affinity, and with ability to be produced (in theory) in unlimited quantities, monoclonal antibodies have a tremendous potential for studies on wheat proteins.

Despite the unusual solubility and alleged poor immune response of gluten proteins, high-resolution techniques, such as immunoblotting and ELISA, are now available, providing an alternative and most useful adjunct to conventional protein chemistry for structural and functional studies on wheat gluten proteins and for rapid and simple approaches to their role in grain technology and human nutrition.

Monoclonal antibodies make it possible to overcome many of the difficulties arising from the use of polyclonal antibodies with wheat proteins: variations in the responses of individual animals, previous dietary exposure of animals in which antisera have been raised, poor specificity, and cross-reactions that confused the interpretation of results. Although polyclonal antibodies may still be recommended for some routine determinations or perhaps for overall quantitation of various antigenic determinants during food processing, monoclonal antibodies are replacing conventional antisera in many fields of immunological research and diagnostic testing.

Immunochemical methods using monoclonal antibodies proved attractive in a number of applications:

1) They have allowed a better identification of toxic components of gluten in celiac disease and a better understanding of the mechanism of toxicity.
2) They proved valuable in the localization of protein components by immunocytochemistry and the screening of expression vector cDNA libraries.
3) Because of the well-defined structural specificity of monoclonal antibodies, some studies on the homologies that exist between different gluten polypeptides and on the configuration of native or denatured antigens have been made possible.

Although much work is still needed to develop tests for specific assays in industry, many indications (including various patent applications) suggest that the 1990s will see the availability of many standardized reagent kits, thus assisting dietary management for celiac patients, quality prediction by breeders, varietal identification or quality assessment by millers and bakers, and monitoring of the composition of various foods (Wrigley et al. 1987b; Skerritt and Hill 1989; Chan et al. 1990; Skerritt et al. 1990c).

Advantages of monoclonal antibodies techniques include: sensitivity (possibility to detect as little as 0.0016% gluten proteins), well-defined structural specificity, speed (superior to electrophoresis or chromatography, longer, however, than NIR reflectance), small-scale and possible screening of a large series of samples, economy, and simplicity (less specialized and costly equipment than HPLC or NIR). However, for the production of monoclonal antibodies, more skilled personnel and much heavier equipment are needed, since cell cultures are involved. The recognition of antibody-producing clones is also a time-consuming task, the cost of which must be considered.

A decisive point in the successful use of monoclonal antibodies depends on the screening of clones. Clones must be selected not only for a specific type of antigen but also for the type of end-product (raw or processed), level of binding affinity and broadness of specificity, assay format (e.g. indirect, competition, or sandwich ELISA; immunocytochemistry), solid phase for antigen or antibody immobilization (e.g. plastic, nylon, or nitrocellulose membrane) as well as the solvent used to extract proteins. Accordingly, nothing is ever gained beforehand when using monoclonal antibodies, however, due to the considerable number of possible epitopes, hope remains, provided the selection of the clone is extremely careful and judicious.

On the other hand, future research into the use of monoclonal antibodies should consider the following aspects:

1) To improve further the understanding of wheat protein structure through the use of antibodies raised against specific peptides, chemically modified proteins, or proteins modified by directed mutagenesis.
2) To investigate the basis of functional properties in situ, by developing antibodies able to discriminate the aggregated/nonaggregated states of wheat

storage proteins or to follow site integrity during aggregate formation in the developing grain.

3) To develop nondestructive seed tests by applying monoclonal antibodies to the screening of thousands of plastic-embedded and abraded half-seeds.

4.2 Nuclear Magnetic Resonance

Developed in the early 1960s in chemistry laboratories, nuclear magnetic resonance has emerged as one of the few and most powerful spectroscopic techniques capable of obtaining useful information on conformation, tridimensional structure, chemical environment, functionality, thermodynamic parameters, and molecular dynamics of proteins and peptides in solution, or on intact complex systems such as wheat gluten. Because the application of high-resolution NMR is not limited to solutions or to crystallized proteins, it may be complementary to circular dichroism, optical rotatory dispersion, crystallography, or X-ray diffraction.

Whereas until very recently, NMR called to mind high cost and outstanding techniques, new generations of commercial apparatuses are now available and can be used for nondestructive routine analyses of various agriculture and food products.

In this chapter on wheat proteins, it has been shown that investigations based on the various NMR methods, especially ^{1}N and ^{13}C relaxation studies, proved attractive in a number of applications:

- Identification of the main motional regimes in dry gluten, in relation to the behavior of individual components (e.g. proteins, lipids).
- Respective effect of lipids and disulfide bonds in the slow and fast molecular motions of the gluten system.
- Description of the changes in rheological properties of hydrated gluten in terms of replacement of labile, nonconvalent protein interactions by permanent disulfide bonds.
- Development of a new model of lipid-protein interactions in wheat gluten and dough, involving rearrangements of phospholipidic membrane structures.
- Development of nondestructive determinations of the protein content in seeds.

Despite their high potential value in exploring the molecular dynamics of protein systems, approaches based on NMR, however, have not been as yet really successful in identifying specific sequences associated with dough rheology or baking quality, or in devising microtests to predict the technological quality in breeding programs.

On the other hand, interpreting NMR experiments still remains an extremely difficult task, as illustrated by the controversial conclusions on the respective contributions of lipids and proteins to gluten NMR signals. Because the different authors have been using various NMR schemes (CP, MAS, SPE, with short or long contact times) on various nuclei (^{13}C high-resolution NMR, ^{1}H relaxation time measurements, etc.), it is always difficult to make general statements on the reported results. Clearly, comments on mobility must be viewed only in the context of each NMR experience.

Because the elucidation of the exact molecular bases of wheat gluten functionality remains the most important challenge to cereal biophysicists in the coming years, future research in the NMR field, in concert with other physicochemical and biochemical studies, should tackle the following aspects:

- The identification of the structural elements which impart dough elasticity or extensibility.
- The determination of the fractions in which the mobility lies and an attempt to account for the differences observed between different cultivars or different gluten origins.
- The elucidation of protein hydration and the further characterization of the protein groups in this respect from comparative studies on dry or hydrated glutens and on purified gliadin or glutenin fractions.
- The elucidation of the dynamics of dough development and the effect of heat and mechanical treatment on the distribution and mobility of protein components through in situ NMR spectroscopy.

References

Ablett S, Barnes DJ, Davies AP, Ingman SJ, Patient DW (1988) C 13 and pulse nuclear magnetic resonance spectroscopy of wheat proteins. J Cereal Sci 7:11–20

Akoka S, Tellier C, Le Roux C, Marion D (1988) A phosphorus magnetic resonance spectroscopy and a differential scanning calorimetry study of the physical properties of N-acylphosphatidyl-ethanol-amines in aqueous dispersions. Chem Phys Lipids 46:43–50

Ariss AJ (1986) Immunocytochemical studies of wheat proteins. Thesis, Hatfield Polytechnic, Hatfield, UK

Asano K, Shinagawa K, Hashimoto N (1983) Characterization of haze-forming proteins of beer and their role in chill haze formation. Rep Res Lab Kirin Brew Co Ltd 26:45–55

Ayob MK, Rittenburg J, Allen JC, Smith CJ (1988) Development of a rapid enzyme-linked immunosorbent assay (ELISA) for gliadin determination in food. Food Hydrocolloids 2:39–49

Baianu IC (1981) Carbon-13 and proton nuclear magnetic resonance studies of wheat proteins. I. Spectral assignments for Flanders gliadins in solution. J Sci Food Agric 32:309–313

Baianu IC, Johnson LF, Waddell DK (1982) High-resolution proton, carbon-13 and nitrogen-15 NMR studies of wheat proteins at high magnetic fields: spectral assignments, changes with concentration and heating treatments of Flinor gliadins in solution – comparison with gluten spectra. J Sci Food Agric 33:373–383

Bebyakin VM, Lutsishina EG (1987a) Heterogeneity of varietal and hybrid populations with respect to gluten quality as determined by NMR. Fiziol Biokhim Kul't Rast 19:369–371 (in Russian)

Bebyakin VM, Lutsishina EG (1987b) Effectiveness of selecting wheat grains for gluten quality by proton NMR relaxation measurement in a system of cyclical and exhaustive crosses. Vestn Skh Nauki 12:25–28 (in Russian)

Belton PS, Shewry PR, Tatham AS (1985) 13 C solid state NMR study of wheat gluten. J Cereal Sci 3:305–317

Belton PS, Duce SL, Tatham AS (1987a) Nuclear magnetic resonance studies of wheat gluten. In: Morton ID (ed) Cereals in a European context. Ellis Horwood Ltd, Bournemouth, UK, pp 489–495

Belton PS, Duce SL, Tatham AS (1987b) Carbon-13 solution state and solid state n.m.r. of wheat gluten. Int J Biol Macromol 9:357–362

Belton PS, Duce SL, Tatham AS (1988a) Proton nuclear magnetic resonance relaxation studies of dry gluten. J Cereal Sci 7:113–122

Belton PS, Duce SL, Colquhoun IJ, Tatham AS (1988b) High-power carbon-13 and proton nuclear magnetic resonance in dry gluten. Magn Reson Chem 26:245–251

Belton PS, Colquhoun IJ, Tatham AS (1989) Nuclear magnetic resonance studies of gluten. Agric Food Chem Consum, Proc Eur Conf Food Chem 5:489–492

Bony M (1990) Détection et dosage du gluten. Biofutur 87:11

Bushuk W, MacRitchie F (1989) Wheat proteins: aspects of structure that determine breadmaking quality. In: Phillips RD, Finley JW (eds) Protein quality and the effects of processing. Marcel Dekker, New York, pp 345–369

Chan HWS, Morgan MRA, Mills ENC (1990) Determining wheat baking quality by glutenin subunit immunoassay. Eur Pat Appl EP 361,838. GB Appl 88/23,027

Ciclitira PJ, Ellis HJ, Evans DJ, Lennox ES (1985) A radio-immunoassay for wheat gliadin to assess the suitability of gluten free foods for patients with celiac disease. Clin Exp Immunol 59:703–708

Daussant J, Bureau D (1988) Immunochemistry of cereal enzymes. In: Pomeranz Y (ed) Advances in cereal science and technology, vol 9. Am Assoc Cereal Chemists, St Paul, MN, USA, pp 47–90

Dawood MR, Howes NK, Bushuk W (1989) Preparation of monoclonal antibodies against specific gliadin proteins and preliminary investigations of their ability to discriminate cereal cultivars. J Cereal Sci 10:105–112

Donovan GR, Skerritt JH, Castle SL (1989) Monoclonal antibodies used to characterize cDNA clones expressing specific wheat endosperm proteins. J Cereal Sci 9:97–111

Du Cros DL, Campbell WP, Skerritt JH, Wrigley CW, Cressey PJ (1984) Characterization of quality-related gluten proteins using chemical methods and monoclonal antibodies. In: Graveland A, Moonen JHE (eds) Proc 2nd Int Workshop on Gluten proteins, TNO, Wageningen, The Netherlands, pp 155–161

Ellis HJ, Freedman AR, Ciclitira PJ (1989) The production and characterization of monoclonal antibodies to wheat gliadin peptides. J Immunol Methods 120:17–22

Ewart JAD (1977) Immunochemistry of wheat proteins. In: Castimpoolas N (ed) Immunological aspects of food. Avi Publ Co, Westport, CT, USA, pp 87–116

Feillet P (1988) Protein and enzyme composition of durum wheat. In: Fabriani G, Lintas C (eds) Durum Wheat: Chemistry and Technology, American Association of Cereal Chemists, St Paul, MN, USA, pp 93–119

Feillet P, Popineau Y (1990) Protein-protein interactions. In: Salovaara (ed) Proc ICC'90, Int Assoc of Cereal Science, Vienna (Abstr), p 36

Freedman AR, Galfre G, Gal E, Ellis HJ, Ciclitira PJ (1987a) Detection of wheat gliadin contamination of gluten-free foods by a monoclonal antibody dot immunobinding assay. Clin Chim Acta 166:323–328

Freedman AR, Galfre G, Gal E, Ellis HJ, Ciclitira PJ (1987b) Monoclonal antibody ELISA to quantitate wheat gliadin contamination of gluten-free foods. J Immunol Methods 98:123–127

Freedman AR, Wieser H, Ellis HJ, Ciclitira PJ (1988) Immunoblotting of gliadins separated by reversed-phase high-performance liquid chromatography: detection with monoclonal antibodies. J Cereal Sci 8:231–238

Friis SU (1988) Enzyme-liked immunosorbent assay for quantitation of cereal proteins toxic in celiac disease. Clin Chim Acta 178:261–270

Fritschy F, Windemann H, Baumgartner E (1985) Quantitative determination of wheat gliadins in foods by enzyme linked immunosorbent assay. Z Lebensm Unters Forsch 181:379–385

Gallant D, Laurière M, Popineau Y, Benhdech H (1989) Use and limits of immunochemistry and cytochemistry in cereal studies. In: Salovaara H (ed) Wheat end-use properties, wheat and flour characterization for specific end-uses, Int Assoc Cereal Science, Vienna, pp 187–207

Grosskreutz JC (1961) A lipoproteic model of wheat gluten structure. Cereal Chem 38:336–349

Guillou-Charpin M, Le Botlan D, Teilier C, Mechin B (1988) Une méthode rapide d'analyse des aliments: la résonance magnétique nucléaire basse résolution. Ind Agric Alim 105:463–471

Howes NK, Lukow OM, Dawood MR, Bushuk W (1989) Rapid detection of the 1 BL/ARS chromosome translocation in hexaploid wheats by monoclonal antibodies. J Cereal Sci 10:1–4

Johnson RB, Labrooy JT, Skerritt JH (1990) Antibody responses reveal differences in oral tolerance to wheat and maize grain protein fractions. Clin Exp Immunol 79:135–140

Kasarda DD, Bernardin JE, Nimmo CC (1976) Wheat proteins. In: Pomeranz Y (ed) Advances in cereal science and technology, vol 1. Am Assoc Cereal Chemists, St Paul, MN, USA, pp 158–236

Kasarda DD, Lafiandra D, Morris R, Shewry PR (1984) Genetic relationships of wheat gliadin proteins. Kulturpflanze 32:41–60

Lecomte JTL, Jones BL, Llinas M (1982) Proton magnetic resonance studies of barley and wheat thionins: structural homology with crambin. Biochemistry 21:4843–4849

Le Grys GA, Booth MR, Al-Baghdadi SM (1981) The physical properties of wheat proteins. In: Pomeranz Y, Munck L (eds) Cereals: a renewable resource. Am Assoc Cereal Chemists, St Paul, MN, pp 243–264

Le Roux C (1987) Contribution à l'étude des interactions lipides-protéines dans les produits céréaliers de cuisson: Propriétés physico-chimiques des lipides endogènes et exogènes dans les pâtes et le gluten de blé tendre. Thèse de Doctorat, Université de Nantes, France

Linskens HF, Jackson JF (eds) (1986) Nuclear magnetic resonance. Mod Meth Plant Analysis – NS Vol 2

Marion D, Le Roux C, Akoka S, Tellier C, Gallant D (1987) Lipid-protein interactions in wheat gluten: a phosphorus nuclear magnetic resonance spectroscopy and freeze-fracture electron microscopy study. J Cereal Sci 5:101–115

Marion D, Le Roux C, Tellier C, Akoka S, Gallant D, Gueguen J, Pipineau Y, Compoint JP (1989) Lipid-protein interactions in wheat gluten: a renewal. Abh Akad Wiss DDR, Abt Math Naturwiss Tech 1:147–152

Mills ENC, Spinks CA, Morgan MRA (1989a) A two-site enzyme-linked immunosorbent assay for wheat gliadins. Food Agric Immunol 1:19–27

Mills ENC, Brett GM, Tatton MJ, Alcocer MJC, Tatham AS, Shewry PR, Morgan MRA (1989b) Monoclonal antibodies as probes of wheat bread-making quality. Agric Food Chem Consum, Proc Eur Conf Food Chem 5:351–354

Mills ENC, Burgess SR, Tatham AS, Shewry PR, Chan HSW, Morgan MRA (1990) Characterization of a panel of monoclonal anti-gliadin antibodies. J Cereal Sci 11:89–101

Moonen JHE, Hemminga MA, Graveland A (1985) Magnetic resonance spectroscopy of wheat proteins: a magic-angle-spinning carbon-13 nuclear magneric resonance and electron spin resonance spin label study. J Cereal Sci 3:319–327

O'Donnell DJ, Ackerman JH, Maciel GE (1981) Comparative study of whole seed protein and starch content via cross polarization – magic angle spinning carbon-13 nuclear magnetic resonance spectroscopy. J Agric Food Chem 29:514–518

Osborne TB (1907) The proteins of the wheat kernel. Carnegie Inst Washington, Washington, DC, No 84

Parker ML, Mills ENC, Morgan MRA (1990) The potential of immuno-probes for locating storage proteins in wheat endosperm and bread. J Sci Food Agric 52:35–45

Reeves CD, Krishnan HB, Okita TW (1986) Gene expression in developing wheat endosperm. Accumulation of gliadin and ADP glucose pyrophosphorylase messenger RNAs and polypeptides. Plant Physiol 82:34–40

Rutar V (1982) A new possibility for nondestructive protein content determination in viable seeds. Appl Spectrosc 36:259–260

Schofield JD, Baianu IC (1982) Solid-state, cross-polarization magic-angle spinning carbon-13 nuclear magnetic resonance and biochemical characterization of wheat proteins. Cereal Chem 59:240–245

Shewry PR, Miflin BJ (1984) Seed storage proteins of economically important cereals. In: Pomeranz Y (ed) Advances in cereal science and technology, vol 7. Am Assoc Cereal Chemists, St Paul, MN, USA, pp 1–83

Skerrit JH (1985a) A sensitive monoclonal-antibody-based test for gluten detection: quantitative immunoassay. J Sci Food Agric 36:987–994

Skerrit JH (1985b) Detection and quantitation of cereal protein in foods using specific enzyme-linked monoclonal antibodies. Food Technol Aust 37:570–572

Skerritt JH (1988) Immunochemistry of cereal storage proteins. In: Pomeranz Y (ed) Advances in cereal science and technology, vol 9. Am Assoc Cereal Chemists, St Paul, MN, USA, pp 263–338

Skerritt JH (1989) Immunodiagnostic approaches to quality and process control in wheat breeding and utilization. In: Pomeranz Y (ed) Wheat is unique: structure, composition, processing, end-use properties, and products. Am Assoc Cereal Chemists, St Paul, MN, USA, pp 131–159

Skerritt JH, Henry RJ (1988) Hydrolysis of barley endosperm storage proteins during malting. II. Quantification by enzyme- and radio-immunoassay. J Cereal Sci 7:265–281

Skerritt JH, Hill AS (1989) Detection of glutens with monoclonal antibody-producing hybridoma cells. Brit. UK Pat Appl GB 2,207,921, 15 Feb 1989, 21 pp

Skerritt JH, Hill AS (1990a) Monoclonal antibody sandwich enzyme immunoassays for determination of gluten in foods. J Agric Food Chem 38:1771–1778

Skerritt JH, Hill AS (1990b) Homologies between grain storage proteins of different cereal species. 2. Effects of assay format and grain extractant on antibody cross-reactivity. J Cereal Sci 11:123–141

Skerritt JH, Lew PY (1990) Homologies between grain storage proteins of different cereal species. 1. Monoclonal antibody reaction with total protein extracts. J Cereal Sci 11:103–121

Skerritt JH, MacRitchie F (1990) Interaction between glutenin polypeptides: simple antibody and HPLC-based methods for prediction of gluten strength. In: Proc ICC'90, Int Assoc Cereal Chemistry, Vienna (Abstr), p 45

Skerritt JH, Martinuzzi O (1986) Effects of solid phase and antigen solvent on the binding and immunoassay of water-insoluble flour proteins. J Immunol Methods 88:217–224

Skerritt JH, Robson LG (1990) Wheat low molecular weight glutenin subunits – structural relationship to other gluten proteins analyzed using specific antibodies. Cereal Chem 67:250–257

Skerritt JH, Smith RA (1985) A sensitive monoclonal-antibody-based test for gluten detection: studies with cooked or processed foods. J Sci Food Agric 36:980–986

Skerritt JH, Underwood PA (1986) Specificity characteristics of monoclonal antibodies to wheat grain storage proteins. Biochim Biophys Acta 874:245–254

Skerritt JH, Smith RA, Wrigley CW, Underwood PA (1984) Monoclonal antibodies to gliadin proteins used to examine cereal grain protein homologies. J Cereal Sci 2:215–224

Skerritt JH, Dilent JA, Wrigley CW (1985) A sensitive monoclonal-antibody-based test for gluten detection: choice of primary and secondary antibodies. J Sci Food Agric 36:995–1003

Skerritt JH, Batey IL, Wrigley CW (1986) New approaches to barley variety identification and quality studies. Proc Conv Inst Brew 19:55–62

Skerritt JH, Martinuzzi O, Wrigley CW (1987) Monoclonal antibodies in agricultural testing: quantitation of specific wheat gliadins affected by sulfur deficiency. Can J Plant Sci 67:121–129

Skerritt JH, Lew PY, Castle SL (1988a) Accumulation of gliadin and glutenin polypeptides during development of normal and sulfur-deficient wheat seed: analysis using specific monoclonal antibodies. J Exp Bot 39:723–737

Skerritt JH, Wrigley CW, Hill AS (1988b) Prospects for the use of monoclonal antibodies in the identification of cereal species, varieties and quality types. In: Konarev VG, Gavrilyuk IP (eds) Biochemical identification of varieties. Inst Plant Industry, Leningrad, USSR, pp 110–123

Skerritt JH, Frend AJ, Robson LG, Greenwell P (1990a) Immunological homologies between wheat gluten and starch granule proteins. J Cereal Sci 12:123–136

Skerritt JH, Devery JM, Hill AS (1990b) Gluten intolerance: chemistry, cellac-toxicity, and detection of prolamins in foods. Cereal Foods World 35:638–644

Skerritt JH, Wrigley CW, Underwood PA (1990c) Cereal identification using antibodies to characteristic protein, and kits containing the antibodies. Pat Specific (Aust) AU 592,987, Appl 85/401, 2 May 1985, 12 pp

Stoyanov S, Tlaskalova-Hogenova H, Kocna P, Kristofova H, Fric P, Hekkens WTJM (1988) Monoclonal antibodies reacting with gliadin as tools for assessing antigenic structures responsible for the exacerbation of celiac disease. Immunol Lett 17:335–338

Tatham AS, Shewry PR, Belton PS (1990) Structural studies of cereal prolamins, including wheat gluten. In: Pomeranz Y (ed) Advances in cereal science and technology, vol 10. Am Assoc Cereal Chemists, St Paul, MN, USA, pp 1–78

Vaag P, Munck L (1987) Immunochemical methods in cereal research and technology. Cereal Chem 64:59–72

Wrigley CW, Bietz JA (1988) Proteins and amino acids. In: Pomeranz Y (ed) Wheat chemistry and technology, vol I, 4th edn. Am Assoc Cereal Chemists, St Paul MN, USA, pp 159–275

Wrigley CW, Batey IL, Campbell WP, Skerritt JH (1987a) Complementing traditional methods of identifying cereal varieties with novel procedures. Seed Sci Technol 15:679–688

Wrigley CW, Skerritt JH, Hill AS (1987b) Test kits for grain proteins. In: Westcott T, Williams Y, Ryken R (eds) Quality control issues in the production of foods and chemicals. Proc 37th Aust Cereal Conf, Royal, Australian Chemical Institute, Melbourne, pp 140–141

Wüthrich K (1976) NMR in biological research: peptides and proteins. North-Holland Publ, Amsterdam

Electrophoretic Analyses of Soybean Seed Proteins

B. R. Hedges, R. G. Palmer, and L. A. Amberger

1 Introduction

The soybean (*Glycine max* L. Merr.) is a crop whose seed is valued as a source of oil and protein. A number of diverse disciplines utilize electrophoresis in some capacity to study the soybean. Starch and polyacrylamide gel electrophoresis (PAGE) are relatively simple technologies that rapidly screen for a large number of genetic markers. Electrophoretic systems detect seed storage proteins and more than 25 enzyme systems encoded by more than 60 genes. Protein electrophoresis has been and will continue to be a powerful tool in soybean research.

One application has been to use isozymes to assess the purity of commercial seed production and for variety identification. Despite the relatively narrow germplasm base of the North American varieties (Delannay et al. 1983), sufficient isozymic variability exists for these applications. The use of protein markers for variety fingerprinting has greater discriminatory power than the use of morphological markers (Cardy and Beversdorf 1984a); Doong and Kiang 1987b; Griffin and Palmer, in press). These same authors note that the traditional variety identification techniques coupled with isozyme analysis are more useful than either method alone.

The commercial value of soybean seed depends on the quality and quantity of oil and protein. Glycinin and β-conglycinin are the major seed storage proteins. Screening of germplasm has identified mutants for the subunits composing these two proteins. Both storage proteins have a low sulfur (S) amino acid content; however, their subunits have a three- to fourfold difference in S-amino acid composition. From populations derived from crosses of seed storage mutants, genotypes were selected that had a 20% higher S-amino acid content (Ogawa et al. 1989). Proteins called protease inhibitors inhibit growth in monogastric animals feeding on unprocessed soybean meal. Polymorphisms exist in the two major classes of inhibitors, the Kunitz trypsin inhibitor and the Bowman-Birk proteinase inhibitor (Orf and Hymowitz 1977; Stahlhut and Hymowitz 1983). The variety Kunitz has a reduced quantity of inhibitors that was achieved by backcrossing and selecting for a null Kunitz inhibitor allele (Bernard et al. 1991).

Soybean oil is found in many food products. During storage, fatty acids are oxidized by lipoxygenases, and the resulting by-products add undesirable flavors to the oil. Elimination of lipoxygenases stabilizes the soybean oil. Three lipoxygenases and corresponding null phenotypes were found during PAGE screenings of soybean accessions (Hildebrand and Hymowitz 1982; Kitamura et al. 1983; Davies and Nielsen 1986). The same PAGE techniques were used to develop germplasm with reduced lipoxygenase activity (Davies and Nielson 1987).

Seed protein profiles have been used in evolutionary studies. Hymowitz and Kaizuma (1981) determined the paths of soybean dissemination from China by analyzing the frequency of β-amylase and Kunitz trypsin inhibitor alleles in soybean populations throughout Asia. Parallel germplasm pools were present in the wild annual soybean (*Glycine soja* Sieb. & Zucc.) (Griffin 1986). The frequency of β-amylase and Kunitz trypsin inhibitor alleles differentiated Japanese cultivars into two clines, each having distinct centers of origin (Hymowitz and Kaizuma 1979).

Numerous studies have used electrophoretic techniques, primarily PAGE, to study seed development. Gayler and Sykes (1981) scanned Coomassie Brilliant Blue R stained gels with a densitometer to measure β-conglycinin subunit accumulation in developing soybean seeds. Synthesis of the α- and α'-subunits was detected an average of 17 days after flowering (DAF), and production peaked 15 – 20 DAF. β-Subunits did not appear until 22 DAF, and the maximum rate of accumulation occurred 25 to 30 DAF. The varying proportions of subunits during seed development resulted in β-conglycinin proteins with a distinctive subunit and amino acid composition.

These examples are by no means a comprehensive review of the utility of electrophoretic separation of soybean seed proteins. The purpose was to illustrate the wide range of disciplines that can employ protein analyses. Numerous protein extraction protocols, buffer types, and running conditions have been utilized, with many laboratories adapting existing technologies to address their research needs. Complex protein mixtures have been resolved in several gel types, including starch, polyacrylamide, and polyacrylamide/starch. The objective of this chapter is to provide basic instructions on electrophoretic techniques in soybean. Many modifications to standard protocols have been made, and where appropriate, these will be discussed.

2 Tissues Utilized as Protein Sources

The soybean seed is an excellent source of storage proteins and enzymes. Proteins can be extracted easily from developing seeds, dry mature seeds, and the cotyledons of germinating seeds. The large seed size enables researchers to remove a portion of the cotyledons and to grow a plant from the remaining embryo and attached cotyledons. Nondestructive sampling procedures make it possible to maintain genotypes of interest.

The developmental stage of the seed in developing pods can be determined from pod size (Meinke et al. 1981). Pods at the appropriate stage can be picked, placed into plastic bags, and stored at $-70\,°C$ until needed. Sampling dry seeds can be destructive or nondestructive. They hypocotyl-radical axis is clearly visible on the seed coat of dry seeds. A razor blade or scalpel will cut off pieces of the cotyledon ends opposite the embryo. Care must be taken so that the cotyledons do not detach because the cotyledons have a tendency to slide past one another.

The cotyledons of germinating seeds are also used as a source of proteins for electrophoretic analysis. Different enzymes are expressed at different stages of ontogeny, so the time of sampling will vary. Often researchers want to maximize the number of isozyme markers, so a compromise is necessary because not all enzymes will be at their optimum expression. Scarification of the testa, especially in lines with hard seed coats, will synchronize germination. A small portion of the seed coat can be removed with a razor blade. Rubbing small seed (such as the wild annual species) between two pieces of sandpaper is less laborious than trying to remove a piece of the seed coat with a razor blade. Excess seed of these lines should be planted so that the required number of seedlings will be at the appropriate developmental stage.

Seed is placed on germination paper or on filter paper in a Petri dish and put into a growth chamber for germination. We have found that 72 to 96 h at 28 to 30 °C will produce a seedling that can be sampled easily and will survive transplanting. Germinating seed in Petri dishes is recommended only if germination time is less than 36 h or if the seedlings do not need to be saved. The ends of the cotyledons or an entire cotyledon can be cut off with a scalpel or razor blade. A faster method uses a micropipetter equipped with a disposable glass micropipette. The two cotyledons are held together, and the micropipet is pushed through the tissue. The tissue plugs are then ejected into a 1.5-ml microcentrifuge tube containing homogenization buffer. The micropipet should be rinsed thoroughly with distilled water to prevent cross-contamination between samples. Immersing the tubes in ice water will minimize protein degradation. The samples can be stored at -20 or $-70\,°C$ until needed.

2.1 Homogenization

The level of technological sophistication associated with homogenization ranges from a hammer and a seed envelope to a grinder with a pestle machined to fit a microcentrifuge tube. Hydrated tissue can be ground in a 1.5-ml microcentrifuge tube with a power-driven pestle. The appropriate homogenization buffer is added before grinding. It is important to keep the tissue cool during grinding because excessive heat degrades protein and can produce artifacts. Mortar and pestles also can be used but are much more time-consuming. After grinding, samples are centrifuged to pellet cellular debris. The supernatant is stored frozen, but long storage periods and thawing the sample a number of times will reduce the protein levels and resolution and increase denaturation. Generally, the sample is ground the day before the gel is run.

Dry seed or seed chips are more difficult to grind. Strategies vary from laboratory to laboratory. The simplest strategy is to add homogenization buffer, allow the tissue to hydrate, and grind (as above). An alternative is to pulverize seed in an envelope with a hammer, crush the seed with pliers, or grind it in a mortar and pestle. The resulting powder is extracted with the appropriate solutions. Subsequent centrifugation will remove cellular remains and sand used for grinding in a mortar.

There are a number of homogenization buffers employed in the analyses of soybean proteins, and buffer composition tends to evolve over time. Almost every laboratory makes minor modifications to "standard" buffers and adds or omits components. In starch gel electrophoresis, the homogenization and extraction buffers are the same because grinding and extraction is a one-step procedure. The three most commonly used buffers are listed in Tables 1, 2, and 3. Our laboratory uses buffer 1 (Table 1) because it is easy and inexpensive to make. For polyacrylamide gel electrophoresis of seed storage proteins, more elaborate extraction protocols are employed. Some laboratories grind the cotyledonary tissue, defat the powder and then extract the protein. For most applications, this adds extra steps and is unnecessary. The seed can be ground in a grinder, crushed, or powdered in a mortar and pestle. Homogenization buffer (Table 3) is added to the powder, and after 30 min, sufficient protein has been extracted. Centrifugation pellets the cellular debris and the supernatant can be used fresh or stored at $-20\,^{\circ}$C. Under nondenaturing electrophoretic

Table 1. Buffers used for starch gel electrophoresis

		References
Homogenization buffers		
1. 16.7% Sucrose	4.2 g	Cardy and Beversdorf (1984b)
8.3% Sodium ascorbate	2.1 g	
Up to 25 ml with water (pH 7.4)		
2. 0.1 M TRIS-HCl	1.21 g	Griffin and Palmer (1987)
4% PVP-40	4.0 g	
0.4 M sucrose	13.7 g	
0.001 M dithiothreitol	0.02 g	
Up to 100 ml with water (pH 7.2)		
3. 0.005 M L-histidine	1.048 g	Bult et al. (1989)
Up to 100 ml with water (pH 7)		
Electrode buffers		Cardy and Beversdorf (1984b)
B. 0.065 M L-histidine	10.1 g	
0.019 M Citric acid	4.0 g	
Up to 1 l with water (pH 5.7)		
D. 0.065 M L-histidine	10.1 g	
0.007 M Citric acid	1.5 g	
Up to 1 l with water (pH 6.5)		
Gel buffers		Cardy and Beversdorf (1984b)
B. 6:1 Dilution of B electrode buffer		
D. 3:1 Dilution of D electrode buffer		

Table 2. Buffers used for polyacrylamide/starch gel electrophoresis[a]

Homogenization buffer	
Same as gel buffer given below	
Gel buffer	
0.005 L-histidine	1.048 g
Up to 1 l with water (pH 7.0)	
Solutions used for gel making	
1. Acrylamide	46.55 g
Bis-acrylamide	2.45 g
Ammonium persulfate	0.70 g
Gel buffer	450.0 ml
2. Starch	14.0 g
Gel buffer	250.0 ml
Electrode buffer	
0.13 M TRIS	15.73 g
0.04 M Citrate	7.68 g
Up to 1 l with water (pH 7)	

[a] Adapted from Kiang et al. (1985) and Bult et al. (1989).

conditions, no sodium dodecylsulfate (SDS) is added to the buffer. In SDS-PAGE, the proteins are denatured by the addition of 0.2% to 4% SDS to the extraction buffer and 0.1% SDS to gel and electrode buffers. The addition of urea aids in the denaturation of proteins (Fontes et al. 1984). Sucrose (15%) or glycerol (10%) is added to increase the density of the protein solution and facilitates loading the samples into the wells in the gel. Tracking dyes such as bromophenol blue can be added to the sample or can be placed in the outer wells with the molecular weight standards.

3 Starch Gel Preparation

A number of companies market hydrolyzed potato starch suitable for making gels. A starch suspension consisting of 64.8 g of starch and 150 ml of the appropriate buffer (Table 1) is agitated on a magnetic stirrer. An additional 400 ml of gel buffer is heated to 85 °C in a microwave oven. If the buffer is overheated, lumps will form in the gel. Add the 400 ml to the starch suspension quickly and swirl the flask to complete the mixing. After mixing, heat the suspension to 85 °C again. Then, the gel is degassed to remove the bubbles introduced during the stirring process by placing the gel under a partial vacuum. When the bubbles have been eliminated, pour the gel into the gel mold. The entire process must be done quickly because the gel cools quickly and becomes too thick to degas or pour.

Table 3. Buffers used for polyacrylamide electrophoresis

		References
Homogenization buffers		
1. 0.1 M Tris-HCl	1.2 g	Griffin and Palmer (1987)
4% PVP	4.0 g	
0.4 M Sucrose	13.7 g	
0.001 M dithiothreitol	0.02 g	
Up to 100 ml with water (pH 7.2)		
2. 0.05 M Tris-HCl	0.61 g	Honeycutt et al. (1989)
0.70 M SDS	20.20 g	
0.15 M β-mercaptoethanol	1.2 g	
Up to 100 ml with water (pH 7.6)		
3. 0.01 M phosphate-citrate	0.14 g	Tihanyi et al. (1989)
0.01 M β-mercaptoethanol	0.08 g	
20% Glycerol	24.3 ml (85% Glycerol)	
Up to 100 ml with water (pH 6.5)		
Stock solutions[a]		
1. Acrylamide	30.0 g	
β-Bis-acrylamide	0.8 g	
Up to 100 ml with distilled water		
2. TRIS	36.3 g	
1 N HCl	48.0 ml	
Up to 100 ml with distilled water (pH 8.9)		
3. SDS	10.0 g	
Up to 100 ml with distilled water		
4. TRIS	5.7 g	
1 N HCl	45.0 ml	
Up to 100 ml with distilled water (pH 6.9)		
5. Ammonium persulfate	1.0 g	
Up to 10 ml with distilled water		
Electrode buffer[a]		
Glycine	28.8 g	
Tris	6.0 g	
SDS (denaturing gels only)	1.0 g	
Up to 1 l with water (pH 8.35)		

[a] Adapted from Davis (1964) and Laemmli (1970).

The gel will set in 30 to 40 min after which it should be covered with plastic wrap to prevent dehydration. Overnight storage at room temperature is recommended because storing gels under refrigeration produces a granular structure. For longer storage, gels can be stored in a plastic bag with 50 ml distilled water to prevent desiccation.

3.1 Gel Loading

A spatula can be used to cut a vertical slit the length of the gel 2 cm from the edge of the gel. Gentle pressure from the palm of the hand will open the slit.

Supernatant from centrifuged samples is absorbed by 2×11 mm wicks cut from filter paper. Wicks are placed vertically along one face of gel with a spacing of 2 to 3 mm. After all wicks have been loaded, gently close the gel, being careful to remove all air bubbles and not to move the wicks. The plastic wrap should be placed back on the top of the gel.

The gel is placed in tanks containing the appropriate buffer (Table 1). Orientate the loaded samples so that the proteins migrate through the gel towards the positive pole. The electrophoretic run should be carried out in a refrigerator, a cold room, or chromatography chamber to prevent the gel from overheating. In addition, a water bag should be placed on top of the gel to provide additional cooling. Gels are run for 5 h at 8.5 W constant power. Some laboratories remove the wicks approximately 45 min into the gel run, but we find that this is unnecessary.

3.2 Gel Slicing

After the run is complete, the gel is trimmed to a 7.5 cm width and removed from the gel mold. The gel is placed on a Plexiglas sheet and the air bubbles are squeezed out from underneath the gel. A 1 mm$\times2$ cm$\times20$ cm Plexiglas strip is placed on each side of the gel. A thin wire (usually a guitar string) held taut in a Π-shaped frame is drawn through the gel. Pairs of strips are added to the stack until the entire gel is processed. The top slice is discarded, and the remaining slices are put into plastic boxes for staining.

4 Polyacrylamide Electrophoresis

Polyacrylamide gel electrophoresis offers some advantages over starch gels. The gels are transparent, thereby allowing densiometric scanning and the quantification of the amount of protein in a band. Enzyme activity stains can be used in conjunction with nondenaturing buffer systems. The addition of SDS to the gel, electrode, and extraction buffers disrupts protein structures. Migration through the gel becomes a function of protein size, permitting the estimation of the protein's molecular weight. For many applications, PAGE is the system of choice.

The system known as disc (discontinuous) gel electrophoresis was introduced by Orstein (1964) and Davis (1964). Two gels, a stacking gel with large pores and a resolving gel with small pores, compose this system. Each gel has a different buffer, and the pH of the stacking gel is 2 to 3 pH units lower than that of the resolving gel. These features enhance protein separation by concentrating the proteins in the stacking gel, and then upon entering the resolving gel, proteins are separated by size and charge. Laemmli (1970) refined this system by adding SDS to disrupt proteins by separating them into polypeptide chains. The addition of urea to the buffers can further improve soybean protein separation (Fontes et al. 1984). Occasionally, continuous PAGE systems (having only one buffer) have been used to separate soybean proteins.

4.1 Polyacrylamide Gel Preparation

One key to successful PAGE is fastidious cleaning of the glass plates, spacers, and combs before the apparatus is assembled. After washing and rinsing with distilled water, all components should be dried. Rinsing with 70% ethanol and wiping with a labwipe will remove any residual grime and fingerprints.

Sometimes acrylamide will leak out of the assembled apparatus. The edges can be sealed by allowing a hot agar solution (10 mg agar/ml water) to run down each side. Often, this is not necessary, so let experience be your guide. Recipes for making stock solutions are presented in Table 3. The stock solutions except for ammonium persulfate can be stored for several weeks in the refrigerator. The quantities of each solution required to make resolving gels with various polyacrylamide percentages are given in Table 4. The acrylamide and ammonium persulfate solutions should be degassed in a suction flask or desiccator to remove dissolved oxygen. The presence of oxygen will inhibit polymerization of the gel. The subsequent addition of TEMED and ammonium persulfate should be done in a manner that will minimize the reintroduction of oxygen. TEMED is added after the solution has been degassed. Ammonium persulfate should be added just before the gel is to be poured.

When all the components have been added, mix by gently swirling the solution. Tip the apparatus and pour the solution down the side while being care-

Table 4. Formulae for making acrylamide gels with different percentages of acrylamide

	5%	10%	15%	20%
Resolving gel				
Stock solution 1[a]	16.7 ml	33.3 ml	50.0 ml	66.7 ml
Stock solution 2	20.0 ml	20.0 ml	20.0 ml	20.0 ml
Stock solution 3[b]	1.0 ml	1.0 ml	1.0 ml	1.0 ml
Final volume	100.0 ml	100.0 ml	100.0 ml	100.0 ml
After degassing add: 50 µl TEMED				
Just before pouring: 500 µl stock solution 5				
Stacking gel				
Stock solution 1[a]	16.7 ml			
Stock solution 3[b]	1.0 ml			
Stock solution 4	12.5 ml			
Final volume	100.0 ml			
After degassing add: 50 µl TEMED				
Just before pouring: 1 ml stock solution 5				

[a] Stock solutions are from Table 3.
[b] Do not add SDS for nondenaturing PAGE.

ful not to trap air bubbles in the gel. Once the gel has been poured, add a layer of distilled water with a Pasteur pipette or syringe to prevent oxygen from entering the gels. After the gel has set, usually 30 min, pour off the water. If the gel sets too quickly, reduce the quantity of ammonium persulfate. The remaining water can be removed with a Pasteur pipette or syringe. To make the stacking gel, repeat the steps just given for the recipe in Table 3. The comb is inserted between the glass plates before the gel is poured. When the stacking gel has set, the gel can be run or stored at room temperature. During storage, the gel should be placed in a plastic bag with enough distilled water to maintain humidity.

To load the gel, first gently remove the comb from the gel. Deformed wells can be straightened with a Pasteur pipette. Fill the wells with electrode buffer (Table 3). The protein samples are denser than the electrode buffer due to the presence of sucrose or glycerol in the sample and will sink when eluted from the pipette tip. Assemble the apparatus and add additional electrode buffer if necessary. We separate proteins on a commercial minigel apparatus at 80 V (constant voltage) for 1 h (or until the protein has moved out of the stacking gel) and then at 140 V (constant voltage) for 2 h (or until the tracking dye runs off the end of the gel).

4.2 Acrylamide/Starch Gel Electrophoresis

The laboratory of Y. T. Kiang at the University of New Hampshire routinely uses a gel matrix consisting of both acrylamide and starch for separating native proteins. These gels can be sliced like starch gels so that multiple enzyme systems can be assayed on one gel.

The protocol (Bult et al. 1989) consists of making a 7% polyacrylamide gel and a 2% starch gel and then mixing the two. Solution 1 (Table 2) is mixed and degassed. Solution 2 (Table 2) is heated to 78–80 °C in a microwave oven. Then, the starch suspension is placed under a partial vacuum to remove air bubbles. The two solutions are mixed quickly, and 1.4 ml of TEMED is added. The gel is poured into a horizontal electrophoresis apparatus. The mixing of the solution and pouring must be carried out quickly to prevent the gel from cooling and setting in the flask (Bult et al. 1989).

5 Principles of Enzyme Staining and Some Considerations

Detection of isozymes within a gel matrix was first developed by Hunter and Markert (1957) and has been termed the zymogram method. Since that time, many research applications and sophisticated methods for detection of specific enzymes have been developed for direct observation of isozymes in situ after electrophoresis of protein samples in a gel matrix. More intricate systems involve the use of coenzymes such as NAD, NADP, or tetrazolium salts. Reduction reactions usually produce a colored, insoluble product at specific activity sites within a gel matrix (Harris and Hopkinson 1976). Almost all types of enzymes can be assayed by electrophoresis followed by isozyme-specific staining methods. Some enzyme-staining techniques are intricate and involve several accessory enzyme reactions to achieve localized color changes (Harris and Hopkinson 1976).

Chromogenic and chemical staining of isozymes are the simplest methods. Either method can be used to detect some enzymatic hydrolyses of a substrate to a colored product. Localized color changes in the gel matrix indicate the presence of specific enzyme activity. An example of a chromogenic/chemical type of reaction is illustrated by the detection of the isozyme acid phosphatase. In this reaction, the α-naphthyl acid phosphate substrate is hydrolyzed and forms a dark orange band in the presence of Fast Garnet, thereby indicating the location of functional isozymes.

The most common staining method involves electron transfer dyes (Harris and Hopkinson 1976). These dyes are usually tetrazolium salts and 3-amino-9-ethyl carbazole which function as electron acceptors in dehydrogenase reactions and in oxidase and peroxidase reactions, respectively (Harris and Hopkinson 1976). Both methods result in the formation of a colored product at the site of enzyme activity. Methyl thiazolyl tetrazolium (MTT) and nitro blue tetrazolium (NBT) are widely used tetrazolium salts in isozyme staining reactions, although NBT is generally less sensitive than MTT. MTT is reduced by an electron donor such as NADH to form a dark blue insoluble formazan band at the region of enzyme activity. Often an intermediate catalyst such as dichlorophenol indophenol (DCPIP) or phenazine methosulfate (PMS) is added to increase the reaction rate. A tetrazolium-PMS staining mixture can be used to detect enzyme reactions that form reduced NAD and NADP coenzymes. Isozyme-staining reactions for malate dehydrogenase, malic enzyme, and isocitrate dehydrogenase proceed at an increased rate in the presence of PMS.

Enzyme-linked staining methods for isozyme detection can be developed by using exogenous enzymes in staining solutions. After a specific isozyme color reaction is established, it may be modified to detect other isozymes with the aid of additional linking enzymes (Harris and Hopkinson 1976). The most common procedure requires the addition of an appropriate coenzyme (linking enzyme) to generate a formazan precipitate at the site of initial enzyme activity. The specificity of each reaction can be directed by the addition of different substrates. Two examples using a glucose-6-phosphate dehydrogenase (G-6-PDH) linking enzyme are the detection of phosphoglucomutase in the presence of glucose-1-phosphate and the detection of phosphoglucoisomerase in the presence of fructose-6-phosphate. Mannose phosphate isomerase isozymes are detected in the presence of mannose-6-phosphate plus the linking enzymes G-6-PDH and phosphoglucoisomerase.

There are practical considerations when developing and using staining techniques for isozymes detection. One of the main considerations depends on reliability and consistency of the results. Agar overlays may be used to improve resolution when detecting enzymes with low activity or to reduce the volume of reaction mixture required for isozyme detection (Brewer and Sing 1970). Reliability of results may vary if endogenous enzymes comigrate with the isozyme being examined; therefore, new staining procedures should be critically reviewed, especially when working with enzyme-linked staining methods. Improper sample preparation and storage may lead to sample degradation and may greatly affect the quality and reliability of the results obtained from isozyme analysis.

5.1 Isozyme-Specific Activity Stains

Staining recipes in Table 5 were adapted from Bult et al. (1989), Cardy and Beversdorf (1984b), and Rennie et al. (1989). The chemical principles were briefly described earlier; therefore, each reaction will not be described. The stains are made in 75 ml of buffer except for agar overlays that use a total volume of 30 ml. The staining solutions with the gel slices are incubated at 37 °C for 60–90 min. Development rates of the zymogram stains vary with specific isozymes; some develop very rapidly and must be cooled after initial development of stain to prevent blurring. Basic zymogram patterns observed for each isozyme in cultivated soybean have been diagrammed by Cardy and Beversdorf (1984a, b), Rennie et al. (1989), and Gorman et al. (1983); (also see Table 6).

5.2 General Protein Staining with Coomassie Brilliant Blue R

Polyacrylamide gel electrophoresis has led to increased resolving power of protein fractions, but gel fixation and staining sensitivity play an important role in the quality of the results. Activity stains can be used in conjunction with PAGE.

Coomassie blue dye is the most extensively used stain for sensitive protein detection in polyacrylamide gel electrophoresis (Chrambach et al. 1967; Weber and Osborn 1969). The gel is transferred directly from the electrophoresis apparatus to the staining solution of 0.1% Coomassie Brilliant Blue R, 7.5% acetic acid and 50% methanol. Minigels take 20 min to stain, whereas larger gels take 1 h or longer. Then pour off the stain and rinse the gel several times with distilled water before adding a destaining solution consisting of 10% methanol and 7.5% acetic acid. The destain will have to be changed every 1–2 h. Labwipes or foam rubber in the destain will absorb the Coomassie blue and this speeds the destaining process. After destaining, the gel can be preserved by drying.

Table 5. Staining recipes for enzyme activity stains[a]

Enzyme	Buffer[b]	Substrate[c]	NAD[c]	NADP[c]	MTT[c]	NBT[c]	PMS[c]	Other[c]
Alcohol dehydrogenase	IV	Ethanol 6 ml	5			5	0.6	
Acid phosphatase	I	α-NAP 150						Fast Garnet 50
Aconitase	II	Aconitate 100		10	10		2	$MgCl_2$ 100 Isocitrate dehydrogenase 40 U
Diaphorase	II	NADH 10			10			DCPIP 0.5
Endopeptidase	III	BANA 10						Black K Salt 10 $MgCl_2$ 100
Isocitrate dehydrogenase[d]								
Solution 1	IV	Isocitrate 100		10	10		2	$MgCl_2$ 100 PVP 300
Solution 2	IV							Agar 150
Leucine aminopeptidase	V[e]	Leucine naphthyl amide 10						Black K Salt 30
Malate dehydrogenase	IV	Malate 100	10			10	1	
Malic enzyme	II	Malate 100		10	10		1	$MgCl_2$ 100
Mannose phosphate isomerase[d]								
Solution 1	II	Mannose-6-phosphate 20	10		5		1	$MgCl_2$ 100 Glucose-6-phosphate dehydrogenase 40 U Phosphoglucoisomerase 100 U
Solution 2	II							Agar 150
6-Phosphogluconate dehydrogenase	VI	Phosphogluconate 15		10	10		1	$MgCl_2$ 20
Phosphoglucoisomerase	II	Fructose-6-phosphate 50	10		5		1.5	$MgCl_2$ 100 Glucose-6-phosphate dehydrogenase 40 U
Phosphoglucomutase	II	Glucose-1-phosphate 250	10		5		2	EDTA 10 Glucose-6-phosphate dehydrogenase 40 U $MgCl_2$ 100
Superoxide dismutase	VI	Riboflavin 1			5			TEMED 10 ml

[a] Recipes were adapted from Cardy and Beversdorf (1984a, b); Bult et al. (1989); Rennie et al. (1989).
[b] Buffer I 0.01 M sodium acetate, pH 5; Buffer II 0.2 M Tris-HCl, pH 8; Buffer III 0.1 M Tris-maleate, pH 5.5; Buffer IV 0.1 M Tris-HCl, pH 9.1; Buffer V 0.01 M sodium acetate, pH 6.2; Buffer VI 0.05 M Tris-HCl, pH 8.
[c] Given in mg quantities unless followed by units of measure.
[d] Agar overlay. Heat solution 2 until the agar is dissolved, cool to 72°C and mix with solution 1 before pouring on the gel.
[e] Soak gel slice in buffer for 30 min before adding stain components.

Table 6. Soybean isozymes and seed storage proteins detected with electrophoresis

Enzyme	Gene symbol	Reference
Acid phosphatase	*Ap*	Gorman and Kiang (1977)
		Hildebrand et al. (1980)
		Cardy and Beversdorf (1984a)
Aconitase	*Aco1, Aco2,*	Doong and Kiang (1987c)
	Aco3, Aco4	Rennie et al. (1987a)
		Griffin and Palmer (1987)
Alcohol dehydrogenase	*Adh1, Adh2*	Gorman and Kiang (1978)
α-Amylase	*Amy1, Amy2*	Gorman and Kiang (1978)
		Kiang (1981)
β-Amylase	*Sp1*	Larsen and Caldwell (1968)
		Hildebrand and Hymowitz (1980)
β-Conglycinin a'-subunit	*Cgy1*	Kitamura et al. (1984)
		Davies et al. (1985)
β-Conglycinin a-subunit	*Cgy2*	Davies et al. (1985)
β-Conglycinin b-subunit	*Cgy3*	Davies et al. (1985)
Diaphorase	*Dia1, Dia2, Dia3*	Gorman and Kiang (1983)
		Cardy and Beversdorf (1984a)
Endopeptidase	*Enp*	Doong and Kiang (1987a)
		Rennie et al. (1987b)
Esterase		Ferrer-Monge (1974)
		Payne and Koszykowski (1978)
Fluorescent esterase	*Fle*	Doong and Kiang (1988)
Glucose-6-phosphate dehydrogenase	*Gpd*	Kiang and Gorman (1983)
Glutamate oxaloacetate transaminase	*Got*	Kiang et al. (1987)
Glutamate dehydrogenase		Yue (1969)
Glycinin	*Gly1, Gly2, Gly3,*	Fontes et al. (1984)
	Gly4, Gly5	Kitamura et al. (1984)
		Cho et al. (1989)
Isocitrate dehydrogenase	*Idh1, Idh2, Idh3*	Cardy and Beversdorf (1984a, b)
		Kiang and Gorman (1985)
Lactate dehydrogenase		Tihanyi et al. (1989)
Leucine aminopeptidase	*Lap1, Lap2*	Kiang et al. (1985)
		Rennie et al. (1989)
Lectin	*Le*	Orf et al. (1978)
Lipoxygenase	*Lx1, Lx2, Lx3*	Hildebrand and Hymowitz (1982)
		Kitamura et al. (1983)
		Kitamura et al. (1985)
Malate dehydrogenase	*Mdh1*	Cardy and Beversdorf (1984a, b)
Malic enzyme		Cardy and Beversdorf (1984a, b)
Mannose-6-phosphate dehydrogenase	*Mpi*	Cardy and Beversdorf (1984a, b)
		Chiang and Kiang (1988)
Peroxidase	*Ep*	Buttery and Buzzell (1968)
6-Phosphogluconate dehydrogenase	*Pgd*	Chiang and Kiang (1987)
Phosphoglucose isomerase	*Pgi1, Pgi2, Pgi3*	Cardy and Beversdorf (1984a, b)
		Chiang et al. (1987)
Phosphoglucomatase	*Pgm1, Pgm2, Pgm3*	Kiang and Gorman (1983)
		Cardy and Beversdorf (1984a)
Superoxide dismutase	*Sod*	Gorman and Kiang (1978)
		Griffin and Palmer (1989)
Trypsin inhibitor	*Ti*	Orf and Hymowitz (1977, 1979)
Urease	*Eu*	Buttery and Buzzell (1971)
		Kloth et al. (1987)

6 Gel Scoring

For genetic studies, the alleles present at each locus are recorded. Alleles with the slowest mobility are designated "a", the second slowest is "b", and so on. Alleles characterized later are lettered consecutively regardless of mobility. For varietal identification, scoring can be simplified because of the homozygosity and uniformity of soybean varieties. Zymogram patterns have been given letter codes (Gorman et al. 1983; Cardy and Beversdorf 1984a). For isozymes such as aconitase that are encoded by multiple loci, this system reduces the amount of record keeping. Table 6 describes the soybean proteins resolved by electrophoresis and gives references for the interpretation of zymogram patterns of the enzymes.

References

Bernard RL, Hymowitz T, Creemens CR (1991) Registration of Kunitz soybean. Crop Sci 31:232–233

Brewer G, Sing C (1970) An introduction to isozyme techniques. Acad Press, New York

Bult CJ, Kiang YT, Chiang YC, Doong HJY, Gorman MB (1989) Electrophoretic methods for soybean genetic studies. Soybean Genet Newslett 16:175–187

Buttery BR, Buzzell RI (1968) Peroxidase activity in seeds of soybean varieties. Crop Sci 8:722–725

Buttery BR, Buzzell RI (1971) Properties and inheritance of urease isoenzymes in soybean seeds. Can J Bot 49:1101–1105

Cardy BJ, Beversdorf WD (1984a) Identification of soybean cultivars using isoenzyme electrophoresis. Seed Sci Technol 12:943–954

Cardy BJ, Beversdorf WD (1984b) A procedure for the starch gel electrophoretic detection of isozymes in soybean (*Glycine max* (L.) Merr.). Dep Crop Sci Tech Bull 119/8401. University of Guelph, Ontario, Canada

Chiang YC, Kiang YT (1987) Inheritance and linkage relationships of 6-phosphogluconate dehydrogenase isozymes in soybean. Genome 29:786–792

Chiang YC, Kiang YT (1988) Genetic analysis of mannose-6-phosphate isomerase in soybeans. Genome 30:808–811

Chiang YC, Gorman MB, Kiang YT (1987) Inheritance and linkage analysis of phosphoglucose isomerase isozymes in soybeans. Biochem Genet 25:893–900

Cho TJ, Davies CS, Nielson NC (1989) Inheritance and organization of glycinin genes in soybean. Plant Cell 1:329–337

Chrambach A, Reisfeld RA, Wyckoff M, Zaccari J (1967) A procedure for rapid and sensitive staining of protein fractionated by polyacrylamide gel electrophoresis. Anal Biochem 20:150–154

Davies CS, Nielsen NC (1986) Genetic analysis of a null-allele for lipoxygenase-2 in soybean. Crop Sci 26:460–463

Davies CS, Nielson NC (1987) Registration of soybean germplasm that lacks lipoxygenase isozymes. Crop Sci 27:370–371

Davies CS, Coates J, Nielsen NC (1985) Inheritance and biochemical analysis of four electrophoretic variants of β-conglycinin from soybean. Theor Appl Genet 71:351–358

Davis BJ (1964) Disc electrophoresis-II: method and application to human serum proteins. Ann NY Acad Sci 121:404–427

Delannay X, Rodgers DM, Palmer RG (1983) Relative genetic contribution among ancestral lines to North American soybean cultivars. Crop Sci 23:944–949

Doong JYH, Kiang YT (1987a) Inheritance of soybean endopeptidase. Biochem Genet 25:847–853

Doong JYH, Kiang YT (1987b) Cultivar identification by isozyme analysis. Soybean Genet Newslett 14:189–226

Doong JYH, Kiang YT (1987c) Inheritance of aconitase isozymes in soybean. Genome 29:713–717

Doong JYH, Kiang YT (1988) Inheritance study on a soybean fluorescent esterase. J Hered 79:399–400

Ferrer-Monge JA (1974) Esterase isozyme patterns in *Glycine max* exposed to gamma radiation. Can J Bot 52:273–275

Fontes EPB, Moreira MA, Davies CS, Nielsen NC (1984) Urea-elicited changes in relative mobility of certain glycinin and β-conglycinin subunits. Plant Physiol 76:840–842

Gayler KR, Sykes GE (1981) β-Conglycinins in developing soybean seeds. Plant Physiol 67:958–961

Gorman MB, Kiang YT (1977) Variety-specific electrophoretic variants of four soybean enzymes. Crop Sci 17:963–965

Gorman MB, Kiang YT (1978) Models for the inheritance of several variant soybean electrophoretic zymograms. J Hered 69:255–258

Gorman MB, Kiang YT, Palmer RG, Chiang YC (1983) Inheritance of soybean electrophoretic variants. Soybean Genet Newsl 10:67–84

Griffin JD (1986) Genetic and germplasm studies with several isoenzyme loci in soybean (*Glycine max* (L.) Merr.) PhD Thesis, Iowa State Univ, Ames (Diss: Abstr 86:27111)

Griffin JD, Palmer RG (1987) Inheritance and linkage studies of five isozyme loci in soybean. Crop Sci 27:885–893

Griffin JD, Palmer RG (1989) Genetic studies of two superoxide dismutase loci soybean. Crop Sci 29:968–971

Griffin JD, Palmer PG Isozyme variability in the USDA soybean germplasm collection. In: Knapp AD (ed) Proc 4th Int Symp. The development and application of new technologies for varietal identification (in press)

Harris H, Hopkinson DA (1976) Handbook of enyzme electrophoresis in human genetics. North Holland, Amsterdam

Hildebrand DF, Hymowitz T (1980) The *Spl* locus in soybean codes for β-amylase. Crop Sci 20:165–168

Hildebrand DF, Hymowitz T (1982) Inheritance of lipoxygenase-1 activity in soybean seeds. Crop Sci 22:851–853

Hildebrand DF, Orf JH, Hymowitz T (1980) Inheritance of an acid phosphatase and its linkage with the Kunitz trypsin inhibitor in seed protein of soybeans. Crop Sci 20:83–85

Honeycutt RJ, Burton JW, Palmer RG, Shoemaker RC (1989) Association of major seed components with a shriveled-seed trait in soybean. Crop Sci 29:804–809

Hunter R, Markert C (1957) Histochemical demonstration of enzymes separated by zone electrophoresis in starch gel. Science 125:1294–1295

Hymowitz T, Kaizuma N (1979) Dissemination of soybeans (*Glycine max*). Seed protein electrophoresis profiles among Japanese cultivars. Econ Bot 33:311–319

Hymowitz T, Kaizuma N (1981) Soybean protein electrophoresis profiles from 15 Asian countries or regions. Hypotheses on paths of dissemination of the soybean from China. Econ Bot 35:10–23

Kiang YT (1981) Inheritance and variation of amylase in cultivated and wild soybeans and their wild relatives. J Hered 72:382–386

Kiang YT, Gorman MB (1983) Soybean. In: Tanksley SD, Orton J (eds) Isozymes in plant genetics and breeding. Elsevier, New York, pp 295–328

Kiang YT, Gorman MB (1985) Inheritance of NADP-active isocitrate dehydrogenase isozymes in soybeans. J Hered 76:279–284

Kiang YT, Gorman MB, Chiang YC (1985) Genetic and linkage analysis of a leucine aminopeptidase in wild and cultivated soybean. Crop Sci 25:319–321

Kiang YT, Chiang YC, Bult CJ (1987) Genetic study of glutamate oxaloacetic transaminase in soybean. Genome 29:370–373

Kitamura K, Davies CS, Kaizuma N, Nielsen NC (1983) Genetic analysis of a null-allele for lipoxygenase-3 in soybean seeds. Crop Sci 23:924–927

Kitamura K, Davies CS, Nielsen NC (1984) Inheritance of alleles for *Cgyl* and *Gy4* storage proteins in soybean. Theor Appl Genet 68:253–257

Kitamura K, Kumagi T, Kikuchi A (1985) Inheritance of lipoxygenase-2 and genetic relationships among genes for lipoxygenase-1, -2, and -3 isozymes in soybean seeds. Jpn J Breed 35:413–420

Kloth RH, Polacco JC, Hymowitz T (1987) The inheritance of a urease-null trait in soybean. Theor Appl Genet 73:410–418

Laemmli VK (1970) Cleavage of structural proteins during the assembly of the head of bacteriophage T4. Nature 227:680–685

Larsen AL, Caldwell BE (1968) Inheritance of certain proteins in soybean seed. Crop Sci 8:474–476

Meinke DW, Chen J, Beachy RN (1981) Expression of storage-protein genes during soybean seed development. Planta 153:130–139

Ogawa T, Tayanna E, Kitamura K, Kaizuma N (1989) Genetic improvement of seed storage proteins using three variant alleles of 7S globulin subunits in soybean (*Glycine max* L.) Jpn J Breed 39:137–147

Orf JH, Hymowitz T (1977) Inheritance of a second trypsin inhibitor variant in seed protein of soybeans. Crop Sci 17:811–813

Orf JH, Hymowitz T (1979) Inheritance of the absence of the Kunitz trypsin inhibitor in seed protein of soybean. Crop Sci 19:107–109

Orf JH, Hymowitz T, Pull SP, Pueppke SG (1978) Inheritance of a soybean seed lectin. Crop Sci 18:899–900

Orstein L (1964) Disc electrophoresis I. Theory. Ann NY Acad Sci 121:321–403

Payne RC, Koszykowski TJ (1978) Esterase differences in seed extracts among soybean cultivars. Crop Sci 18:557–559

Rennie BD, Beversdorf WD, Buzzell RI (1987a) Genetic and linkage analysis of an aconitate hydratase variant in soybean. J Hered 78:323–326

Rennie BD, Beversdorf WD, Buzzell RI (1987b) Inheritance and linkage analysis of two endopeptidase variants in soybeans. J Hered 78:327–328

Rennie BD, Thorpe ML, Beversdorf WD (1989) A procedural manual for the detection and identification of soybean (*Glycine max* (L.) Merr.) isozymes using the starch gel electrophoretic system. Dep Crop Sci Tech Bull, University of Guelph, Ontario, Canada

Stahlhut RW, Hymowitz T (1983) Variation in low molecular weight proteinase inhibitors of soybean. Crop Sci 23:766–769

Tihanyi K, Fontabot A, Thirion JP (1989) Soybean L-(+)-lactate dehydrogenases: purification, characterization, and resolution of subunit structure. Arch Biochem Biophys 274:626–632

Weber K, Osborn M (1969) The reliability of molecular weight determinations by dodecyl sulfate-polyacrylamide gel electrophoresis. J Biol Chem 244:4406–4412

Yue SB (1969) Isoenzymes of glutamate dehydrogenase in plants. Plant Physiol 44:453–457

Analysis of Storage of Proteins in Rice Seeds

D. S. LUTHE

1 Introduction

For several years my laboratory has been interested in characterizing rice seed proteins and studying their synthesis and deposition during seed development. Much of the work has been based on earlier work by Juliano's group (Tecson et al. 1971; Juliano and Boulter 1976); Villareal and Juliano 1978) at the International Rice Research Institute in the Philippines, and on work doen with rice (Zhao et al. 1983) and oat seed proteins (Peterson 1978; Brinegar and Peterson 1982a, b). In addition, others have contributed to our knowledge of rice storage proteins by (1) studying the immunological relationships among rice seed proteins and those of other seeds (Robert et al. 1985; Okita et al. 1988); (2) determining how the storage proteins are processed (Krishnan and Okita 1986; Sarker et al. 1986) and deposited in protein bodies in the endosperm (Tanaka et al. 1980; Yamagata et al. 1982); and (3) isolating and sequencing genes of the major rice storage protein classes (Higuchi and Fukazawa 1987; Takaiwa et al. 1987; Wang et al. 1987; Kim and Okita 1988a, b; Masumura et al. 1989a). This chapter will describe the techniques that have been used in my laboratory (Luthe 1983; Wen and Luthe 1985) to isolate the subunits of the major rice storage protein, glutelin, and methods used for in vivo labeling and electrophoretic analysis of rice and proteins.

2 Seed Proteins

Seed storage proteins are an abundant group of proteins that accumulate in the seed druing embryogenesis and are degraded during germination to provide nitrogen and energy sources for the germinating embryo. Their synthesis begins after fertilization and they accumulate during seed maturation. In dicotyledonous plants storage proteins are deposited in protein bodies in the cotyledons, whereas in monocotyledonous plants they are primarily present in protein bodies in the endosperm.

Seed storage proteins have traditionally been classified into four solubility fractions: albumins, globulins, prolamins, and glutelins (Osborne 1910). The albumins are water-soluble proteins that usually consist of enzymes present in the seed. The globulins are salt-soluble proteins that comprise the most abundant storage protein class in many dicotyledonous seeds. They are especially prevalent in legume seeds and have been classified as being either legumin-like proteins with a sedimentation coefficient of 11 to 12S or vicilin-like proteins with a sedimentation coefficient of 7S (Derbyshire et al. 1976). The prolamins are alcohol-soluble proteins of heterogeneous sizes that make up the major storage protein class in the cereals (see recent review by Shotwell and Larkins 1989). The glutelins are soluble in dilute acid or alkali and are generally considered to be the proteins remaining after all of the other fractions have been extracted. Glutelins are also prevalent in wheat and other cereal seeds.

2.1 Isolation of the Four Storage Protein Fractions

Because much of the early research on rice seed proteins was done prior to the development of high resolution sodium dodecyl sulfate polyacrylamide gel electrophoresis (SDS-PAGE), we decided to try several alternative methods to isolate the major storage protein fractions from rice seeds. The methods currently being used for the sequential extraction of rice (*Oryza sativa* L, cv. Lebonnet) seed proteins is shown in Fig. 1. The extraction buffers are listed in Table 1. To prevent deleterious effects of extremes in pH, a solution containing the chaotropic detergent, sodium dodecyl sulfate (SDS), was used to extract the glutelin fraction rather than dilute solutions of acid or alkali. In order to diminish cross-contamination among solubility fractions, the extraction of each fraction was repeated three times. Each protein fraction was precipitated from the supernatant fractions as described in Fig. 1.

2.2 Electrophoretic Analysis of Storage Protein Fractions

For electrophoretic analysis the protein samples were dissolved in sample buffer (Laemmli 1970) containing 1 mM phenylmethanesulfonyl fluoride (PMSF) and heated in a boiling water bath for 5 min prior to electrophoresis. The storage protein fractions are analyzed by SDS-PAGE using the discontinuous buf-

Table 1. Solutions for rice seed protein extraction

Extraction buffer	Composition
Albumin	10 mM Tris·HCl (pH 7.5) and 1 mM PMSF[a]
Globulin	0.5 M NaCl, 10 mM Tris·HCl (pH 7.5) and 1 mM PMSF
Prolamin	70% Ethanol, 10 mM Tris·HCl (pH 7.5) and 1 mM PMSF
Glutelin	0.5% (W/V) Sodium dodecyl sulfate (SDS), 10 mM Tris·HCl (pH 8.5) and 1 mM PMSF

[a] PMSF = phenylmethylsulfonyl fluoride, a protease inhibitor. Stock solutions of PMSF were prepared in 95% ethanol and alqiuots were added to the extraction buffers immediately prior to use.

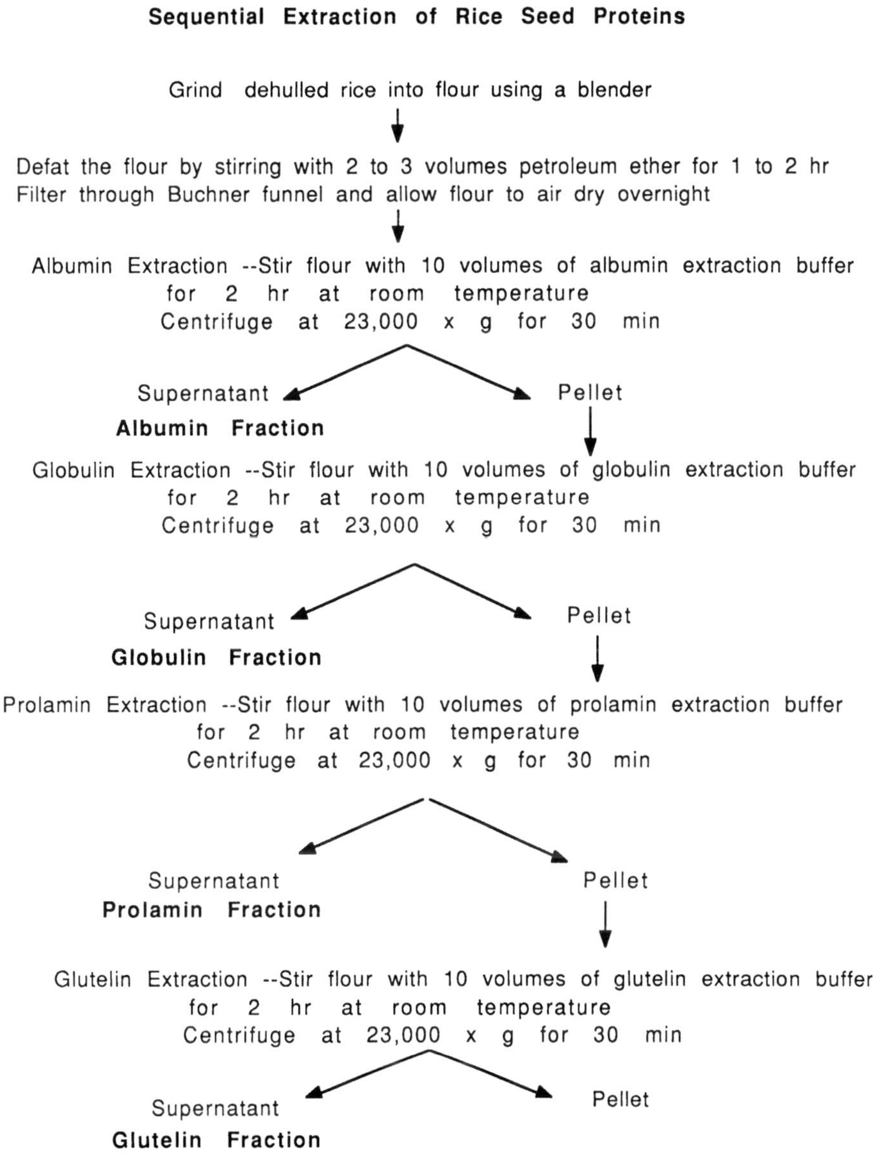

Sequential Extraction of Rice Seed Proteins

Grind dehulled rice into flour using a blender

↓

Defat the flour by stirring with 2 to 3 volumes petroleum ether for 1 to 2 hr
Filter through Buchner funnel and allow flour to air dry overnight

↓

Albumin Extraction --Stir flour with 10 volumes of albumin extraction buffer
for 2 hr at room temperature
Centrifuge at 23,000 x g for 30 min

Supernatant ← → Pellet
Albumin Fraction

Globulin Extraction --Stir flour with 10 volumes of globulin extraction buffer
for 2 hr at room temperature
Centrifuge at 23,000 x g for 30 min

Supernatant ← → Pellet
Globulin Fraction

Prolamin Extraction --Stir flour with 10 volumes of prolamin extraction buffer
for 2 hr at room temperature
Centrifuge at 23,000 x g for 30 min

Supernatant ← → Pellet
Prolamin Fraction

Glutelin Extraction --Stir flour with 10 volumes of glutelin extraction buffer
for 2 hr at room temperature
Centrifuge at 23,000 x g for 30 min

Supernatant ← → Pellet
Glutelin Fraction

Fig. 1. Flow chart describing the sequential extraction of rice storage protein fractions. The proteins are isolated from the supernatant fractions of albumin, globulin, and prolamin by acetone precipitation. Two volumes of ice-cold acetone are added to the supernatants followed by precipitation overnight at $-20\,^\circ$C. Samples are centrifuged and allowed to air dry. The glutelin-containing supernatant is dialyzed against distilled water at $4\,^\circ$C overnight to remove residual SDS and to precipitate the glutelin. Following dialysis the suspension is centrifuged at $4000\times$g for 30 min and the pellet is lyophilized

fer system of Laemmli (1970) with the following modification. We have determined that the proteins were better resolved in the gradient gel system that in gels containing a single acrylamide concentration. A 5% to 20% gradient of acrylamide was poured into a vertical slab gel (14.5×17×0.15 cm) (Gibco-BRL Model V16) using a gradient-forming apparatus (Gibco-BRL, Model 150). A stacking gel (5% acrylamide) was then poured on top of the gradient gel.

Samples were applied to the gel and electrophoresed at 20 mA until the tracking dye move out of the stacking gel, then the current was increased to 40 mA until the tracking dye moved to the bottom of the gel. Gels were fixed and stained with 0.2% (w/v) Coomassie brilliant blue R-250 in 50% methanol (v/v), 10% acetic acid and destained with 5% (v/v) methanol, 7.5% (v/v) acetic acid.

Figure 2 shows the electrophoretic analysis of the four rice storage protein fractions. Despite attempts to prevent it, cross-contamination of solubility fractions did occur. Several polypeptides with apparent MW of 23 and 53 kD were present in both the albumin and globulin fractions, and a group of low MW proteins (approximately 13 to 16 kD) were present in all fractions. A polypeptide with apparent MW of 23 kD was the most prevalent protein in the globulin fraction and is probably similar to the 26-kD salt-soluble protein previously reported (Tanaka et al. 1980; Yamagata et al. 1982). The two major polypeptides in the prolamin fraction where those with apparent MW of 15.6 and 14.6 kD, which may correspond to the prolamin polypeptides of 13 and

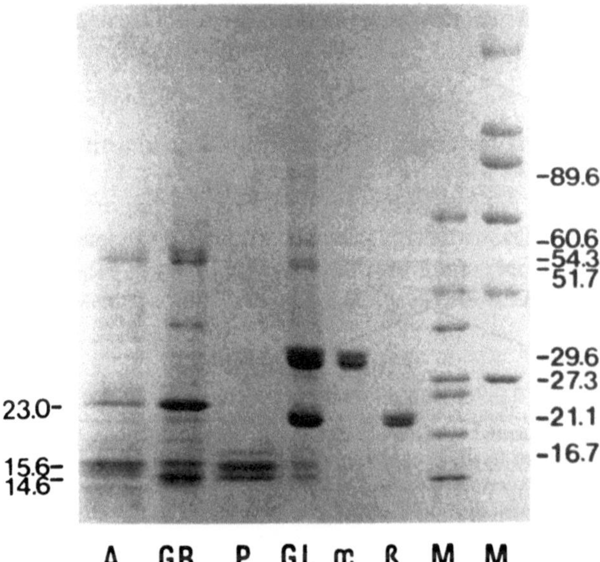

Fig. 2. SDS-PAGE of rice storage protein fractions. *Lanes: A* albumin; *GB* globulin; *P* prolamin; *GL* glutelin; *α* purified α-subunit; *ß* purified ß-subunit; *M* molecular weight markers. *Numbers* in the margins refer to the apparent molecular weight in kilodaltons (kD) of the major polypeptides

16 kD that were recently reported by Hibino et al. (1989). The glutelin fraction was comprised of two major groups of polypeptides: 29 and 21 kD groups. Since glutelin is the major storage protein fraction in rice, comprising about 80% of the total protein, this fraction was selected for further study.

3 Characterization of Rice Glutelin

The electrophoretic pattern of rice glutelin was very similar to that of oat globulin, the major oat seed storage protein (Peterson 1978; Brinegar and Peterson 1982a, b) and the legumin-like proteins found in soybean and pea (Derbyshire et al. 1976; Shotwell and Larkins 1989). These legumin-like proteins have a sedimentation coefficient in the range of 11 to 12 S, are comprised of six acidic (α) and six basic (β) subunits joined by disulfide bonds, and have a molecular weight of about 360 kD. The α-peptides range in size from approximately 30 to 40 kD, and the β-peptides have molecular weights of about 20 kD (Derbyshire et al. 1976). The α- and β-peptides are synthesized as a precursor with a molecular weight of approximately 60 kD which contains a signal sequence required for movement into the protein bodies and a tandem arrangement of one α- and one β-subunit connected by several amino acids (Shotwell and Larkins 1989). Following translation the precursor is cleaved resulting in a dipeptide containing one α-subunit and one β-subunit joined by a disulfide bond (Shotwell and Larkins 1989). Six of these dipeptides assemble to form the 11 or 12 S holoprotein (Shotwell and Larkins 1989). Two characteristics of rice glutelin lead us to believe that it might be similar to the legumin-like proteins: the apparent molecular weights of the major polypeptides, and the fact that it was synthesized as a precursor, which was posttranslationally cleaved to form the α- and β-subunits (Yamagata et al. 1982; Luthe 1983). Because of these similarities we decided to determine some of the other biochemical properties of rice glutelin.

3.1 Two-Dimensional Electrophoresis

Two-dimensional electrophoresis (O'Farrell 1975)[1] consisting of isoelectric-focusing (IEF − separation according to charge) in the first dimension and SDS-PAGE (separation according to size) in the second dimension was conducted to determine the relative charges on the α- and β-subunits of glutelin. This information should indicate (1) if rice glutelin was comprised of acidic

---

[1] Before conducting two-dimensional electrophoresis one should carefully read the paper by O'Farrell (1975) and O'Farrell et al. 1977). In addition, the book, *Methods in Molecular Biology*, Vol. 1, *Proteins*, edited by J. M. Walker (Hurmana) has helpful hints on many types of electrophoresis.

and basic subunits like the other legumin-like proteins and (2) if the charge/size properties of the protein could be used to design a protocol for purifying the α- and β-subunits.

When conventional two-dimensional electrophoresis (O'Farrell 1975) was conducted the β-subunits of glutelin, which we believed to be basic proteins, were not well resolved. The lack of resolution was the result of using IEF in the first dimension. There are several reasons why IEF is not suitable for the resolution of basic proteins. First, the interaction of urea with the basic ampholines destablizes the pH gradient in the basic region of the gel after prolonged exposure to the electric current. Consequently, some of the basic proteins to not migrate into the gel, and those that do are streaked. Second, IEF gels are prerun to form a pH gradient and the proteins are electrophoresed until they reach a position in the gel where the pH is equivalent to their pI; if the gradient has been disrupted the basic proteins will be unable to focus. Third, since electrophoresis is toward the anode the basic proteins are the last to enter the gel resulting in poor resolution. To alleviate this problem nonequilibrium pH gradient electrophoresis (NEPHGE, O'Farrell 1977) was used for the first dimension instead of IEF. For NEPHGE the gels are not prerun, so no preformed pH gradient is established. The time of electrophoresis is shortened so the proteins are not electrophoresed until they reach their equilibrium position. Finally, electrophoresis is toward the cathode with the acidic reservoir on top and the basic reservoir on the bottom, which allows the basic proteins to enter the gel first and improves resolution. Because the proteins are not electrophoresed until they reach an equilibrium position as in IEF, their absolute pIs cannot be determined, but NEPHGE can be used to determine whether the polypeptides are acidic or basic.

NEPHGE was conducted in vertical tube gels (13×0.3 cm) containing 9 M urea, 5% acrylamide (32:1 weight ratio of acrylamide:bis-acrylamide), 2% ampholine mixture (see below), 0.025% ammonium persulfate, and 0.07% TEMED (N,N,N',N', tetramethylethylenediamine). The ampholines[2] in the mixture were in the following pH ranges: pH 4−6 (0.07%), pH 5−7 (0.14%), pH 6−8 (0.14%), pH 9−11 (0.2%) and pH 3.5−10 (1.6%). Lyophilized protein samples were dissolved in lysis buffer containing 9.5 M urea, 2% Nonidet P-40, 5% β-mercaptoethanol, and 2% ampholines, pH 5−7 (1.6%) and pH 3−10 (0.4%). The sample (30 to 160 µg in about 20 µl) was loaded on the gel and overlayered with a solution containing 8 M urea and 1% ampholines (pH 5−7 − 0.8% and pH 3.5−10 − 0.2%). Gels were placed in a tube gel adaptor (Gibco-BRL) for the vertical electrophoresis apparatus. The upper reservoir (anode) contained 0.01 M H_3PO_4 and the lower reservoir (cathode) contained 0.02 M NaOH. The NaOH must be degassed prior to use. NEPHGE was run at 400 V for 4 to 6 h (1 600−2 400 volt-hours). Following electrophoresis the gels were stored at −70 °C for electrophoresis in the second dimension.

Before electrophoresis in the second dimension the tube gels were equili-

---

[2] The most consistent results are obtained when LKB Ampholines are used. Ampholines can now be obtained from Pharmacia/LKB Biotechnology Inc., Piscataway, New Jersey, USA.

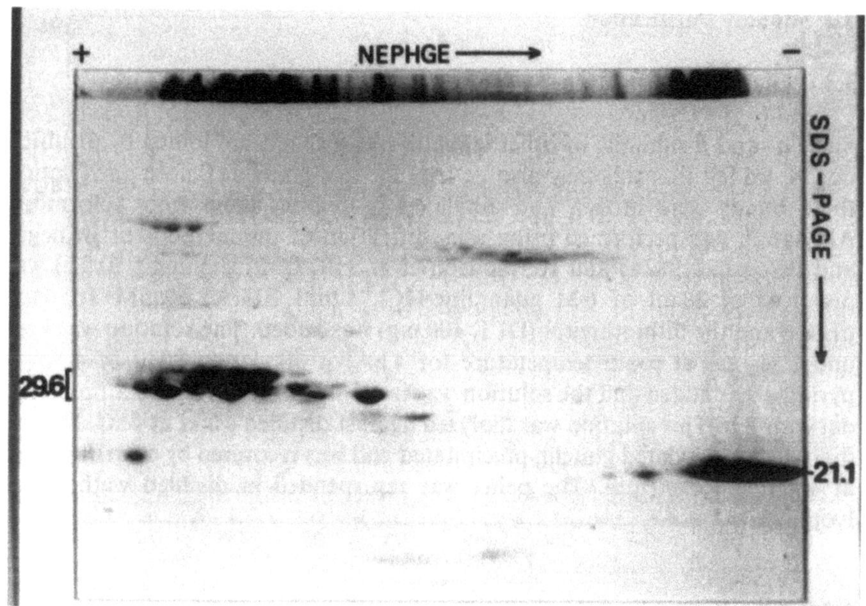

Fig. 3. Two-dimensional electrophoresis of rice glutelin. Nonequilibrium pH gradient electrophoresis (NEPHGE) was conducted in the first dimension. After electrophoresis for 2800 V-h, the final pH range was 4.1 to 9.1. Electrophoresis in the second dimension was SDS-PAGE. *Numbers* in the margin refer to the average apparent molecular weight of the α-subunit (29.6 kD) and the β-subunit (21.1 kD). (Wen and Luthe 1985)

brated in 5 ml of SDS sample buffer (Laemmli 1970) for 1.5 h at room temperature. The IEF gels were immediately loaded into 1% agarose which covered the stacking gel portion of the SDS-polyacrylamide gel (5 to 20% acrylamide gradient). SDS-PAGE was conducted at 20 mA constant current for about 5 h. The gel was stained by shaking it in a solution containing 0.005% Coomassie brilliant blue R-250, 50% ethanol, and 7% acetic acid for 36 h. The gel was stained and rehydrated in the same solution without the ethanol (O'Farrell 1975). This procedure eliminates the staining of the ampholines which migrate to the bottom of the gel.

Two-dimensional electrophoresis indicated that the α- and β-subunits of glutelin were acidic and basic, respectively. Each subunit consisted of several charge and size variants (Fig. 3). These properties were consistent with those of other legumin-like proteins. Because of the differences in charge, chromatofocusing was used for purification of the α-subunits and cation exchange chromatography was used for purification of the β-subunit (Wen and Luthe 1985).

3.2 Subunit Purification

3.2.1 Glutelin Purification

Since α- and β-subunits of other legumin-like proteins are joined by disulfide bonds, we felt that this may also be true for rice glutelin. Before purification these bonds were broken and alkylated to prevent them from reforming. Alkylation was performed using a modification of the methods of Brinegar and Peterson (1982a) and Hermodson et al. (1977). Rice glutelin (0.8 g) was dissolved in 20 ml of 6 M guanidine-HCl, 1 mM EDTA, 50 mM Tris-Hcl, pH 8.6 and the dithiothreitol (DTT, 400 mg) was added. The solution was kept under N_2 gas at room temperature for 4 h. For alkylation 80 µl of 4-vinyl-pyridine was added and the solution was incubated at room temperature in the dark for 2 h. This solution was dialyzed against distilled water at 4 °C. During dialysis, the alkylated glutelin precipitated and was recovered by centrifugation at 4000×g for 30 min. The pellet was resuspended in distilled wather and lyophilzed.

3.2.2 Glutelin Acidic Subunit Purification

Chromatofocusing[3] is a column chromatographic technique similar to IEF — it separates proteins according to their isoelectric point (pI). This technique was used to purify the acidic subunits of glutelin. Two chromatofocusing columns were used during the purification procedure. The first column used a gradient buffer system from pH 8.3 to 5. In this pH range the basic subunits should be positively charged and will be excluded from the column. The matrix of the first column was Polybuffer Exchanger PBE 94 (Pharmacia). The column (1×40 cm) was equilibrated with about 20 column volumes of start buffer containing 9 M urea, 25 mM Tris-acetate, pH 8.3. Because of the insolubility of rice glutelin, it was essential to keep 9 M urea in all of the column buffers. Alkylated glutelin (130 mg) in 10 ml of start buffer was loaded on the column and the column was eluted with 12.5 vol of a buffer containing 30% of 1 : 10 diluted polybuffer 96 (Pharmacia, pH range 9 to 6), 70% of 1 : 10 diluted polybuffer 74 (pH range 7 to 4) in 9 M urea. The pH of the elution buffer was adjusted to pH 5 with acetic acid prior to elution of the column. Fractions (2 ml) were collected and the absorbance at 280 nm was measured. SDS-PAGE was used to determine the protein composition of the fractions. Figure 4 shows the results from the first chromatofocusing column. Because the column was overloaded, fraction I contained some α-polypeptides, but it also contained several β-polypeptides and what appeared to be a prolamin polypeptide. Most of the α-polypeptides eluted in fractions II, III, and IV, which were in the pH range 7.5 to 6. Since these fractions still contained some contamination

---

[3] See Pharmacia Handbook, *Chromatofocusing with Polybuffer and PBE.*

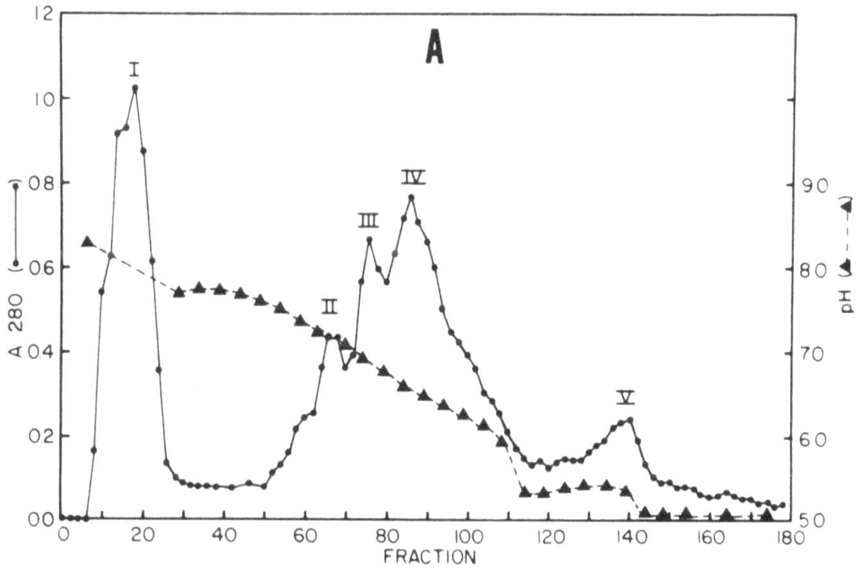

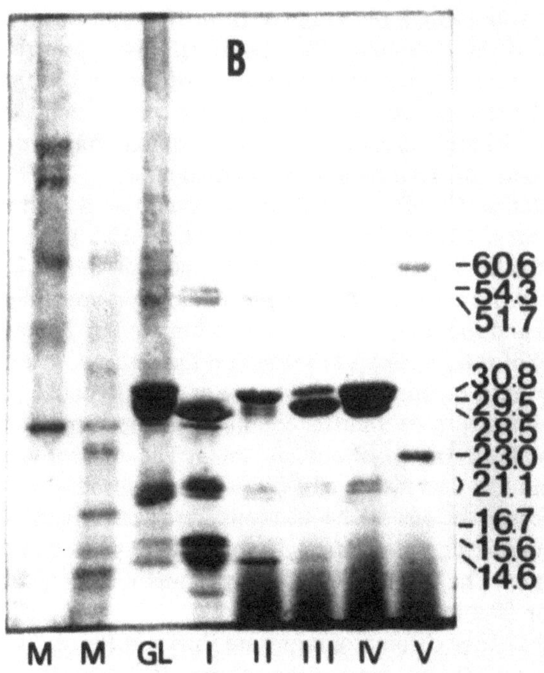

Fig. 4. Elution of glutelin subunits from an acidic chromatofocusing column (A). After chromatography the peak fractions were pooled and designated fractions *I–V* and analyzed by SDS-PAGE (B). *Lanes: M* molecular weight markers; *GL* unpurified glutelin; and *I–V* peak fractions from the column. *Numbers* in right-hand margin refer to the apparent molecular weights of major polypeptides in kD

β-polypeptides, they were pooled and applied to a second chromatofocusing column.

The second chromatofocusing column (1×36 cm) contained the same matrix as the first, but the samples were eluted with a pH gradient of pH 8 to 6. Pooled fractions from the first column were dissolved in 5 ml of start buffer and applied to the column. The eluting buffer was 9 M urea in $1:13$ diluted polybuffer 96, pH 6. Fractions were collected and analyzed as described above. In this case, all the β-subunits were excluded from the column and were present in fraction 12 (Fig. 5). The α-polypeptides were eluted in fractions A1, A2, and A3, with the 30.8, 29.5, and 28.5 kD polypeptides eluting first, second, and third, respectively. After a more detailed analysis of the protein compositions of the fractions, fractions from 131 to 210 were pooled. Because fraction A1 contained some 14.6 kD polypeptide, this fraction was not collected. The results from this column indicate that with sufficient care it may be possible to isolate separate charge variants of the α-polypeptides using chromatofocusing.

3.2.3 Glutelin Basic Subunit Purification

Unfortunately, chromatofocusing using a pH gradient from 10.5 to 8 was not suitable for the purification of the β-subunit (Wen 1984). In this system the pH gradient declined rapidly, probably due to the presence of the urea in the elution buffer; consequently, the β-subunit could not be separated from the 14.6 and 15.6 kD polypeptides. Therefore, cation exchange chromatography on CM-Sepharose CL-6B was used for purification of the β-subunit.

The column of CM-Sepharose CL-6B (1×15 cm) was equilibrated with a start buffer containing 9 M urea, 50 mM β-mercaptoethanol, 10 mM glycine-NaOH, pH 8.5. At this pH the basic polypeptides of the β-subunit should be positively charged and bind to the column, whereas the acidic α-polypeptides should be negatively charged and should not bind to the column. After 200 mg of alkylated glutelin in 10 ml of start buffer was loaded on the column, it was washed with the start buffer until no further absorbance at 280 nm could be measured. The column was eluted with start buffer containing a linear gradient of NaCl from 0 to 0.3 M. The flow rate was about 22.5 ml/h. The effluent was collected in 3 ml fractions and the absorbance at 280 nm of each fraction was measured. The polypeptide compositions of the fractions were determined by SDS-PAGE. The fractions containing the purified β-subunit were pooled and dialyzed against distilled water. The precipitated β-subunit was recovered by centrifugation at $4000 \times g$ for 20 min and lypophilized.

The results of the cation exchange chromatography are shown in Fig. 6. The acidic polypeptides (see fractions) did not bind to the column. Polypeptides of 51.7, 15.6, and 14.6 kD eluted with the β-subunits in the B1 fraction (fractions $11-26$), but β-polypeptides of acceptable purity were isolated from fraction B2 (fractions $28-90$). SDS-PAGE of the purified α- and β-subunits is shown in Fig. 2.

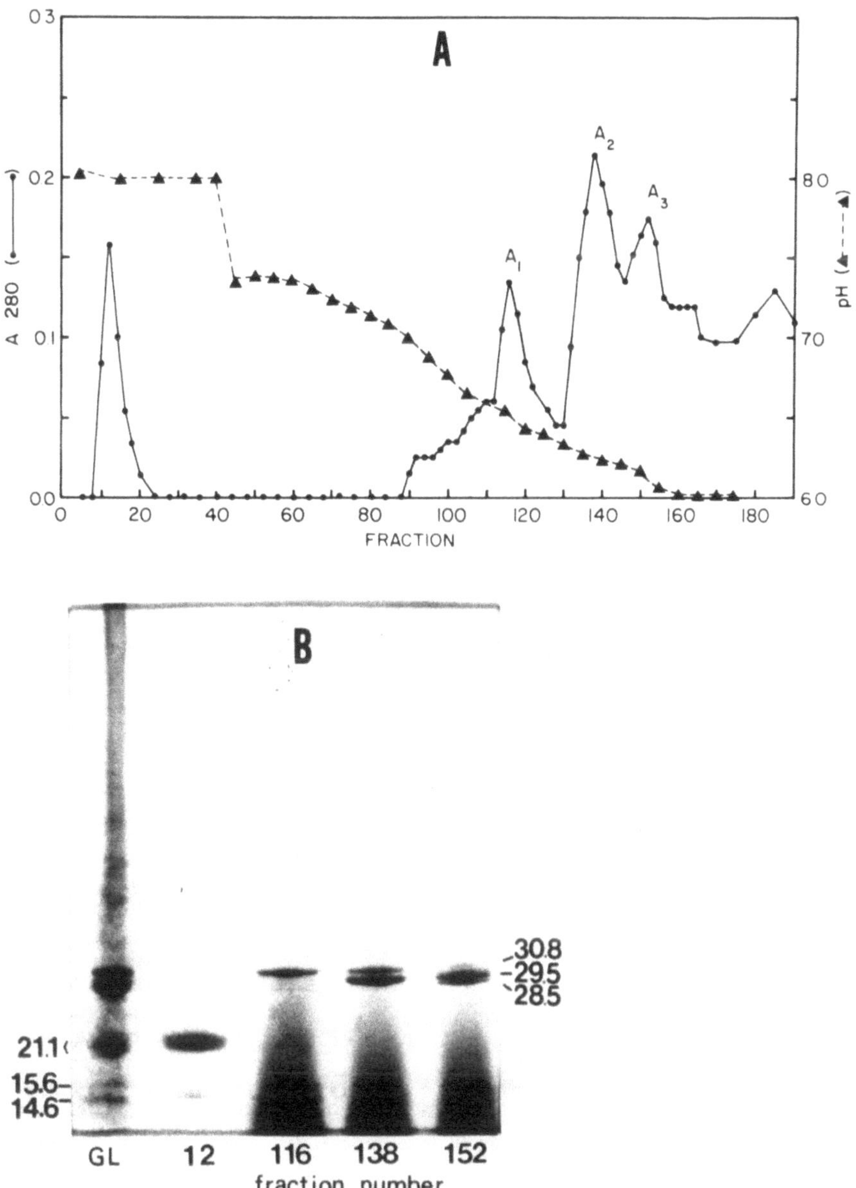

Fig. 5. Chromatofocusing of the α-subunit enriched fractions (II, III, and IV) recovered from the column described in Fig. 4. **A** Peak fractions were pooled and designated A_1, A_2, and A_3. **B** Selected fractions from the column were analyzed by SDS-PAGE. *GL* Unpurified glutelin; *numbers* in the margin refer to the apparent molecular weight in kD of major polypeptides. (Wen and Luthe 1985)

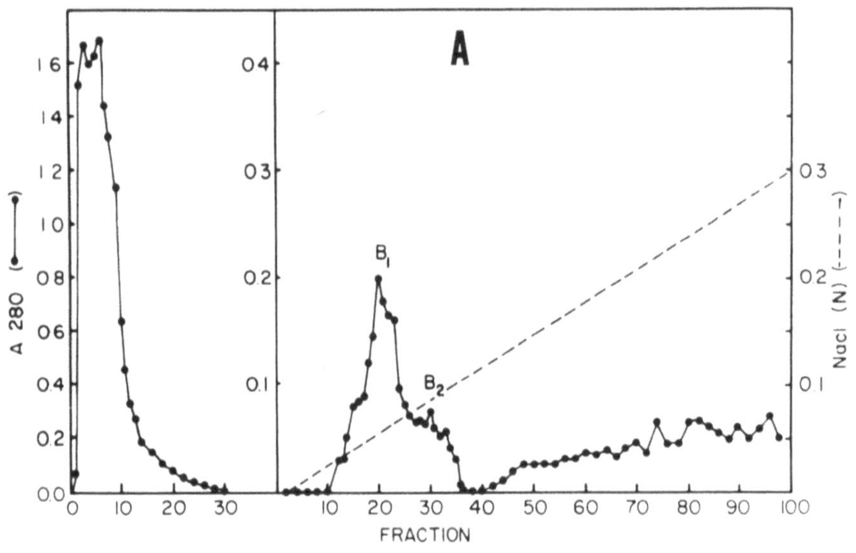

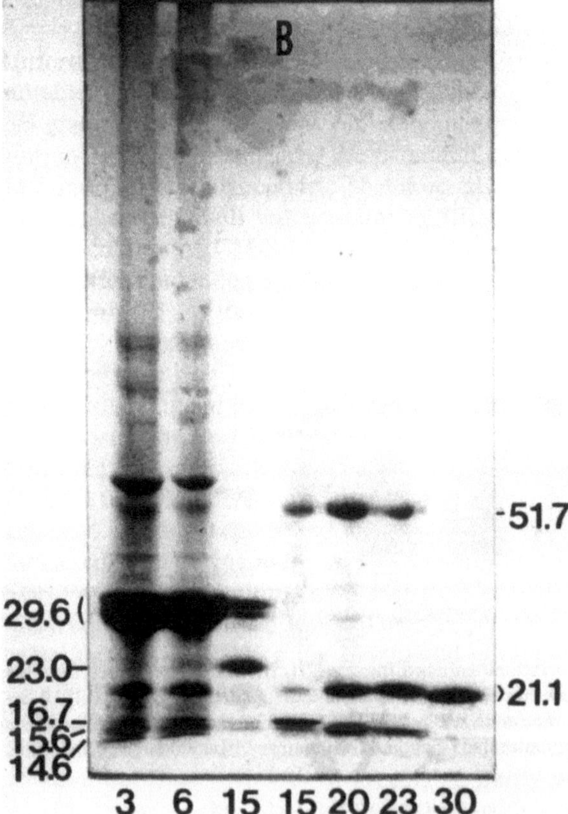

Fig. 6. A Purification of the β-subunit of rice glutelin by cation exchange chromatography on CM-Separarose CL-6B. The *left panel* represents protein that did not bind to the column. Peaks B_1 and B_2 eluted in the NaCl gradient (0 to 0.3 M). **B** SDS-PAGE analysis of selected fractions from the column. The first three fractions (*3, 6, 15*) represent the unbound protein. The remaining fractions (*15, 20, 23, 30*) are samples taken from the B_1 and B_2 peaks. *Numbers* in the margins refer to the apparent molecular weights of the major polypeptides in kD

3.2.4 Determination of Isoelectric Points (pI)

Results from the two-dimensional gel analysis, column chromatography, and SDS-PAGE indicated that both the α- and β-subunits were comprised of several polypeptides with slight differences in size and charge. This was not surprising because most storage proteins are encoded by small multigene families (Shotwell and Larkins 1989). Although we had difficulty focusing the basic polypeptides in the two-dimensional analysis, IEF on horizontal gels was attempted to determine the number of charge variants and pI values of the α- and β-subunits.

An LKB 2117 Multiphor system was used for IEF. Slab gels ($8 \times 11.5 \times 0.2$ cm) contained 5% acrylamide (32:1 acrylamide: bis-acrylamide), 10% glycerol, 2% ampholines (pH 6−4, 0.07%; pH 5−7, 0.14%; pH 6−8; 0.14%; pH 9−11, 0.20%; pH 3.5−10, 1.6%), 6 M urea, 0.025% ammonium persulfate, and 0.07% TEMED (Brinegar and Peterson 1982a). The gel mixture was degassed for about 2 min before the ammonium persulfate and TEMED were added. The mixture was immediately loaded into the mold and overlayered with distilled water. The anode and cathode wicks were soaked in 1 M H_3PO_4 and 1 M NaOH, respectively. The lyophilized samples were dissolved in a buffer containing 9.5 M urea, 2% Nonidet P-40, 5% β-mercaptoethanol, and 2% ampholines (pH 5−7, 1.6%, pH 3−10, 0.4%) (O'Farrell 1975), absorbed into filter paper strips, and applied to the gel surface near the anode wick. IEF was run at 300 V for 20 min, 600 V for 10 min, 650 V for 10 min, 700 V for 10 min, 800 V for 30 min, and 900 V for 30 min. The IEF gel was cooled by circulating H_2O at 10 °C. The sample strips were removed after the gel had run for 45 min. IEF was terminated when the colored protein marker, cytochrome C, pH 10.2 (a component of the standard IsoGel pI marker kit; FMS Corp.) reached a well-focused and stable position. Following IEF, the pH gradient of the gel was measured using a surface pH electrode (Bradley-James Corp.). Then the gel was fixed with 10% TCA containing 3% sulfosalicylic acid for at least 30 min and stained with 0.12% Coomassie brilliant blue R250 in 25% ethanol and 8% acetic acid.

Figure 7 shows the IEF analysis of oat globulin, rice glutelin, and the purified α- and β-subunits. The α-subunit of rice glutelin had several polypeptides with pI values in the range of pH 6.5 to 7.5; the β-subunit had several bands in the range of pH 9.4 to 10.3. The higher isoelectric point of 10.3 was determined by extrapolation. Because of the insolubility of rice glutelin it was necessary to run the IEF in the presence of 6 M urea; therefore, the pI values of the polypeptides were shifted upward (O'Farrell 1975). The pI values of the marker proteins (Fig. 7) were also shifted up by approximately 0.40 pH units when the IEF was conducted in the presence of urea. In addition, the PIs of the α- and β-subunits of oat globulin were slightly higher than those obtained when they were focused in the absence of urea (Brinegar and Peterson 1982a). From the two-dimensional gel analysis we determined that there was approximately 16 variants of the α-polypeptides. The β-polypeptides could not be clearly resolved on the two-dimensional gels, but about nine polypeptides could be distinguished on the horizontal IEF gel.

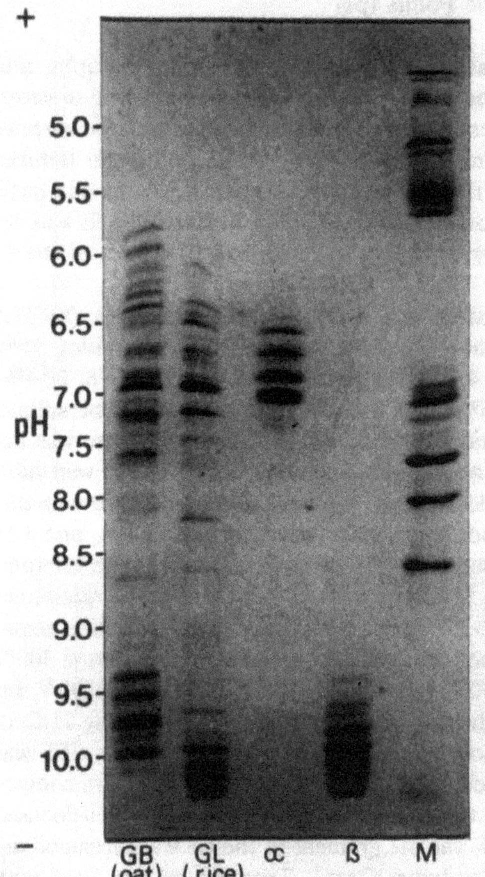

Fig. 7. Horizontal IEF of oat globulin (*GB*), rice glutelin (*GL*), and the purified α- and β-subunits of rice glutelin. *Lane M* contains proteins with known pI values: whale myoglobin (major band, pH 8.2; minor band, pH 7.7), horse myoglobin (major band, pH 7.4; minor band, pH 7), carbonic anhydrase (pH 6.1), b-lactogloglovin (major band, pH 5.4; minor band, pH 5.5), ovalbumin (pH 4.8), glucose oxidase (pH 4.2), and amyloglucosidase (pH 3.6). (Wen and Luthe 1985)

We have raised polyclonal antibodies against the purified α- and β-subunits that were specific when used to probe a Western blot containing rice glutelin. Each antibody reacted with the appropriate subunit and the 53-kD precursor protein (data not shown). Other workers have simplified the purification procedure by successfully isolating the α- and β-subunits directly from polyacrylamide gels (Krishnan and Okita 1986).

4 Analysis of Storage Protein Deposition During Rice Seed Development

In addition to determining the properties of rice glutelin, in vivo labeling has been used to determine when rice seeds were most active in protein synthesis and when specific storage fractions were being synthesized.

4.1 Plant Material and Total Protein Analysis

For these studies long-grain white rice (*Oryza sativa*, cv-Lebonnet) was grown in a paddy on the Mississippi Agricultural and Forestry Experiment Station Farm, Mississippi State, Ms. Individual panicles were tagged when anthesis occurred mid-panicle. For total seed protein anaylsis panicles were harvested at 2-day intervals during the maturation period of approximately 30 days. The caryopses were removed from the lemna and palea (dehulled), and either frozen and stored at $-70\,^{\circ}$C or lyophilized. Total protein was extracted from the seeds by homogenizing the seeds in SDS-PAGE sample buffer (Laemmli 1970) containing PMSF as described above. Seeds were homogenized using a Tekmar Tissumizer or a mortar and pestle. For rice the ratio of sample buffer to seed number was approximately 0.25 ml/seed, but this will vary depending on the type of seed being analyzed. Following homogenization the extracts were boiled for 5 min and centrifuged at $13\,000\times g$ in a microcentrifuge. The supernatant containing the seed protein was stored at $-20\,^{\circ}$C. Seed proteins were analyzed by SDS-PAGE as described above. SDS-PAGE analysis of rice seed proteins (Fig. 8A) indicated that most of the protein deposition occurred between 8 and 10 DPA (days postanthesis). There was no great increase in storage protein accumulation after about 12 DPA. There appeared to be coordinate expression of the α- and β-subunits of glutelin, the 25 kD globulin polypeptide, and the 14.6 and 15.6 kD polypeptides.

4.2 In Vivo Labeling of Rice Seed Proteins

In order to determine when rice seeds were most active in protein synthesis panicles were harvested at various intervals during the grain-filling period and the proteins were labeled in vivo with either ^{3}H-leucine or ^{35}S-sulfate. In early experiments ^{3}H-leucine was used to label the seed proteins and the efficiency of incorporation (incorporation into protein/uptake) was about 15-fold higher than that of ^{35}S-sulfate. The lower efficiency of ^{35}S-sulfate incorporation into protein probably occurs because the sulfate must first be converted into methionine and cysteine prior to incorporation into protein. Despite this disadvantage, there are several advantages to the use of ^{35}S-sulfate: it is less expensive than ^{3}H-leucine; and it emits β-particles with higher energy than ^{3}H, which makes it easier to detect the labeled proteins by fluorography (Bonner and Laskey (1974).

For the experiments reported here the panicles were harvested at 2-day intervals during the grain-filling period. After harvesting they were quickly transported to the laboratory in water. For in vivo labeling three panicles were cut while the stem was held under water, which prevents a vacuum from forming in the vascular tissue and allows radioisotopic tracers to be transported to the caryopsis. The panicles were placed in sterile test tubes containing 0.5 mCi of ^{35}S-sulfate in 0.5 ml of sterile distilled H_2O. After this solution was taken up (about 30 min) it was followed with sterile distilled H_2O for 2 h. The seeds

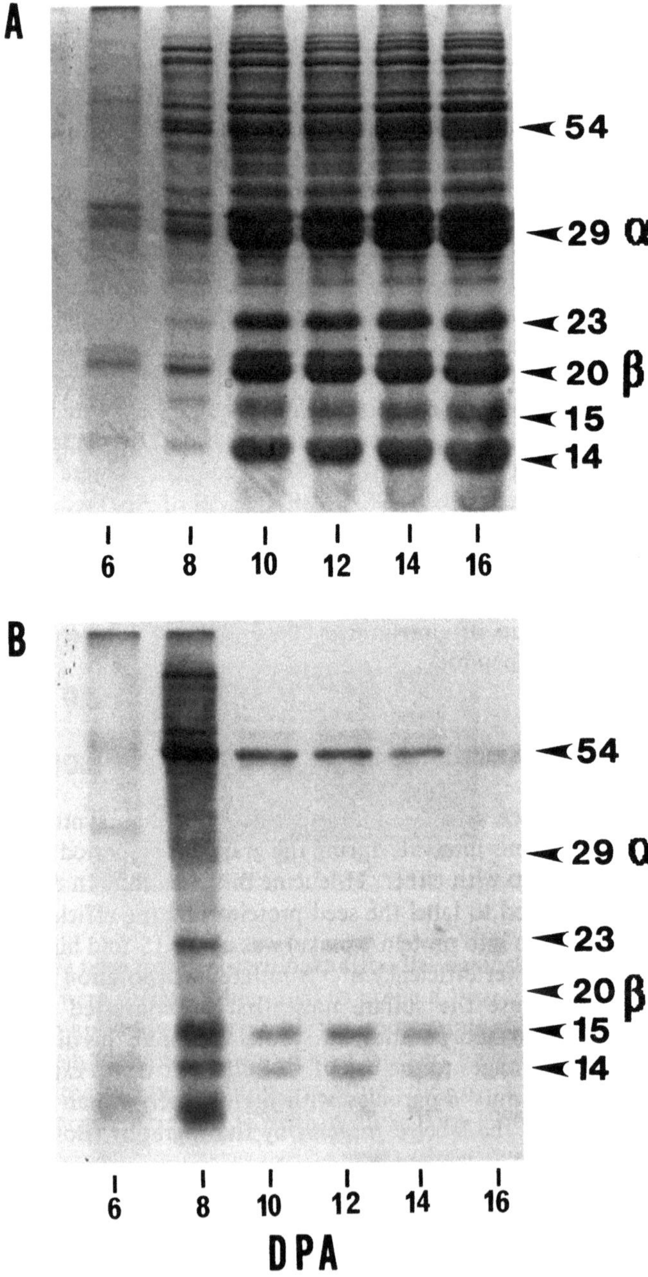

Fig. 8. A Coomassie blue-stained SDS-PAGE analysis of protein extracts from rice seeds harvested at various days postanthesis (DPA). **B** Fluorography of SDS-PAGE analysis of protein extracts from developing rice seeds labeled with ^{35}S-sulfate. Equivalent volumes, not cpm, were loaded on the gel; consequently, the data is represented on a per seed basis. *Numbers* in the margins refer to the apparent molecular weight of major polypeptides in kD. (Luthe 1983)

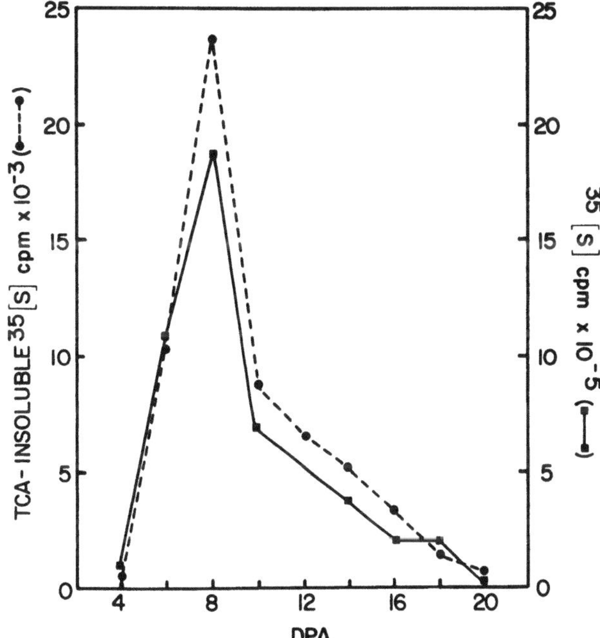

Fig. 9. Uptake of ^{35}S-sulfate and incorporation into TCA-insoluble material at various days postanthesis (DPA) during rice embryogenesis. (Luthe 1983)

were dehulled, frozen in liquid N_2, and lyophilized. Total protein was extracted from the seeds using SDS-PAGE sample buffer as described above. The amount of radioactivity incorporated into protein was determined by spotting aliquots of the samples (four aliquots/sample) on Whatman 3-MM filter paper disks. One set of samples (two disks) was counted to determine ^{35}S uptake; the other set was washed in trichloroacetic acid (TCA) to determine incorporation of lable into protein. Samples were washed in TCA (10%) for 15 min, washed in acetone for 5 min, and dried. The amount of radioactivity was determined by scintillation spectroscopy in a toluene-based fluid. When aliquots of equivalent volumes are used, the data can be expressed as cmp/seed. The results of this experiment are shown in Fig. 9 and indicated that maximum incorporation of amino acids into TCA-insoluble material occurred at 8 DPA. These data support that of Fig. 8 A indicating a massive increase in storage protein accumulation between 8 and 10 DPA. This labeling technique has also been used to determine when oat seeds are most active in protein synthesis (Luthe 1987).

4.3 Patterns of Proteins Synthesized in Vivo During Seed Development

To determine which storage proteins are synthesized at a particular developmental stage, they must be labled in vivo at that stage. The radioactive proteins are then analyzed by SDS-PAGE and fluorography. For the experiments described here, the seed protein samples were labeled and extracted as described above. CPM data from the TCA-precipitable fraction were used to determine the amount of sample to be applied to the gel. If one wants to determine quantitative differences in protein synthesis per seed, then equivalent volumes of samples should be analyzed; if one wants to observe qualitative differences, then equivalent cpm per lane should be used. To minimize the amount of time required for fluorography, the maximum number of cpm possible should be loaded on each lane. We have found that about 50000 to 100000 cpm per lane provides sufficient radioactivity to expose the film in about 3 to 4 days.

Following electrophoresis, the gels were stained with the Commassie blue solution and destained. Staining allows one to assess the quality of the gel prior to fluorography. Fluorography is a technique which facilitates the detection of radioactive substances which emit low energy β-particles such as ^{3}H, ^{14}C, and ^{35}S. In this technique the gels are impregnated with a fluor or scintillator, which intereacts with the radioactive substance producing light and exposing the X-ray film. For the data presented here a modification of the Bonner and Laskey (1974) procedure was used. This procedure uses the water-soluble fluor, sodium salicylate (Chamberlin 1979). After destaining (destaining does not need to be complete if the gel is to be analyzed by fluorography), the gel is soaked in 20 vol of H_2O for 30 min, and then soaked in 10 vol of 1 M sodium salicylate for 30 min. The gel is then placed on Whatman 3 MM paper moistened with H_2O and dried under vacuum using a commercially available gel drier. Although the salicylate method uses less hazardous chemicals, the protein bands are generally more diffuse than those obtained using the Bonner and Laskey method. Many fluorography reagents, both aqueous and nonaqueous, are now commercially available. We have had success using Resolution (EM Corporation, Chesnut Hill, MA).

The results from the fluorography of labeled rice protein are shown in Fig. 8 B. These data confirmed that rice seeds were most active in protein synthesis at 8 DPA and indicated that the major proteins synthesized were 56, 25, 16, and 13 kD polypeptides. The small polypeptide (less than 14 kD) that was labled 6 and 8 DPA, may correspond to the 10 kD rice prolamin which is rich in methionine and cysteine (Hibino et al. 1989). The pattern of proteins synthesized each day during embryogenesis was quite different than the pattern of protein that accumulated (Fig. 8 A). There were no proteins with apparent molecular weights similar to the α- and β-subunits of glutelin. Since it is known that other legumin-like proteins are synthesized from a larger precursor which is posttranslationally processed (Shotwell and Larkins 1989), it is possible that this may also be the case for rice glutelin. A pulse-chase experiment was conducted (Luthe 1983) to determine precursor-product relationships among the polypeptides. For this experiment, 14 panicles, harvested 8 DPA, were labeled with 3 mCi of ^{35}S-sulfate for 1 h. The panicles were then transferred to growth medium without hormone (Murashige and Skoog 1962), two panicles were removed at 3, 6, 12, 24, 28 and 48 h, seeds were dehulled, stored, and homogenized as described above. The results of this experiment confirmed

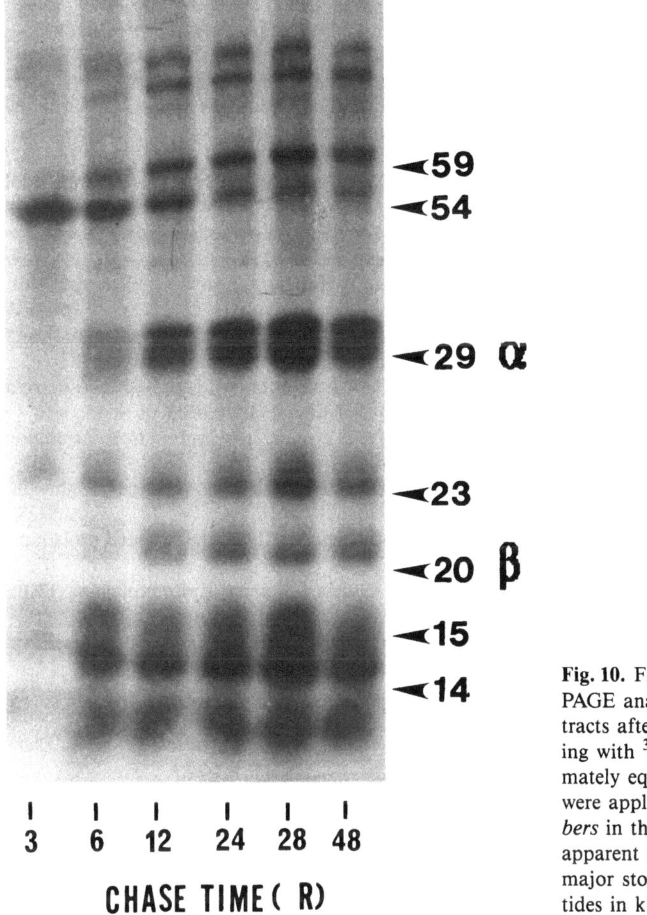

Fig. 10. Fluorography of SDS-PAGE analysis of rice seed extracts after a pulse-chase labeling with ^{35}S-sulfate. Approximately equal numbers of cpm were applied to each lane. *Numbers* in the margin refer to the apparent molecular weights of major storage portein polypeptides in kD

those of Yamagata et al. (1982) indicating that a 56 kD protein was converted to the α- and β-subunits (Fig. 10). Others (Krishnan and Okita 1986; Sarker et al. 1986) have used antibodies to confirm that the 56 kD polypeptide was the precursor of the two smaller groups of polypeptides.

Acknowledgments. I would like to acknowledge Mr. Tan Nan Wen for his careful and patient effort in purifying the glutelin subunits. This work was supported by the Mississippi Agricultural and Forestry Experiment Station (MAFES) Project 1614-000. This is MAFES publication number BC-7619.

References

Bonner WM, Laskey RA (1974) A film detection method for tritium-labelled proteins and nucleic acid in polyacrylamide gels. Eur J Biochem 48:83–88

Brinegar AC, Peterson DM (1982a) Separation and characterization of oat globulin polypeptides. Arch Biochem Biophys 219:71–79

Brinegar AC, Peterson DM (1982b) Synthesis of oat globulin precursors. Plant Physiol 170:1767–1769

Chamberlin JP (1979) Fluorographic detection of radioactivity in polyacrylamide gels with the water-soluble fluor, sodium salicylate. Anal Biochem 98:132–135

Derbyshire E, Wright DJ, Boulter D (1976) Legumin and vicilin, storage proteins of legume seeds. Phytochemistry 15:3–24

Hermodson M, Schnier G, Kurachi K (1977) Isolation, crystallization, and primary amino acid sequence of human platelet factor 4. J Biol Chem 252:6267–6279

Hibino T, Kidzu K, Masumura T, Ohtsuki K, Tanaka K, Kawabata M, Fujii S, (1989) Amino acid composition of rice prolamin polypeptides. Agric Biol Chem 53:513–518

Higuchi W, Fikazawa C (1987) A rice glutelin and a soybean glycinin have evolved from a common ancestral gene. Gene 55:245–253

Juliano BO, Boulter D (1976) Extraction and composition of rice endosperm glutelin. Phytochemistry 15:1601–1606

Kim W-T, Okita TW (1988a) Structure, expression, and heterogeneity of the rice seed prolamins. Plant Physiol 88:49–655

Kim W-T, Okita TW (1988b) Nucleotide and primary sequence of a major rice prolamine. FEBS Lett 231:308–310

Krishnan HB, Okita TW (1986) Structural relationship among the rice glutelin polypeptides. Plant Physiol 81:748–753

Laemmli UK (1970) Cleavage of structural proteins during the assembly of the head of bacteriophage T4. Nature 227:680–685

Luthe DS (1983) Storage protein accumulation in developing rice (*Oryza sativa* L.) seeds. Plant Sci Lett 32:147–158

Luthe DS (1987) Storage protein synthesis during oat (*Avena sativa* L.) seed development. Plant Physiol 84:337–340

Masumura T, Kidzu K, Sugiyama Y, Mitsukawa N, Hibino T, Tanaka K, Fujii S (1989a) Nucleotide sequence of a cDNA encoding a major rice glutelin. Plant Mol Biol 12:723–725

Masumura T, Shibata D, Hibino T, Kato T, Kawabe K, Takeba G, Tanaka K, Fujii S (1989b) cDNA cloning of an mRNA encoding a sulfur-rich 10 kD prolamin polypeptide in rice seeds. Plant Mol Biol 12:123–130

Murashige T, Skoog F (1962) A revised medium for rapid growth and bioassays with tobacco tissue cultures. Physiol Plant 15:473–497

O'Farrell PH (1975) High resolution two-dimensional electrophoresis of proteins. J Biol Chem 250:4007–4021

O'Farrell PZ, Goodman HM, O'Farrell PH (1977) High resolution two-dimensional electrophoresis of basic as well as acidic proteins. Cell 12:1133–1142

Okita TW, Krishnan HB, Kim WT (1988) Immunological relationships among the major seed proteins of the cereals. Plant Sci 57:103–111

Osborne TB (1910) Die Pflanzenproteine. Ergeb Physiol 10:47–215

Peterson DM (1978) Subunit structure and composition of oat seed globulin. Plant Physiol 62:506–509

Robert LS, Nozzolillo C, Altosaar I (1985) Homology between rice glutelin and oat 12 S globulin. Biochim Biophys Acta 829:19–26

Sarker SC, Ogawa M, Takahasi M, Asada K (1986) The processing of a 57 kD precursor peptide to subunits of rice glutelin. Plant Cell Physiol 27:1579–1586

Shotwell MA, Larkins BA (1989) The biochemistry and molecular biology of seed storage proteins. In: Stumpf PK, Conn EE (eds) The Biochemistry of Plants, vol 15. Acad Press, New York, pp 297–345

Takaiwa F, Kikuchi S, Oono K (1987) A rice glutelin gene family — a major types of protein bodies in the rice endosperm. Agric Biol Chem 32:76–80

Tanaka K, Sugimoto T, Ogawa M, Kasai Z (1980) Isolation and characterization of two types of protein bodies in sice endosperm. Agric Biol Chem 44:163–169

Tecson EMS, Esmana BV, Lontok LP, Juliano BO (1971) Studies on the extraction and composition of rice endosperm glutelin and prolamin. Cereal Chem 48:186–181

Villareal RM, Juliano BO (1978) Properties of glutelin from mature and developing rice grain. Phytochemistry 17:177–182

Wang C-S, Shastri K, Wen L, Huang J-K, Sonthayanon B, Muthukrishnan S, Reeck GR (1987) Heterogeneity in cDNA clones encoding rice glutelin. FEBS Lett 222:135–138

Wen T-N (1984) Biochemical characterization of the glutelin storage protein of rice. MS Thesis, Mississippi State University Library

Wen T-N, Luthe DS (1985) Biochemical characterization of rice glutelin. Plant Physiol 78:172–177

Yamagata H, Sugimoto T, Tanaka K, Kasai Z (1982) Biosynthesis of storage proteins in developing rice seeds. Plant Physiol 70:1094–1100

Zhao W-M, Gatehouse JA, D Boulter (1983) The purification and partial amino acid sequence of a polypeptide from the glutelin fraction of rice grains: homology to pea legumin. FEBS Lett 162:96–102

Protein and Fat Determination in Corn

D. BULLOCK and K. MOORE

1 Introduction

Determination of protein and fat content are fundamental to the analyses of corn (*Zea mays* L.). For protein the major determination methods are Kjeldahl, Dumas, and near-infrared reflectance spectroscopy (NIRS). For fat the most common determination techniques are solvent extraction, nuclear magnetic resonance spectroscopy (NMR), and NIRS. Standard reference procedures for protein and fat analysis of corn grain may be found in AOAC (1984a, b, 1989), AACC (1983a, b, c), and SAM (1990). This chapter describes the techniques and the steps necessary to ensure reliable results.

2 Protein Determination

2.1 The Dumas Method

The Dumas method is a combustion technique (Dumas 1831). The manual Dumas method is accurate, but slow, and requires homogeneous samples and ample technical skill. This method is not used widely except when the sample is expensive or in limited quantity, but at the same time is reasonably homogeneous.

Automated Dumas instruments have been introduced and accepted (AOAC 1984a). Coleman Inc., LECO Inc., and Perkin-Elmer Inc. all offer instruments. The early models are limited to about 30–40 samples per day and required ample technical skill to maintain them. Newer models have been shown to require much less technical skill and are accurate and rapid (Sweeney and Rexroad 1987). The LECO FP-228, for example, can analyze a complete sample in about 3 min, producing data not significantly different from AOAC Kjeldahl methods. Similarly, the LECO CHN 600 Analyzer analyzes a sample within about 4 min (McGeehan and Naylor 1988). Despite these improvements the automated Dumas instruments are not used widely.

2.2 The Kjeldahl Method

The Kjeldahl method (Kjeldahl 1883), in one form or another, is the standard protein determination procedure for laboratories dealing in limited numbers of samples. The Kjeldahl procedure involves several steps. A ground sample is submersed in concentrated H_2SO_4 and usually combined with catalysts and salts. The mixture is digested at 325 to 375 °C which results in a conversion of organic and inorganic forms of N to NH_4^+. The NH_4^+-N is converted to NH_3 by distillation of the digest with alkali. The NH_3 is collected in a boric acid solution and the N content is quantified. Many variations in this basic scheme exist. Most are accurate if done correctly.

The Kjeldahl procedures developed prior to 1960 are now known as the macro-Kjeldahl methods since they require relatively large digestion flasks (350 to 800 ml). The macro-Kjeldahl methods are accurate, but expensive and require a lot of space. Procedures using 30 to 50 ml digestion flasks or tubes are known as semimicro-Kjeldahl procedures. Perradeo et al. (1983) calculated that the semimicro-Kjeldahl procedure used one tenth of the reagents and was one-ninth the cost of the macro-Kjeldahl procedure. Semimicro-Kjeldahl procedures are now preferred to macromethods in most laboratories, so discussion will be limited to the former. Excellent reviews of the macro-Kjeldahl procedures are presented by Kirk (1950) and Bremner (1965).

2.2.1 Sample Size and Grind

Semimicro-Kjeldahl methods call for 50 to 200 mg of grain. This small sample size requires that the samples be finely ground so that they are homogeneous. Sample size requirement is a function of the grind. For samples ground to pass a 40 mesh, a sample size of not less than 100 mg should be used. If samples are ground to pass an 80 mesh, then 50 mg samples are acceptable (Nelson and Sommers 1980).

2.2.2 Pretreatments

The standard Kjeldahl method (i.e., no pretreatment) recovers an unknown and variable amount of the N from compounds containing $N-N$ and $N-O$ bonds. For tissues high in NO_3^- (>1000 ppm) this is a serious problem, but corn grain contains very little NO_3^- so standard Kjeldahl methods suffice. Tissues high in NO_3^- can be analyzed and NO_3^- recovered via a pretreatment of the sample with salicyclic acid-H_2SO_4 (du Preez and Bate 1989 a, b).

2.2.3 Oxidizing Agents

Hydrogen peroxide is used widely as an oxidizing agent, but not accepted universally. Oxidizing agents speed digestion time, but can lead to N_2 loss

(Nelson and Sommers 1980). Strong oxidizing agents like MnO_4^- and ClO_4^- should be avoided. Hydrogen peroxide will decrease digestion but may increase clearing time and thus offer limited net savings (Florence and Milner 1979). But even more important is the argument of Hambleton and Noel (1975), who cautioned against the use of H_2O_2 due to excessive foaming and sample loss as well as the argument of Nelson and Sommers (1973) that the use of H_2O_2 without salts or catalysts results in incomplete N recovery.

Despite these arguments many procedures still advocate the use of H_2O_2. Hach et al. (1985) reported a H_2O_2 digestion procedure which was up to 25 times faster than conventional procedures, but even they recognize that many analysts report occasional low N values from standard test samples when using various peroxy methods. In a later paper (Hach et al. 1987) they reported a system using a Vigreux fractionating head fit to a 100 ml volumetric flask. This appears to allow for very short digestion periods, does not require distillation, and prevents N loss. The resulting digest is also suitable for direct colorimetric analysis of many elements in addition to Kjeldahl N. The procedure is attractive, but the necessity for individual vacuum lines and additional equipment will probably hamper acceptance.

It is worth noting that the use of H_2O_2 pretreatments may increase the speed of digestion of the grain sample, but H_2O_2 does not improve reproducibility nor decrease variance. With that in mind and the previously noted potential problems, it is difficult to present a convincing argument for the mandatory use of H_2O_2 for corn grain analysis.

2.2.4 Acid

Concentrated H_2SO_4 is the most commonly used acid for sample digest in the Kjeldahl procedure. A portion of the H_2SO_4 is lost by volatilization and an even larger portion is consumed during the oxidation of the grain. A minimum concentrated H_2SO_4: sample size ratio (w/w) of 40:1 should be used (Nelson and Sommers 1980). If too little acid is used, the sample may solidify during digestion and substantial amounts of N will be lost by volatilization. Mixtures of H_2SO_4 and either H_3PO_4 or $HClO_4$ have been proposed, but not widely adopted (Kirk 1950; Skjemstad and Reeve 1976). Nelson and Sommers (1980) indicate that the H_3PO_4 methods produce substantial bumping and spattering and that the clearing time is much longer than originally reported.

2.2.5 Salts

Salts are added to most Kjeldahl digestions to raise the boiling point of the H_2SO_4, which decreases digestion time and allows for more complete recovery of N. The most common salt used in Kjeldahl procedures is K_2SO_4, but other salts such as K_2HPO_4 and Na_2SO_4 are also used. Note that Na_2SO_4 should be used with caution since it can increase spattering during the digestion step.

The addition of 0.33 to 0.5 g K_2SO_4 per ml H_2SO_4 is optimum (Nelson and Summers 1980). Higher $K_2SO_4 : H_2SO_4$ ratios can be used and have been advocated because they increased digestion temperature and reduced digestion time, however, problems arise, especially as the ratio exceeds 0.8 g K_2SO_4/ml H_2SO_4 (Bremner and Mulvany 1982). High salt ratios produce mixtures which tend to bump and splatter during the digestion process and may solidify during cooling or even during the digestion process.

Solidification during the digestion results in the loss of N via volatilization, while solidification during cooling results in a salt mass which is difficult to dissolve prior to NH_4^+ quantification.

Even without solidification during heating, N loss can occur if the temperature of the digest exceeds 400 °C. This occurs when more than 1.3 to 1.4 g K_2SO_4/ml of H_2SO_4 is used (Bremner and Mulvany 1982). High salt concentrations also result in frothing of the digest which increases the clearing time. Kjeldahl procedures using sealed tubes which allow for temperature as high as 470 °C without loss of N have been proposed (White and Long 1951), but have not been widely used for grain analysis. The sealed tubes also prevent the loss of N through bumping, decomposition of NH_4HSO_4, and entry of atmospheric NH_3 into the digest.

Issac and Johnson (1976) reported that if H_2SeO_3 is used as a catalyst then salts are not needed, but Nelson and Sommers (1973) reported that $H_2SO_4 - H_2SeO_3$ mixtures in the absence of salts recover less total N than obtained with the standard Kjeldahl procedures involving salts and catalysts.

2.2.6 Catalysts

Catalysts affect the rate of digestion only when the salt concentration is low (Bremner and Mulvaney 1982). At least 40 different metals have been examined as potential catalysts in the Kjeldahl process. Mercuric oxide alone or in combination with $CuSO_4$ or elemental Se, or $CuSO_4$ in combination with elemental Se, are the most common catalysts (Jones 1987). Note that $CuSO_4$ alone will not suffice; it must be combined with either HgO or elemental Se (Jones 1987).

Most procedures recommend HgO at a rate of about 5% (w/w) of the K_2SO_4 used. If HgO is used as a catalyst, then $Hg - NH_4^+$ complexes form upon addition of the alkaline material during the digestion. Thus steps must be taken to destroy the $Hg - NH_4^+$ complexes and to precipitate the Hg, e.g. addition of Na_2S or $Na_2S_2O_3$ to form HgS or the addition of Zn dust to form metallic Hg.

There is evidence suggesting that catalysts are not necessary in all situations. Studies involving soil have shown that Kjeldahl analysis is possible without the use of a catalyst, but it requires the use of a salt concentration of 1 g K_2SO_4/ml H_2SO_4 (Bremner and Mulvaney 1982). The use of such a high salt concentration has inherent solidification problems which, under most conditions, are unacceptable.

2.2.7 Digestion Time

The various Kjeldahl procedures are not identical in the digestion time required for the quantitative recovery of N. All agree that the digest must be clear, however, the boiling period after clearing differs. For plant materials 0 to 16 h have been recommended. Jones (1987) suggests that as a general rule "the digestion time after clearing should be 2 to 3 times that required to reach clearing." Undoubtedly, some period of boiling is required after clearing since only about 92% – 93% of the organic N in plant material is converted to NH_4^+ at the time of clearing (Nelson and Sommers 1980). Most methods suggest a boiling period of at least 60 min.

2.2.8 Tubes and Heating Blocks

The introduction of small Kjeldahl flasks resulted in substantial savings compared to the macro-Kjeldahl procedures. Nelson and Sommers (1973) reported additional savings with a semimicro-Kjeldahl procedure for plant material using Pyrex Folin-Wu nonprotein nitrogen tubes heated in a aluminium block. Numerous procedures have been published dealing with tube digestion methods (Nelson and Sommers 1980; Campbell 1986) and firms are now offering tube digestion systems (Technicon Instruments Corp., Tarrytown, NY; Tecator, Inc. Herndon, VA). However, since commercial systems are expensive, interest is high for custom-made system which usually cost far less. In our laboratory we have had local machinists assemble heating blocks complete with heat strips and temperature controls for about one-fifth the cost and with three times the capacity (120 vs 40 tubes) of commercial systems. Thus, the block and tube digestion procedures are widely used because they are simple, rapid, and require little space.

Tubes and blocks can cause excessive foaming during digestion (Cataldo et al. 1974; Issac and Jonhson 1976). This can be controlled by first preheating the sample with the H_2SO_4 and salt catalyst for about 10 min or until the initial fuming subsides.The tubes can then be cooled and the catalyst and salt mixture can be added followed by a second digestion period (Campbell 1986).

Heat digestion can be replaced by microwave digestion (He et al. 1990). Microwave systems specifically designed for Kjeldahl digestion are available (CEM Corp., Mathew, NC and Prolabo, Paris France). The digestion time is very fast (about 30 min), but the small size of the microwave compartments limits digestion to only one sample at a time. Larger systems have been reported (Vittori Antisari and Sequi 1988; He et al. 1990), but even these systems are limited to 12 an 5 samples, respectively. Thus, the microwave procedure is adequate for grain, but is not necessarily more convenient or rapid (especially for large samples) than conventional digestion with an Al block due to the time required for loading and unloading.

2.2.9 Ammonium Quantification of Digest

Determination of the ammonium content of the Kjeldahl digest can be determined by alkaline distillation and titration, colorimetric procedures, or ammonia-sensing electrodes. If done correctly and then standardized, the different methods give similar results.

2.2.9.1 Alkaline Distillation

Distillation under alkaline conditions is the standard method for quantification of ammonium N in Kjeldahl digests. In the original Kjeldahl method the NH_3 released during distillation was collected in a measured volume of standard H_2SO_4 and then quantified by titration of the excess acid with a standard alkali such as NaOH and an indicator. The method was accurate, but required two standard reagents. Winkler (1913) modified the procedure so that the NH_3 is distilled into H_3BO_3 and then titrated with standard H_2SO_4. Standard HCL is now more commonly used for the titration. The advantages of the Winkler modification is that neither the volume nor the strength of the H_3BO_3 solution need to be known accurately and that an excess of H_3BO_3 can be used to ensure complete absorption of NH_3 so the potential for NH_3 saturation of the receiving H_3BO_3 is small. When distilling, it is important than the end of the condenser tip be below the surface of the boric acid solution. Failure to do so can result in loss of approximately 3% of the distilled ammonia (Bremner and Breitenbeck 1983).

2.2.9.2 Colorimetric Methods

The colorimetric methods are attractive because they lend themselves to automation. In the colorimetric procedures the NH_4^+ in the Kjeldahl is separated from other components of the digest by distillation or dialysis. In some procedures the NH_4^+ is then treated with a mixture of Na salicylate and a chlorine source (Wall et al. 1975), dichloroisocyanurate (Crook and Simpson 1971), or more commonly, phenol-hypochlorite reagents to produce a colored indophenol complex via the Berthelot (1859) reaction (Smith 1980). The amount of NH_4^+ is then quantified by the absorptions of wavelengths ranging from 620 to 660 nm. Under proper conditions the color intensity produced by the Berthelot reaction obeys Beer's law (Kirk 1950) which allows for tremendous sensitivity.

Another colorimetric method is the Hach system (Hach Co., Loveland, CO) which uses a H_2SO_4/H_2O_2 digestion procedure without salt or catalysts followed by colorimetric determination with an improved Nesslerization method. The Hach method has been shown to give results similar to the Kjeldahl procedure (Watkins et al. 1987), but is still not used widely.

2.2.9.3 NH_3-Sensing Electrodes

Ammonia gas-sensing electrodes have also been introduced for the determination of NH_4^+ in Kjeldahl digests and are attractive since they do not need distillation of the digest. The NH_3-sensing electrodes are rapid, accurate, and capable of detecting nitrogen concentrations in digests containing very small amounts of N with a net fourfold saving in analytic time per sample (Powers et al. 1981).

3 Fat Determination

The majority of the fat contained in corn grain is found in the scutellum portion of the germ as microscopic droplets of oil.

Traditionally, oil content has been measured as the amount of lipid extracted from ground grain by an organic solvent.

3.1 Solvent Extraction

In industry the germ is first separated from the rest of the kernel and the oil is then removed by various combinations of heat, pressure, and solvents. In most laboratories oil is removed from ground corn grain by some sort of solvent extraction procedure. Solvent extraction methods include the Soxhlet, Butt-type, Goldfisch (Labconco Inc., Kansas City, MO), and Soxtec (Tecator Inc., Herndon, VA) procedures. The first three procedures are similar in that they involve dripping the organic solvent onto the corn sample while the Soxtec procedure involves submersion of the corn sample into the organic solvent.

3.1.1 Organic Solvents

In industry, hexane is the most commonly used organic solvent, but petroleum ether, diethyl ether, and carbon tetrachloride are also used (Weber 1987). The yields of extractable oil obtained with diethyl ether and carbon tetrachloride are similar to one another, but slightly greater than that obtained with other organic solvents (SAM 1990). The AOAC (1984b) and AACC (1983c) methods call for diethyl ether, while the SAM (1990) procedurecalls for carbon tetrachloride.

3.1.2 Sample Size and Moisture Level

A sample size should be used which will yield between 100 to 250 mg of oil. For most corn grain this is about 3 to 5 g. In most procedures corn is dried to an equilibrium moisture point prior to solvent extraction. This is not necessary for carbon tetrachloride and probably not necessary for diethyl ether. Recent data indicate that for carbon tetrachloride extraction corn grain moisture can vary between 10% and 20% without significantly affecting the extractable oil yield (SAM 1990). If drying is deemed necessary, it is acceptable to dry the sample in either a vacuum oven for about 5 h at $95-100\,^{\circ}C$ under pressure not greater than 100 mm Hg or for about 24 h over H_2SO_4 under pressure not greater than 10 mm Hg. In order to avoid fat oxidation, it is critical to avoid exceeding either the temperature or pressure limits and to extract samples within 16 h of being ground (SAM 1990).

3.1.3 Soxtec

The Soxtec extraction system (Tecator Herndon, VA) is similar to the Soxhlet, Butt-type, and Goldfisch methods in that it utilizes an organic solvent extraction; however, Soxtec is much faster. The newest versions of the Soxtec systems are the HT2 and HT6 which allow for simultaneous extraction of two or six samples, respectively. The Soxtec method is based on the automted Soxhlet extraction procedure developed and patented by Randall (1974; US patent No. 3798133). The original procedure gave results similar to both the Soxhlet and Goldfisch procedures (Randall 1974). In the Soxhlet procedure solvent is dripped through the sample rapidly (150 drops/min). The Soxtec procedure is reported to reduce extraction time to less than 20% of the time necessary for Soxhlet and also allows for recovery of 60% to 70% of the extraction solvent (Bhatty 1985). A wide range of organic solvents can be used with the Soxtec system.

A more recent comparison of the Soxtec and Goldfisch systems showed that the Soxtec system consistently extracted substantially less oil and produced a larger standard deviation than the Goldfisch system, but corn was not among the grains tested (Bhatty 1985). Performance was improved substantially by using a smaller particle size. For example, when using a wheat grain particle size of 1 mm, the Soxtec and Goldfisch procedures reported an oil content of $1.25\% \pm 0.02\%$ and $1.62\% \pm 0.05\%$, respectively, but when the particle size was reduced to 0.5 mm the results were $1.71\% \pm 0.02\%$ and $1.73\% \pm 0.05\%$, respectively. Bhatty (1985) also reported that for soybean the Soxtec method gave results similar to the Goldfisch method only if immersion time was increased to several hours. Thus, for some grains the required extraction time for Soxtec is similar to that for Goldfisch and Soxhlet. This essentially eliminates the major advantage that Soxtec system has over the other two extraction system. The necessity for an extended extraction time has not been reported for corn.

3.2 Nonextraction Methods

3.2.1 Nuclear Magnetic Resonance Spectroscopy (NMR)

When referring to fat analysis via NMR we are usually referring to wide-line NMR, but other forms such as pulsed NMR, transient NMR, and ^{13}C NMR have been examined.

Wide-line NMR is nondestructive and allows single seeds to be analyzed for fat content without affecting germination. Therefore, wide-line NMR is used extensively by plant breeders interested in breeding for oil characteristics.

Conway (1960) was the first to demonstrate that wide-line NMR could detect oil content in ground corn grain. Later, Conway and Smith (1963) showed that wide-line NMR was accurate for 25-g samples of whole seeds of 18 different species with an excellent correlation between wide-line NMR reading and oil content (r = +0.99). Furthermore, the technology could even use single corn seeds with an error of $\pm 3\%$ and the procedure did not hurt germination. Similar correlations (r = +0.99) between wide-line NMR readings of single corn seeds and oil content or ground corn grain and oil content were reported by Alexander et al. (1967).

The wide-line NMR procedure measures the resonance of hydrogen nuclei, thus excessive moisture ($>4.5\%$) in the grain does interfere with the system (Alexander et al. 1967). Wide-line NMR is not capable of differentiating oil components.

Rutar (1989) suggests that magic angle sample spinning (MASS) NMR is the preferred technique for oil analysis of seeds and that it makes other versions of NMR obsolete. MASS NMR is nondestructive and can use single seeds. In MASS NMR the sample is rotated quickly about an axis which makes an angle of $54°44'$ (the magic angle) with an external magnetic field. The net result is a reduction in the broadening of the signal and better resolution. MASS NMR is capable of detecting and quantifying many of the various fatty acids which are of prime interest in oil quality. Linolenic acid levels are easily quantified. Linolenic acid levels are also quantifiable but with less accuracy and requiring more work. It should be noted that the kernels found in the middle of the ear will have a larger percent oil content than kernels from either the tip or the base (Lambert et al 1967). Weber (1987) suggests that kernels from the middle of the ear should be taken for oil analysis.

4 Near-Infrared Reflectance Spectroscopy (NIRS)

Use of NIRS for determining the compositon of grain has become increasingly common. NIRS has several advantages which have enhanced its acceptance: it requires minimal sample preparation, allows simultaneous analysis of several constituents, is nondestructive to the sample, and is fast (0.5 to 3 min/sample)

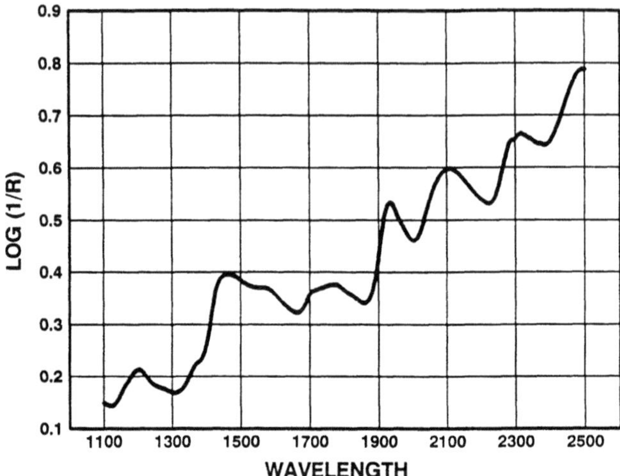

Fig. 1. Near-infrared reflectance spectrum of a ground corn sample

(Melchinger et al. 1986; Norris 1989a, b). NIRS is used for both fat and protein determination in corn. NIRS is used commonly in laboratories which handle very large numbers of samples.

NIRS is an empirical technique and calibrations must be developed from a set of samples of known chemical composition. NIRS analysis is based on the principle that each of the chemical constituents in a sample has a unique NIRS absorbance spectrum which can be used to differentiate it from other constituents. However, because of the complexity of sample spectra (Fig. 1), mathematical and statistical techniques nust be applied to extract useful information. NIRS is therefore by definition a chemometric technique (Norris 1983a, b; Barton 1987).

As early as 1974 Hymowitz and coworkers developed and reported successful NIRS calibrations for predicting protein and oil concentrations in corn and other grains. They reported a correlation of 0.994 between Kjeldahl protein and NIRS protein values for corn and a correlation of 0.993 between corn oil concentrations determined by NMR and NIRS. However, NIRS analysis was not used widely fo the analysis of grain until the advent of more advanced commercial NIRS spectrophotometers in the late 1970s and was further spurred by the rapid development of microprocessor technology during the 1980s which simplified the calibration process (Butler 1983; Osborne and Fearn 1986). The American Association of Cereal Chemists has developed and official method for protein determination in grain using NIRS (AACC 1983a). This method received final approval in 1982 and was revised in 1986 and 1989. The Association of Official Analytical Chemists has recently published a method for determining protein concentration in feeds (AOAC 1989).

4.1 NIRS Instrumentation

There are two general types of NIRS instruments available for grain analysis; scanning monochromators, which generate a continuous spectrum usually from 1100 to 2500 nm, and filter instruments. In the latter type, the filters may be fixed, thus generating discrete wavelength bands, or tilted during analysis to generate a small but continuous spectrum. For the purpose of determining protein concentration in grain, the less expensive and relatively more durable fixed-filter instruments are adequate (Osborne and Fearn 1986; Norris 1989a).

Near-infrared transmittance instruments have also been used to determine grain composition. These instruments generally utilize the region of the spectrum between 800 and 1100 nm and are capable of measuring transmittance through 0.3 cm of ground grain and 2 cm of whole-grain samples (Norris 1983b).

4.2 Sample Preparation

In most cases, grain is ground to a fine powder (100–500 µm) and packed into a sample holder for NIRS analysis. Some instruments have special sample holders which allow the analysis of whole-grain samples. However, results of NIRS analyses of whole-grain samples of corn have been less successful than for other grains with more uniform kernel characteristics (Tkachuk 1981, 1987).

Consistent sample preparation is critical for both the accuracy and precision of NIRS analyses (Williams 1975). Samples should be ground to pass through a 1-mm screen (AACC 1983a, b; AOAC 1989). The grinding method can have a significant impact on the results of NIRS analyses. Use of a cyclone mill is preferred to a shear mill because it produces a more uniform particle size (Williams 1984; Abrahms 1989). Duration of grinding and mill temperature have also been demonstrated to bias NIRS results (Hymowitz et al. 1974). Regardless of the method used to grind samples, it is very important that the same procedures employed to prepare the samples used in calibration, should also be employed to prepare samples of unknown composition.

High-moisture corn samples (>25% moisture) should be dried and allowed to equilibrate at ambient temperature and humidity prior to grinding. It is generally recommended that such samples be dried in a forced-draft oven at 60 °C for 24 h (AOAC 1989). However, some researchers have successfully employed microwave ovens for drying high-moisture samples. Again, it is important that regardless of the method used to prepare high-moisture samples for analysis, it must be consistent between calibration and unknown samples.

Samples should be mixed thoroughly before loading into the sample cell (Williams 1975). The specifics of loading samples cells varies from one manufacturer to the next. However, in all cases, it is critical that the sample be loaded so that it is pressed uniformly against the lens of the sample cell and that no air pockets are present. Windham et al. (1989) recommend the placement of four random portions into each quadrant of the sample cell.

4.3 Sample Analysis by NIRS

Most NIRS instruments require a warm-up period of at least 15 min before they can be used for analyses (AOAC 1989). The manufacturer's recommended diagnostic procedures must be followed to ensure that the instrument will perform satisfactorily. For most instruments this involves assessing noise levels by taking multiple readings from a ceramic standard and evaluating their differences. For scanning monochromators, use of a polystyrene standard is recommended for checking wavelength accuracy (AOAC 1989).

When the instrument is ready, place a loaded sample cell into the instrument's sample holder and start the analysis. With most instruments the process is automated and the result is displayed on a control panel or is transmitted to a computer. When collecting spectral data for calibration development, the resulting reflectance measurements are stored in a data file for later analysis.

4.4 Calibration Development and Validation

There are a number of mathematical techniques which are used to develop calibrations for NIRS. These range from simple linear regression to relatively complex multivariate techniques (Norris 1983 a, b; Hruschka 1987; Martens and Naes 1987). The most common technique for developing NIRS calibrations to determine protein and oil concentrations in corn grain, and the one to be discussed here, is multiple linear regression (Osborne and Fearn 1986). Calibrations developed by multiple linear regression are of the form: $Y = a + b_1 X_1 + b_2 X_2 + \ldots + b_k X_k$; where: X_i = the reflectance (R) usually expressed as (log $1/R$) or some mathematical treatment thereof at wavelength i, b_i is the respective regression coefficient, and a is the intercept.

Several software packages are available for calibration development and are generally available from NIRS instrument manufacturers. The USDA has developed a series of public software programs for calibration development which are described in USDA Agriculture Handbook No. 643.

4.4.1 Calibration Sample Selection

A minimum of 40 to 50 samples should be selected as a calibration set. The samples should represent the population of samples which is to be analyzed using the calibration (AACC 1983a, c; AOAC 1989). The sample set should contin a wide range of concentrations for each constituent to be determined and if possible should be evenly distributed throughout that range (Osborne and Fearn 1986).

If the population of the samples to be analyzed using the calibration is well-defined, it is considered a closed or finite population. Selecting calibration samples at random from a finite sample polulation usually results in acceptable calibrations (Osborne and Fearn 1986). Sample populations which are not well-defined are considered open or infinite. Selecting calibration samples for an infinite population is more difficult since is not possible to foresee and therefore account for all potential sources of variability in the calibration. Therefore, a minimum of 150 calibration samples are recommended when developing calibrations for infinite populations. Selection of calibration samples from large sample populations can be aided by using programs which identify unique samples on the basis of spectral characteristics (Honigs et al. 1985; Windham et al. 1989).

In addition to the calibration sample set, another set of at least 20 samples should be selected as a validation set. These samples are used to evaluate the performance of the calibration equations on an independent sample set and should be selected using the same criteria as the calibration samples. The mean constituent concentrations and standard deviations of the validation and calibration sample sets should be within 20% of each other (Windham et al. 1989).

4.4.2 Reference Method Analysis

The accuracy and precision of NIRS results depend upon the accuracy and precision of the reference method. In general, the results of NIRS analyses cannot be expected to be better than those obtained by the reference method. Therefore, it is critical that reference chemistry be performed with great care and that any errors associated with the method be minimized (Barton 1989; Windham et al. 1989). Kjeldahl protein is the most widely used reference method for NIRS protein calibrations and solvent extraction is the standard reference method for fat. However, any of the other methods for determining protein or fat described above would also be apropriate reference methods.

4.4.3 Developing the Calibration Equation

There are a number of algorithms available for selecting the wavelengths to be used in calibrations developed using multiple linear regression (Hruschka 1987). Some of the more common methods are stepwise multiple regression, backward elimination, and all possible combinations. Shenk et al. (1979) developed a procedure they termed modified stepwise regression which has been incorporated into public software developed by USDA (Shenk 1989). Any of the above methods of wavelength selection are capable of identifying appropriate wavelengths for calibration and in most cases the method used will be determined by the software supplied with the spectrophotometer.

Usually, several potential calibration equations will be developed by the calibration software. The different equations may reflect differences in the number of selected wavelengths and/or various mathematical treatments of the spectral data. The residuals between actual and NIRS values should be examined for each equation. Any sample with a large residual should be closely examined. Such outliers are usually identified by their large t-value and usually occur due to inaccurate reference method analysis or subsampling error (AOAC 1989). Samples identified as outliers should be reanalyzed to determine and correct the source of error. Once this is accomplished, a new set of calibrations should be developed.

The standard error of calibration (SEC) is the standard deviation of errors about the regression. The lower the SEC, the better the equation fits the calibration data. The SEC should not exceed two times the standard error of the laboratory (SEL), which is the standard deviation of repeated reference method analyses (Barton et al. 1989). The coefficient of determination (R^2) is the proportion of the variation in the reference method data explained by the calibration equation. The R^2 should normally exceed 0.90 for good calibrations (Windham et al. 1989). However, if the range of constituent values in the calibration sample population is relatively narrow, a lower value of R^2 may be acceptable (AOAC 1989). Some other criteria to consider are that no two wavelength selected for an equation should be less than 40 nm apart and that no wavelength should have an F-value less than 10 for its regression coefficient (Windham et al. 1989).

4.4.4 Validating the Calibration Equation

Once the initial calibration equations are obtained they should be validated using the validation sample set. This involves determining constituent concentrations by NIRS and regressing these values against actual values as determined by the reference method. The slope of this regression should not differ from 1 by more than 5% (0.95 – 1.05) and the intercept (bias) should not be statistically different from zero. The standard error of validation (SEV) is the standard deviation of differences between NIRS and reference vallues and is indicative of how well the calibration equation performs on an independent set of samples. The lower the SEV, the better the calibration equation performs. Calibration equations with the lowest SEC values will not necessarily have the lowest SEV values. The SEC decreases with each additional wavelength included in a regression equation. The SEV, however, will decrease with each additional term until overfitting occurs. At this point, the SEV will start to increase as additional wavelengths are added. For a given mathematical treatment, the calibration equation with the lowest SEV should provide the best performance for routine sample analysis. A calibration equation with both excellent calibration and validation statistics should be selected for best overall performance (AOAC 1989). The performane of NIRS calibrations should be monitored overtime to ensure the quality of analyses. Detailed procedures for monitoring the performance of NIRS calibrations are described by Shenk et al. (1989).

References

AACC (1983a) Method 39–10. Near infrared reflectance method for protein determination. In: Christensen EA (ed) Approved methods of the AACC. American Association of Cereal Chemists, St. Paul, MN

AACC (1983b) Method 46–13. Crude protein micro Kjeldahl method. In: Christensen EA (ed) Approved methods of the AACC. American Association of Cereal Chemists, St. Paul, MN

AACC (1983c) Method 30–20. Crude fat in grain and stock feeds. In: Christensen EA (ed) Approved methods of the AACC. American Association of Cereal Chemists, St. Paul, MN

Abrahms F (1989) Sample preparation. In: Marten G, Shenk J, Barton F (eds) Near infrared reflectance spectroscopy (NIRS): analysis of forage quality. US Department of Agriculture, Agriculture Handbook No 643, p 23

Alexander DE, Silvela L, Collins FI, Rodgers RC (1967) Analysis of oil content of maize by wide-line NMR. J Am Oil Chem 44:555–558

AOAC (1984a) Protein (crude) in animal feed. In: Williams S (ed) Official methods of analysis of the Association of Official Analytical Chemists

AOAC (1984b) Fat (crude) or ether extract in animal feed. In: Williams S (ed) Official methods of analysis of the Association of Official Analytical Chemists

AOAC (1989) Fiber (acid detergent) and protein (crude) in animal feed and forages. Near-infrared reflectance spectroscopic method. J Assoc Off Anal Chem 72:182–184

Barton FE II (1987) Analytical application to fibrous foods and commodities. In: Williams P, Norris K (eds) Near-infrared technology in the agricultural and food industries. American Association of Cereal Chemists, St. Paul, MN, pp 169–183

Barton FE II (1989) Appendix 2. Considerations of chemical analyses. In: Marten G, Shenk J, Barton F (eds) Near infrared reflectance spectroscopy (NIRS): analysis of forage quality. US Department of Agriculture, Agriculture Handbook No 643, pp 68–82

Berthelot MP (1859) Violt d'aniline. Repert Chim Appl 1:284

Bhatty RS (1985) Comparison of the Soxtec and Goldfisch systems for determination of oil in grain species. Can Inst Food Sci Technol J 18:181–184

Bremner JM (1965) Total nitrogen. In: Black CA, Evans DD, White JL, Ensminger LE, Clarke FE (eds) Methods of soil analysis, part 2. Am Soc Agron, Madison, WI, Agronomy 9:1149–1178

Bremner JJ, Breitenbeck GA (1983) A simple method for determination of ammonium in semimicro-Kjeldahl analysis of soil and plant materials using a block digester. Commun Soil Sci Plant Anal 14:905–913

Bremner JM, Mulvaney CS (1982) Nitrogen – total. In: Page AL, Miller RH, Keeney DR (eds) Methods of soil analysis, part 2. Chemical and microbiological properties. Am Soc Agron, Madison, WI, Agronomy 9:595–624

Butler LA (1983) The history and background of NIR. Cereal Foods World 28:238–240

Campbell DC (1986) Micro-Kjeldahl analysis using 40-tube block digester and steam distillation. J Assoc Off Anal Chem 69:1013–1016

Cataldo DA, Schroder LE, Youngs VL (1974) Analysis by digestion and colorimetric assay of total nitrogen in plant tissues high in nitrate. Crop Sci 14:854–856

Conway TF (1960) Proceedings of a symposium on high-oil corn. Dept Agron, University of Illinois, pp 29–32

Conway TF, Smith VR (1963) In: Ferraro JR, Ziomek JS (eds) Developments in applied spectroscopy, vol 2. Plenum, New York, pp 115–127

Crooke WM, Simpson WE (1971) Determination of ammonium in Kjeldahl digests of crops by an automated procedure. J Sci Food Agric 22:9–10

Dumas JBA (1831) Procedes de l'analyse organique. Ann Chim Phys 47:198–213

Florence E, Milner DF (1979) Routine determination of nitrogen by Kjeldahl digestion without use of catalyst. Analyst 104:378–381

Hach CC, Brayton SV, Kopelove AB (1985) A powerful Kjeldahl nitrogen method using peroxymonosulfuric acid. J Agric Food Chem 33:1117–1123

Hach CC, Bowden BK, Kopelove AB, Brayton SV (1987) More powerful peroxide Kjeldahl diges-
    tion method. J Assoc Off Anal Chem 70:783–787
Hambleton LC, Noel RJ (1975) Protein analysis of feeds, using a block digester. J Assoc Off Anal
    Chem 58:143–145
He XT, Mulvaney RL, Banwart WL (1990) A rapid method for total nitrogen analysis using micro-
    wave digestion. Soil Sci Soc Am J 54:1625–1629
Honigs DE, Hieftje GM, Mark HL, Hirschfeld TB (1985) Unique-sample selection via near-infra-
    red spectral subtraction. Anal Chem 57:2299–2303
Hruschka WR (1987) Data analysis: wavelength selection methods. In: Williams P, Norris K (eds)
    Near-infrared technology in the agricultural and food industries. American Association of Ce-
    real Chemists, St. Paul, MN, pp 35–55
Hymowitz T, Dudley JW, Collins FI, Brown CI (1974) Estimations of protein and oil concentra-
    tion in corn, soybean, and oat seed by near-infrared light reflectance. Crop Sci 14:713–715
Issac RA, Johnson WC (1976) Determination of total nitrogen in plant tissue using a block digest-
    er. J Assoc Off Anal Chem 59:98–100
Jones JB Jr (1987) Kjeldahl nitrogen determination – what's in a name. J Plant Nutr
    10:1675–1682
Kirk PL (1950) Kjeldahl method for total nitrogen. Anal Chem 22:354–358
Kjeldahl J (1883) Neue Methode zur Bestimmung des Stickstoffs in organischen Körpern. Z Anal
    Chem 22:366–382
Lambert RJ, Alexander DE, Rodgers RC (1967) Effects of kernel position on oil content in corn
    (*Zea mays* L.). Crop Sci 7:143–144
Martens H, Naes T (1987) Multivariate calibration by data compression. In: Williams P, Norris
    K (eds) Near-infrared technology in the agricultural and food industires. American Associa-
    tion of Cereal Chemists, St. Paul, MN, pp 57–87
McGeehan SL, Naylor DV (1988) Automated instrumental analysis of carbon and nitrogen in
    plant and soil samples. Commun Soil Sci Plant Anal 19:493–505
Melchinger AE, Schmidt GA, Geiger HH (1986) Evaluation of near infra-red reflectance spectros-
    copy for predicting grain and stover quality traits in maize. Plant Breed 97:20–29
Nelson DW, Sommers LE (1973) Determination of total nitrogen in plant material. Agron J
    65:109–112
Nelson DW, Sommers LE (1980) Total nitrogen analysis of soil and plant tissues. J Assoc Off Anal
    Chem 63:770–778
Morris KH (1983a) Extracting information from spectrophotometric curves. Predicting chemical
    composition from visible and near-infrared spectra. In: Martens H, Russwurm H jr (eds) Food
    research and data analysis. Applied Science Publ, New York
Norris KH (1983b) Multivariate analysis of raw materials. In: Shemilt T (ed) Chemistry and world
    food supplies: the new frontiers. Pergamon Press, New York
Morris KH (1989a) Definition of NIRS analysis. In: Martens G, Shenk J, Barton F (eds) Near
    infrared reflectance spectroscopy (NIRS): analysis of forage quality. US Department of Agri-
    culture, Agriculture Handbook No 643, p 6
Norris KH (1986b) NIRS instrumentation. In: Marten G, Shenk J, Barton F (eds) Near infrared
    reflectance spectroscopy (NIRS): analysis of forage quality. US Department of Agriculture,
    Agriculture Handbook No 643, pp 12–17
Orthoefer FT, Sinram RD (1987) Corn oil: composition, processing, and utilization. In: Watson
    SA, Ramstead PE (ed) Corn chemistry and technology. American Association of Cereal
    Chemists, St. Paul, MN, pp 535–552
Osborne BG, Fearn T (1986) Near infrared spectroscopy in food analysis. Longman Scientific &
    Technical, Essex
Perradeo J, Dorsant H, Cuesta A, Laredo MAC (1983) Determinacion de nitrogeno en varias
    fuentes alimenticias, utilizando los metodes de macro y micro-Kjeldahl. Rev Inst Colomb
    Agropecu 18:233–239
Powers RF, Van Gent DL, Townsend RF (1981) Ammonia electrode analysis of nitrogen in micro-
    Kjeldahl digest of forest vegetation. Common Soil Sci Plant Anal 12:19–30
Preez DR du, Bate GC (1989a) A sample method for the quantitative recovery of nitrate-N during
    Kjeldahl analysis of dry soil and plant samples. Commun Soil Sci Plant Anal 20:345–357

Preez DR du, Bate GC (1989b) Recovery of nitrate-N in dry soil and plant samples by the standard unmodified Kjeldahl procedure. Commun Soil Sci Plant Anal 20:1915–1931

Randall EL (1974) Improved method for fat and oil analysis by a new process of extraction. J Assoc Off Anal Chem 57:1165–1168

Rutar V (1989) Magic angle sample spinning NMR spectroscopy of liquids as a nondestructive method for studies of plant seeds. J Agric Food Chem 37:67–70

SAM (1990) Method A-6. Corn analysis (crude fat). In: Bernetti R (ed) Standard analytical methods of the member companies of the corn industries research foundation. Corn Refiners Assoc, Washington

Shenk JS (1989) Public software. In: Martin G, Shenk J, Barton F (eds) Near infrared reflectance spectroscopy (NIRS): analysis of forage quality. US Department of Agriculture, Agriculture Handbook No 643, pp 18–21

Shenk JS, Westerhaus MO, Hoover MR (1979) Analysis of forages by infrared reflectance. J Dairy Sci 62:807–812

Shenk JS, Westerhaus MO, Abrams SM (1989) Supplement 2. Protocol for NIRS calibration: monitoring analysis results and recalibration. In: Marten G, Shenk J, Barton F (eds) Near infrared reflectance spectroscopy (NIRS): analysis of forage quality. US Department of Agriculture, Agriculture Handbook No 643, pp 104–110

Skjemstad JO, Reeve R (1976) The determination of nitrogen in soils by rapid high-temperature Kjeldahl digestion and autoanalysis. Commun Soil Sci Plant Anal 7:229–239

Smith VR (1980) A phenol-hypochlorite manual determination of ammonium-nitrogen in Kjeldahl digest of plant tissue. Commun Soil Sci Plant Anal 11:709–722

Sweeney RA, Rexroad PR (1987) Comparison of LECO FP-228 "nitrogen determinator" with AOAC copper catalyst Kjeldahl method for crude protein. J Assoc Off Anal Chem 70:1028–1030

Tkachuk R (1981) Protein analysis of whole wheat kernels by near infrared reflectance. Cereal Foods World 26:584–587

Tkachuk R (1987) Analysis of whole grains by near-infrared reflectance. In: Williams P, Norris K (eds) Near-infrared technology in the agricultural and food industries. American Association of Cereal Chemists, St. Paul, MN, pp 233–240

Vittori Antisari L, Sequi P (1988) Comparison of total nitrogen by four procedures and sequential determination of exchangeable ammonium, organic nitrogen, and fixed ammonium in soil. Soil Sci Soc Am J 52:1020–1023

Wall LL, Gehrke CW, Neuner TE, Cathey RD, Rexroad PR (1975) Total protein nitrogen: evaluation and comparison of four different methods. J Assoc Off Anal Chem 58:811–817

Watkins KL, Trygvel LV, Drause GF (1987) Total nitrogen determination of various sample types: a comparison of the Hach, Kjeltec, and Kjeldahl methods. J Assoc Off Anal Chem 70:410–412

Weber EJ (1987) Lipids of the kernel. In: Watson SA (ed) Corn chemistry and technology. American Association of Cereal Chemists, St. Paul, MN, pp 311–349

White LM, Long MC (1951) Kjeldahl microdigestion in sealed tubes at 470°C. Anal Chem 23:363–365

Williams PC (1975) Application of near infrared reflectance spectroscopy to analysis of cereal grains and oilseeds. Cereal Chem 52:561–576

Williams PC (1984) A study of grinders used for sample preparation in laboratory analysis of grains. Cereal Foods World 29:770–775

Windham WR, Mertens DR, Barton FE II (1989) Supplement 1. Protocol for NIRS calibration: sample selection and equation development and validation. In: Marten G, Shenk J, Barton F (eds) Near infrared reflectance spectroscopy (NIRS): analysis of forage quality. US Department of Agriculture, Agriculture Handbook No 643, pp 96–103

Winkler LW (1913) Beitrag zur titrimetrischen Bestimmung des Ammoniaks. Z Angew Chem 26:231–232

Analysis of Cereal Starches

W. R. Morrison

1 Introduction

1.1 Nature of Cereal Starches

The starchy endosperm of the mature cereal caryopsis contains 60% – 70% starch which is the stable reserve polysaccharide of the seed, but there are also small amounts of starch in the pericarp of developing grain and in the embryo, scutellum of germinating grain which are transient in nature. This chapter is concerned only with endosperm starches. The views expressed below and the recommended analytical methods differ appreciably from many of those in the standard texts (Whistler 1964; Radley 1968; Ullmann 1973; Banks and Greenwood 1975; Radley 1976; Whistler et al. 1984) – this reflects our improved understanding of cereal starches and new developments in analytical methodology.

Starch granules are formed within amyloplasts, but the amyloplast envelope is rarely seen in electron micrographs of the mature endosperm, and membrane proteins and lipids are not recovered with the isolated starch granules. However, the starch granules are intimately associated with cytoplasmic and storage proteins in the dried-out endosperm, and these proteins, together with various lipids, will contaminate isolated starches unless suitable precautions are taken (Sect. 2).

In every cereal the starch granules cover a range of sizes about the mean value (Sect. 4.2 and Table 1). In the *Triticeae* size distributions are normally bimodal, whereas in other cereals they are unimodal. In rice and oats the granules often occur as compound granules, comprised of ca. 50 – 80 single granules, which can be isolated intact with care.

In normal cereal starches the polysaccharides are ca. 70% amylopectin (AP) and 30% amylose (AM). In the diploid cereals where there are also stable mutations affecting starch composition (maize, sorghum, rice, and barley), AM contents are commonly 0% – 5% in waxy starches while in the high-AM types they are reported to be as high as 70%. AM is an essentially linear α-(1→4)-glucan with a mean degree of polymerization (DP_n) of 900 – 1300 (Morrison and Karkalas 1990). Over 50% of the molecules are unbranched, and the rest have widely separated α-(1→6)-branch points giving an overall average of 3 – 5 branches/molecule and a mean chain length (CL) of 250 – 370.

Table 1. Typical dimensions of waxy and normal cereal starch granules measured with a 256-channel Coulter Multisizer

Starch	Diameter		Volume		Percent by number	Percent by volume
	Mean (μm)	Range (μm)	Mean (μm^3)	Range (μm^3)		
Wheat A-granules	14 – 17	8 – 30	1 700 – 3 000	270 – 14 140	6 – 10	65 – 80
Wheat B-granules	4 – 5	1 – 10	55 – 80	< 1 – 525	90 – 94	20 – 35
Barley A-granules	10 – 14	5 – 25	700 – 1 700	65 – 8 180	8 – 26	88 – 96
Barley B-granules	2 – 3	< 1 – 6	10 – 27	< 1 – 120	74 – 92	4 – 13
Oats	5 – 6	< 2 – 20	90 – 150	< 4 – 4 200	100	100
Rice	5 – 6	1 – 10	88 – 170	< 1 – 525	100	100
Maize	8 – 1	1 – 20	500 – 900	< 1 – 4 200	100	100

AP in cereals has a DP_n of $4.7 - 12.8 \times 10^3$ with extensive chain branching so that average CL $= 11 - 29$ (Morrison and Karkalas 1990). Models for the structure of AP have been proposed by many authors and are discussed at length by Manners (1989a, b). There are three types of anomalous AP in cereal starches that need to be considered (Morrison and Karkalas 1990). The first has longer than normal external chains with enhanced iodine-binding capacity (IBC) which give false high values for AM determined by iodometric methods (Sects. 3.2.1 and 3.2.4). The second type has a high IBC like the first type, but has a MW similar to AM so that it also gives a false value for AM by GPC (gel permeation chromotography) of native starches (Sect. 3.2.2). Both types occur in *amylose extender* and some other mutants of maize (Baba et al. 1987; South et al. 1991). The third type of anomalous AP has long internal chains with limited branching, hence, it has an enhanced IBC which gives false high values for AM by iodometric methods; it occurs in normal maize and in high-AM rice starches (Hizukuri 1986; Takeda et al. 1987, 1988).

Cereal starches are possibly unique in having monoacyl lipids throughout the starch granules in amounts proportional to AM content (Morrison 1988a; South et al. 1991). However, hydrated starch granules can absorb monoacyl lipids from the other (nonstarch) lipids in the endosperm, and these so-called surface lipids are considered to be artifacts (Morrison 1981).

All cereal starches have small amounts of protein ($<0.15\%$) inside the granules and very variable quantities on their surfaces. The major integral protein (ca. 59 kD) has been identified as granule-bound starch (amylose) synthase (Greenwell and Schofield 1986; Goldner and Boyer 1989) and amounts are proportional to AM content. When the starch granules are carefully purified to remove residues of storage proteins, small amounts of low molecular weight (8 – 30 kD) surface proteins often remain. The surface protein friabilin (14 – 15kD) is associated with soft endosperm texture in wheat and rye (Greenwell and Schofield 1986; Morrison et al. 1992).

1.2 Principles for Starch Isolation and Analysis

Starch should be isolated quantitatively since inadvertent fractionation will give wrong granule size distributions and (sometimes) misleading chemical analyses. There is no advantage in using direct starch extraction with a solvent such as aqueous dimethyl sulfoxide because only limited analyses are then possible, and incomplete extraction gives a fraction with an atypical composition (Morrison and Karkalas 1990).

It is also necessary to avoid enzymatic and mechanical damage to the granules, and for this reason grain is normally softened by steeping in water or dilute $NaHSO_3$. Damaged granules are particularly susceptible to attack by amylases, and they swell excessively in cold water with loss of crystalline order and associated physical properties (Sect. 4). Mercuric chloride (10^{-2} M) may be used to inhibit enzymes during starch isolation (Adkins and Greenwood 1966) but it is very toxic and its use can be avoided. If thiomersal (= thimerosal, also very toxic) is used to inhibit microbial growth during steeping, all residues must be thoroughly removed since they interfere with iodometric determination of AM (Sects. 3.2, 3.4). Proteolytic enzymes are sometimes used to assist in removing storage proteins and improve recovery of very small starch granules (Table 1), but it is essential that they should first be treated to remove all traces of amylases (McDonald and Stark 1988). Discoloration caused by polyphenol oxidases present in some proteases may be prevented by adding some $NaHSO_3$. Steeping in water at temperatures up to 53 °C speeds up grain softening and protease action (if used), but at temperatures within $5-10$ °C of the starch gelatinization temperature (GT) considerable annealing of crystallites occurs and the GT of the native granules can be increased by $10-15$ °C. In practice these problems do not arise if clean grain is used, and if all operations are conducted at $0-4$ °C.

2 Isolation of Starches

2.1 Isolation from Single Kernels (SK Method)

Sufficient starch can be isolated from one to three grains of wheat or barley for many of the analyses described below. The method (South and Morrison 1990) is invaluable when studying F_2 generation progeny and other samples where an individual grain is likely to be different from its neighbors. Although developed for wheat and barley, the method may be used with slight modifications for other cereals.

Remove the husk from covered grain such as barley. Dissect the germ from the caryopsis with a scalpel, and crush the grain between smooth-jaw pliers. Transfer all fragments to a 15-ml Uni-Form homogenizer (Jencons Scientific Ltd.), add 2 ml water and steep at 4 °C overnight. Grind the softened tissue with several aliquots of water, transferring the starchy suspension to two 10-ml tubes, centrifuge at 3000 g for 10 min to recover the solids, and recombine into a single tube.

Suspend the solids in water (<1 ml) and layer carefully above 0.8 ml of 80% (w/v) CsCl in two 1.5-ml Eppendorf microcentrifuge tubes, centrifuge at 14000 g for 7 min (work in a cold room at 4°C). Stir the supernatant with a needle to release trapped starch granules in the supernatant and repeat the centrifuging. Discard the supernatant and wash/centrifuge the starch several times with water to remove CsCl, combining the two parts into a single 10-ml screw-cap tube. Add sodium laurate (25 mg) and water (5 ml), seal the tube, and shake horizontally for 30 min at room temperature. Recover and wash the starch as before, then pass as a slurry through Miracloth (60-μm aperture) (Plastok UK, Ltd.) held in a 13-mm-diameter Swinnex filter holder (Millipore UK, Ltd.) to remove fragments of cell debris. Finally, wash/centrifuge the starch twice with acetone to remove water and air-dry overnight. Washing with laurate is an optional step to remove surface proteins and lipids; residual laurate does not interfere with lipid analysis (Sect. 3).

2.2 Isolation from Flour, with CsCl Treatment

This is a scaled-up version of the SK method to give sufficient starch for analysis of granule surface proteins (Sulaiman and Morrison 1990). Alternatively, the SK method can be used (15–20 kernels/5–6 tubes) for the initial extraction. The flour could be prewashed with diethyl ether to minimize risk of surface lipid contamination, but polar alcoholic solvents should not be used.

Flour (100–150 mg) is slurried in water (1 ml), carefully layered above 80% (w/v) CsCl (8 ml) in a tube, and centrifuged at 18000 g for 45 min. The sediment of starch is recovered and may be purified further by repeating the CsCl step twice. The starch is then washed with water to remove CsCl, and with acetone to remove water, then air-dried as before. There is no point in passing the starch through Miracloth (60-μm aperture) unless the starting material is cracked whole grain.

2.3 Isolation from Grain, Larger Scale Method

The procedure given below contains features from methods used by Morrison et al. (1984), Soulaka and Morrison (1985), Morrison and Gadan (1987), McDonald and Stark (1988), and South et al. (1991). Methods should be tailored to suit each particular cereal; stages (i) and (iv) greatly reduce lipid artifacts in maize starch, and stage (iv) is now recommended for all cereals.

(i) Manually degerm maize kernels, *or* (ii) crack wheat or barley grains by passing between smooth rolls with a suitable gap *or* (iii) break up rice or maize kernels by coarse grinding in a coffee grinder or mill; (iv) remove most nonstarch lipids by extracting several times with 2–5 vol diethyl ether/methanol (4:1 by vol) and air-dry solids; (v) steep for 18–24 h in 0.02 M HCl at 2–4°C to inactivate amylases and other enzymes, then adjust to pH 6–7 with NaOH, *or* (vi) steep in 0.5% (w/v) NaHSO$_3$ for 18–24 h at 30–35°C to soften endosperm particles, *or* (vii) suspend the solids in Tris-HCl buffer (pH 7.6, 10 ml/g grain) containing 0.5% (w/v) NaHSO$_3$ and 0.01% thiomersal, and digest with amylase-free protease (Sigma Type XIV, 5 mg/g grain, or other equivalent enzyme), at 2–4°C for 18–24 h to degrade proteins, *or* (viii) steep in water at 2–4°C for 18–24 h; (ix) gently grind the material from stages (v), (vi), (vii), or (viii) to release starch granules and pass the starch slurry through a 75-μm mesh sieve; (x) centrifuge for 10–15 min at 3000 g to recover starch, discard the supernatant liquor, and scrape off the dark proteinaceous layer on top of the starch; (xi) repeatedly wash the starch by stirring in water, centrifuging, and scraping as indicated above till judged free from contaminating protein and solutes; (xii) slurry in several volumes of acetone and filter or centrifuge to recover the starch, and air-dry overnight.

2.4 Size Fractionation of Starch Granules

Starch granules that are approximately spherical sediment according to Stoke's Law, and can be separated by sedimenting in a column of water for various times (Decker and Höller 1962; Morrison and Gadan 1987). In the *Triticeae* the large A-type granules are somewhat flattened and sediment at faster rates than predicted; typical conditions to separate large A-type and small B-type granules of wheat starch would be 60–90 min sedimentation time in an 18-cm column of water, depending on average granule dimensions. Longer times are required to separate the smaller A- and B-type granules of barley starches, and in every case it is essential to check the separations by microscopic examination or by Coulter size analysis (Sect. 4.2).

3 Chemical Analysis

3.1 Total Polysaccharides (α-Glucan)

The best methods for determining starch content depend on quantitative conversion of starch to glucose, using enzymes that hydrolyze only α-(1→4) and α-(1→6)-glucosidic linkages. They also have the added advantage that they can be used to measure any intermediate oligosaccharides derived from starch. There are numerous other less specific and less sensitive methods which are not discussed here. The procedure given below was developed by Karkalas (1985); several similar methods have been published but they are neither simpler nor more sensitive. Methods involving acid hydrolysis of starch are not recommended since there are inevitable losses of glucose.

1) Glucose oxidase-peroxidase (GOP) reagent. Dissolve in water and make up to 500 ml: $Na_2HPO_4 \cdot 12\,H_2O$ (11.5 g), KH_2PO_4 (2.5 g), phenol (500 mg), 4-aminophenazone (75 mg), peroxidase (Sigma P-8125, from horseradish, 3500 purpurogallin units = ca. 40 mg), and glucose oxidase (Sigma G-6500, from *Aspergillus niger*, 3000 units = ca. 3 ml). Filter through a glass microfiber filter (Whatman GF/A), store at 4 °C protected from light, and use within 1 month.

2) Amyloglucosidase solution (AMG). Dissolve amyloglucosidase (Boehringer Mannheim 208 469, from *Aspergillus niger*, 30 mg lyophilized product) in 15 ml buffer (pH 4.6) containing citric acid monohydrate (460 mg) and sodium citrate dihydrate (840 mg), and dilute to 100 ml with water. Prepare a fresh solution daily.

To determine starch, weigh samples (50–100 mg starch, correct to 0.1 mg) into 10-ml screw-cap tubes, add water (2 ml) and α-amylase solution (Sigma A-3403, from *Bacillus licheniformis*, 0.2 ml) and vortex mix. Add more water (6 ml), seal the tubes, and mix well by inversion, then heat at 85 °C – shake frequently for 3 min, then at intervals for 30 min. Cool to 20 °C and make up to 100 ml. Prepare blank enzyme solution (0.2 ml α-amylase/100 ml water) and a standard solution of glucose (100 mg glucose/1 l of water). To sample and blank tubes (1 ml aliquots) add AMG solution (1 ml), mix gently and incubate at 60 °C for 30 min. Cool to 20 °C, add 8 ml water, mix, and filter through glass microfiber filters (Whatman GF/A, 7 cm diam.), or centrifuge at 3000 g for 5 min, and determine glucose in 1-ml aliquots.

To determine glucose, take aliquots (1 ml) containing 10 to 160 µg glucose/ml and add 5 ml GOP reagent; use 1 ml water + 5 ml GOP as reference. Heat in a water bath at 35 °C for 40 min, protected from light, cool to ambient temperature for 10 min in the dark, then measure absorbance at 505 nm within 30 min against the reference solution. After correcting for the enzyme blank, calculate glucose content, then starch (anhydrous) = 0.9×glucose. If soluble dextrins or oligosaccharides are being determined, the hydrolysis with bacterial α-amylase may be omitted and the blanks changed accordingly. If the samples are alkaline eluates from GPC columns (Sects. 2.2, 2.3) they should be neutralized first. If the samples are in DMSO (dimethyl sulfoxide) solution, they should be diluted so that the DMSO concentration is less than 20% (by vol).

3.2 Amylose and Other Polysaccharides

3.2.1 Colorimetric Determination of Amylose

The long linear α-(1→4) chains in AM form blue inclusion complexes with polyiodide ions with λ_{max} at 620–650 nm, while the short chains in AP form pink to purple complexes which absorb at 530–540 nm but have little absorptivity at 620–650 nm. Thus, AM content can be determined colorimetrically in the presence of normal types of AP, but high values are obtained if anomalous AP is present (Sect. 1.1). Since the stoichiometry of the complexes is variable, it is essential to control ion concentrations, pH (weakly acidic), and temperature (Morrison and Laignelet 1983), and to ensure that starches are not contaminated with substances that react with I_2 or KI (e.g. $NaHSO_3$, $HgCl_2$, thiomersal).

Weigh (correct to 0.1 mg) starch (75–80 mg) or high-AM starch (20–40 mg) into 10-ml screw-cap tubes fitted with PTFE-faced liners in the caps. Add urea-dimethyl sulfoxide solution (UDMSO = 1 vol 6 M urea: 9 vol DMSO; 10 ml) rapidly from a dispenser, cap the tubes, and vortex mix immediately until the starch is completely dispersed. Place the tubes in boiling water, vortex mix frequently for 5–15 min, then maintain at 100 °C for 90 min in an oven. Cool and check for the absence of clear gel (undissolved starch) by inverting the tubes.

To determine apparent amylose (i.e., with some interference from native starch lipids), transfer an aliquot (1 ml) to a dry volumetric flask (100 ml), *rapidly* add water (90–95 ml) and *mix immediately*, than add I_2/KI solution (2 mg I_2, 20 mg KI/ml, use 2 ml), mix well and make up to 100 ml – *the total time taken from the addition of water should not exceed 60 s*, otherwise AM will begin to retrograde. Hold at 20 °C for 15 min to allow full color development, then measure absorbance at 635 nm, preferably in a cell holder thermostatically controlled to 20 °C.

To determine total AM (in lipid-free starch) transfer aliquots (1 ml) of starch-UDMSO solution to 10-ml screw-cap tubes, add cold ethanol (9 ml), cap the tubes, and mix well. Allow the starch precipitate to form for 15 min, then centrifuge at 2000 g for 15 min to pack the precipitate, pour off the supernatant and leave the tubes upside down to drain for a few minutes. Add UDMSO (1 ml) cautiously and vortex mix gently to redissolve the starch. Heat the tubes at 100 °C for 15–30 min to complete dissolution, cool, quantitatively wash out the contents of the tubes with water into 100 ml volumetric flasks and immediately add I_2/KI reagent, as before.

Positive displacement volumetric dispensers and pipettes should be used with the viscous DMSO solution, otherwise gravimetric checks will be necessary. Blue Value (BV) is the absorbance at 635 nm of 10 mg *anhydrous* starch (or α-glucan) in 100 ml diluted I_2/KI solution at 20 °C. If the temperature deviates much from 20 °C a correction can be applied to BV (Morrison and Laignelet 1983). Then:

$$AM\ (\%) = 28.414\ BV - 6.218.$$

3.2.2 Gel Permeation Chromatography (GPC) of Native Starches

GPC of waxy and normal cereal starches gives AM contents similar to those obtained colorimetrically (Sect. 3.2), but GPC will give slightly lower AM contents (which are more correct) in the presence of the high MW types of AP and very high results in the presence of low MW AP (Sect. 1.1) which elutes with AM. These anomalous results can be largely eliminated by GPC of debranched starches (Sect. 3.4). Numerous systems for GPC of native starches are possible (Praznik 1985, 1986), and only the general principles are described here.

The starch should be free from lipid – this can be achieved by dissolving in DMSO or UDMSO and precipitating with ethanol (Sect. 3.2). The amorphous starch is then dissolved in the eluting solvent which may be water, or *dilute* buffered solutions (Praznik 1986). Retrogradation (recrystallization) of AM can be prevented by eluting with 0.01–0.05 M KOH, ammonium bicarbonate buffer (which can be evaporated with the water if fractions are to be recovered by preparative GPC), or dilute aqueous DMSO.

The column (typically $2\,m \times 16\,mm$ diam.) should be packed with a gel to fractionate in the nominal MW range $1 \times 10^5 - 2 \times 10^7$, but if alkaline eluting solvents are used only the comparatively stable cross-linked gels can be used (e.g., Sepharose CL-2B). If the eluate is collected in fractions any nonspecific method can be used to measure carbohydrate, e.g. phenol-sulfuric acid (Dubois et al. 1956) or anthrone-sulfuric acid (Koehler 1952; Ough 1964). However, continuous monitoring of the eluate is preferable. This can be done using a postcolumn derivatizer (autoanalyzer) with phenol-sulfuric acid (Robyt and Bemis 1967) or cysteine-sulfuric acid (Sargeant 1982; Morrison et al. 1984) which give a direct weight response. Refractive index (RI) detectors are nondestructive and respond to the weight of eluted carbohydrate but cannot be used with DMSO solutions which have a high RI or with alkali which corrodes the optical surfaces. Laser light-scattering detectors give a molar response which decreases with lower polysaccharide molecular weights, and they have been used with HPLC systems to determine number-average molecular-weights and DP_n (Hizukuri and Takagi 1984; Takagi and Hizukuri 1984; Yu and Rollings 1987). Postcolumn reaction with I_2/KI reagent *in parallel* with total carbohydrate monitoring can be very useful, especially if full spectral scans (450–700 nm) are carried out on collected fractions. If the final concentrations are glucan = $<1.6\,mg$, $I_2 = 1.6\,mg$, and KI = 20 mg/100 ml, CL can be estimated (Morrison and Karkalas 1990):

$$Cl = 3290/(635 - \lambda_{max})\ .$$

This method is useful for estimating CL and DP_n if $\lambda_{max} < 590$ nm.

The continuous enzymatic quantification of eluted α-glucan, as glucose, has been described recently (Karkalas and Tester 1992).

3.2.3 GPC of Debranched Starches

Enzymatic hydrolysis of α-(1→6)-branch points gives the component linear α-(1→4)-glucan chains of AM and AP which may be separated by GPC on columns fractionating in the nominal molecular weight range $1 \times 10^4 - 1 \times 10^6$, e.g. Sepharose CL-6B. The tendency for linear glucans to retrograde is greatest when CL = 75 – 100 (Pfannemüller et al. 1971; Kodama et al. 1978; Gidley et al. 1986; Pfannemüller 1986), and eluting solvents should therefore contain DMSO or 0.01 – 0.05 M KOH, otherwise very misleading results will be obtained. Isoamylase (from *B. amyloderamosa*, *Pseudomonas* spp. or *Cytophaga* spp.) will completely debranch AP but will not release short maltosyl (G_2) or maltotriosyl (G_3) stubs from AM and β-limit dextrins (Sect. 3.4). For this reason it is advisable to use pullulanase, which does release G_2 and G_3 stubs (Mercier and Kainuma 1975), as a supplementary enzyme or as an alternative. Isoamylase will act in solutions containing $< 40\%$ DMSO, and pullulanase in solutions containing $< 20\%$ DMSO.

Starch in UDMSO is precipitated with ethanol to remove lipid (Sect. 3.2) and redissolved in anhydrous DMSO (5 mg/0.5 ml). Acetate buffer (pH 3.8, 0.01 M, 3.5 ml) is added, then 10 μl of isoamylase solution (Hayashibara Biochemical Labs, Okayama, Japan) containing 500 units of activity is added and the mixture incubated at 30 °C for 24 h. Denature the enzyme and ensure dissolution of the α-glucans by adding 0.25 M KOH (1 ml), giving a solution (0.05 M KOH) suitable for GPC.

Alternatively, resuspend lipid-free waxy or normal starch in acetate buffer (2 ml), boil briefly and cool, then add 10 μl of isoamylase to the cloudy mixture which will soon clear, and incubate at 25 °C for 24 h. Denature the enzyme by boiling the mixture (with added buffer suitable for subsequent GPC, if desired) and centrifuge to obtain a clear supernatant for GPC.

3.2.4 Other Methods

There are numerous other methods which may be used to characterize the structures of AM and AP in considerable detail. These are discussed in reviews by Manners (1989a, b) and by Morrison and Karkalas (1990) who give references to the original papers. Briefly, AM content can be determined iodometrically by potentiometric or amperometric titration, subject to all the caveats affecting the colorimetric assay (Sect. 3.2.1). There are chemical and enzymatic methods to determine the internal, external, and average CL of AP, and by measuring reducing endgroups before and after debranching, it it possible to calculate the number of branch points per molecule. Exhaustive digestion of AP or AM with pure β-amylase gives β-limit dextrins used to determine the molar ratios of A- and B-type external chains in AP and linear and branched molecules in AM.

Starch granules that are mechanically damaged swell spontaneously in cold water with loss of birefringence and other indices of internal molecular order (Sect. 4), and they are much more susceptible to enzymatic digestion. They are readily observed under the light microscope by their ability to absorb dye from Congo Red solution (0.1%), becoming red. Damaged starch content of small samples ($< 50 - 100$ mg) can be quantified by an enzymatic method (Karkalas et al. 1992) that correlates well with standard methods used with wheat flour (Farrand 1964; AACC 1984). In the new method damaged starch is selectivity and quantitatively hydrolyzed to glucose which is determined colorimetrically as in Sect. 3.1, whereas the older methods measure reducing power of variable proportions of maltose, maltotriose, and glucose.

3.3 Lipid Content

Although lipid generally comprises <1% of cereal starches, it is of considerable biochemical interest and affects many of the technological properties of cereal starches (Morrison 1988a, b, 1989). A simple determination of ether-extractable crude fat is merely a measure of contamination since the true starch lipids are not extracted with normal lipid solvent systems (Morrison 1981).

3.3.1 Total Hydrolysate Lipid

This method (Morrison et al. 1980) destroys polysaccharide and protein, and gives quantitative recoveries of saturated fatty acids (FA) which are converted to methyl esters (FAME) for quantification by gas chromatography (GC). Some losses of unsaturated FA are unavoidable, especially if very small samples are used or if lipid levels are very low (e.g. in waxy starches). Nonlipid artifacts of the acid hydrolysis, which interfere with gravimetric methods, are removed by thin-layer chromatography (TLC).

Weigh replicate samples (400–600 mg, correct to 0.1 mg) into 10 ml screw-cap tubes with PTFE faced liners for the caps, add methanol (2 ml) containing heptadecanoate (17:0 FFA or FAME) as internal standard (100–200 µg for waxy starches, 500–1000 µg for normal and high AM starches). *Flush thoroughly with oxygen-free nitrogen* and add, under nitrogen, conc. HCl (12 M, 1.8 ml). Close the tubes and vortex mix immediately. Heat the tubes in boiling water with regular inversion mixing for 60 min. *It is essential to cautiously slacken the caps to vent off accumulated gas pressure several times*, especially during the first 20–30 min, otherwise the tubes may burst. After hydrolysis, cool the tubes, add chloroform (2 ml), mix, and centrifuge. As the solvent-water interface is not always visible, withdraw about 1 ml chloroform from the bottom of the tube with a disposable Pasteur pipette, transfer to another tube, remove the solvent with a stream of nitrogen and add, under nitrogen, methanolysis reagent (14% BF_3 or 5% v/v H_2SO_4 in methanol, 2 ml). Cap the tube, heat in boiling water for 10 min, then cool and extract the FAME by adding pentane (2 ml) followed by water (1–2 ml), mixing and centrifuging.

Recover the supernatant pentane layer, remove the solvent under nitrogen, then apply the impure FAME redissolved in chloroform (ca. 100 µl) as a band (ca. 3 cm wide) to TLC plates coated with activated silica gel (e.g. Merck Silica gel G) and develop 8–14 cm with toluene. Allow the plates to dry, spray lightly with 2′,7′-dichlorofluorescein (0.01% in methanol) to reveal the FAME band ($R_f = 0.4$). Scrape this band into a small funnel plugged at the end of the stem with solvent-washed glass wool, and elute the FAME off the microcolumn of silica with dry diethyl ether (2–3 ml). Evaporate the ether, redissolve the FAME in 100 µl ether, and use aliquots (1 µl) for quantitative gas chromatography (GC). A GC fitted with a column containing a polar packing suitable for 16:0–18:3 FAME and a flame ionization detector is recommended, and corrections should be applied when calculating results as mg FAME/100 g starch (Morrison et al. 1980).

With experience recoveries of linoleate (18:2) and linolenate (18:3) will be 90%–95%, but recoveries of saturated FAME (16:0, 18:0, and 17:0 internal standard) will always be quantitative and can be compared directly with those from solvent-extracted lipid (Sect. 3.3.2).

3.3.2 Total Extractable Lipid

Normal lipid solvents extract little or none of the true starch lipids, which require hot alcohol-water mixtures to give controlled swelling of the starch granules and solubilization of lipid (Morrison and Coventry 1985).

Weigh replicate samples (300–600 mg, correct to 0.5 mg) into 10 ml screw-cap tubes, flush with oxygen-free nitrogen, and add propanol-water (3 vol propan-1-ol or propan-2-ol : 1 vol water, $\ll$ 1.6 ml/100 mg starch) and seal the tubes. Shake well to disperse the starch and heat in boiling water (with occasional shaking) for 2 h. Cool the tubes and centrifuge briefly with the *minimum* force required to lightly pack the swollen starch, cautiously withdraw most of the clear supernatant with a Pasteur pipette and transfer to a 25-ml volumetric flask. Add the original volume of fresh propanol-water to the sample, close the tube, redisperse the starch, and repeat the extraction twice (1×2 h, 1×1 h). Make the combined extracts up to 25 ml with propanol-water. Evaporate aliquots (2–5 ml) containing 17 : 0 FAME internal standard (50–100 µg) to dryness under nitrogen in a boiling water bath, add BF_3-methanol (1 ml) and heat to convert starch lipids to FAME for quantitative GC as before. Since FAME prepared this way are very clean, the TLC step to remove nonlipid artifacts (Sect. 3.3.1) should be omitted.

There are negligible losses of unsaturated FAME by this method, and the percentage composition will therefore be correct and will probably show more 18 : 2 and 18 : 3 than in the hydrolysate lipid FAME. However, incomplete extraction may occur and results expressed as mg FAME/100 g starch will then be low. In this situation the 16 : 0 content of the hydrolysate lipid should be taken as the reference value.

3.3.3 Determination of Lipid Classes

In pure cereal starches the lipids are comprised of free fatty acids (FFA) and lysophospholipids (LPL) with negligible amounts of other lipids (Morrison 1988a). LPL can be calculated from the phosphorus content (Sect. 3.5) of an aliquot (200–500 µl) of the total lipid extract, or they can be determined (as FAME) together with FFA by TLC and GC, as described below. In the *Triticeae* the starch lipids should be > 90% LPL, while in other cereals there are roughly equal amounts of FFA and LPL Morrison et al. 1984; Morrison 1988a).

Remove propanol-water from an aliquot (5–10 ml) of total extractable lipid (Sect. 3.3.2) by heating at 90–100 °C, flushing well with nitrogen, and redissolve in a small volume of chloroform (note that there is no 17 : 0 internal standard present at this point). Apply the lipid as a narrow band (3–4 cm wide) on an activated TLC plate coated with silica gel, and develop with diethyl ether-acetic acid (99 : 1, by vol) to 4–5 cm. Air-dry the plates and spray with 2′,7′-dichlorofluorescein followed by sufficient NH_4OH (8 M) to neutralize the acetic acid on the plate which would otherwise quench fluorescence. Scrape off the FFA band (near the solvent front) and the LPL band (at the origin) into tubes containing 17 : 0 FFA or FAME internal standard (use 20%–25% of anticipated weight of total FAME, preferably > 50 µg), and convert the lipids to FAME for GC with BF_3-methanol as before (Sect. 3.3.2). A few simple conversion factors are applicable to the lipids from nonwaxy starches, based on the fact that the lipids have well-defined FA compositions and nearly constant proportions of LPL classes (Morrison et al. 1975; Morrison 1988a):

$$FFA = 0.95 \times FAME$$
$$LPL = 1.76 \times FAME$$
$$LPL = 16.4 \times P \text{ (\textit{Triticeae}, maize)}$$
$$LPL = 16.2 \times P \text{ (rice)}$$

total lipid $= 1.7 \times FAME$ (*Triticeae* only).

3.4 Protein Content

3.4.1 Total Protein Content

Most authors use the Kjeldahl nitrogen content and a conversion factor to obtain the protein content of cereal starches, but this ignores several points. Firstly, most of the LPL contain nitrogen; secondly, conversion of choline-N (in lysophosphatidylcholine) into ammonia is not quantitative unless conditions are carefully controlled; and, thirdly, there can be more N in lipid than in protein of a purified nonwaxy cereal starch.

The Kjeldahl method ideally requires $0.5-1$ g starch for the initial digestion, which should be at 425 °C using mercuric oxide or a copper selenite catalyst. The small amount of ammonia can then be determined by distillation and colorimetric assay (Osborne and Voogt 1978). For very small samples of starch complete combustion and determination of nitrogen oxides by a chemiluminescence method is much more reliable (Sulaiman and Morrison 1990). Since 90% of the LPL have an $N:P$ molar ratio of $1:0$, a correction factor for lipid-N can be obtained using the weight of lipid phosphorus (or, as an approximation, starch phosphorus) (Sulaiman and Morrison 1990):

$$N\,(LPL) = 0.4 \times P\,(LPL) \ .$$

Then,

$$\text{Total protein} = 6.25\,\{N\,(Kjeldahl) - N\,(LPL)\} \ .$$

3.4.2 Granule Surface Proteins

Methods for the study of starch granule surface proteins are still being developed. The method used to isolate the starch is critically important, and should avoid proteases and protein denaturing substances, yet should remove endosperm storage proteins quantitatively. Methods based on CsCl purification (Sects. 2.1, 2.2) are useful for wheats, but sometimes do not give completely clean starches with very hard wheats and barleys. Some authors prefer to isolate starch by aqueous washing and to suffer considerable losses of starch in order to obtain relatively pure preparations (Greenwell and Schofield 1986; Goldner and Boyer 1989).

Surface proteins can be desorbed from starch granules by extracting with 1% – 2% (w/v) sodium dodecyl sulfate (SDS) at 20 °C. Electrophoresis is done in the presence of SDS on polyacrylamide gels with a gradient from the top to the bottom of 10%→20% (Greenwell and Schofield 1986; Goldner and Boyer 1989; Sulaiman and Morrison 1990). If total surface proteins are being determined directly on SDS extracts, or by the difference on starch before and after SDS extraction, it should be remembered that SDS also extracts some LPL (South and Morrison 1990).

3.4.3 Granule-Bound (Integral) Proteins

The integral proteins of the granule are best studied on starches that have been rigorously purified by a method that will remove surface proteins, e.g. isolation using protease and/or extraction with SDS at 20 °C. Extraction of integral proteins requires SDS at elevated temperatures which are a compromise to obtain reasonable yields without solubilizing too much polysaccharide, which will interfere with electrophoresis, and without causing sufficient swelling to absorb all of the SDS solution! In practice starch (100 – 300 mg) is usually extracted with 2% SDS (2 – 3 ml) at 50 °C for 30 min (Greenwell and Schofield 1986; Goldner and Boyer 1989; Sulaiman and Morrison 1990), but for starches with a high GT it may be better to use 5 – 10 °C below GT to obtain sufficient granule swelling. Electrophoresis is done using the same gels as for surface proteins, and if hot SDS extraction is used to obtain surface plus integral proteins together, the principal integral protein at 56 – 59 kD acts as an internal standard to compare yields of surface proteins.

3.5 Phosphorus Content

3.5.1 Total Phosphorus in Starch and Lipids

In cereal starches most of the phosphorus is in LPL, and there are only very small amounts of hexose-phosphate residues in the polysaccharides. Inorganic phosphate should be negligible, unless phosphate buffer was used during starch isolation! Thus, total LPL can be measured conveniently by determining phosphorus in aliquots of total starch lipid (Sect. 3.3.2), and total starch phosphorus will give only a slight overestimate (<5 mg P/100 g starch). The usual factor to convert weight of phosphorus to LPL is 16.4 (Sect. 3.3.3). The method given below (Morrison 1964) is for $1 - 10$ μg P. In practice starch samples should not exceed 10 mg otherwise too much H_2O_2 will be required (giving high blanks) and if too much SO_2 is lost, the final acidity will be too low for controlled color development.

Weigh starch samples (3 – 5 mg, correct to 0.02 mg) into graduated borosilicate glass tubes fitted with glass stoppers, or evaporate aliquots of lipid extract (Sect. 3.3.2; containing 2 – 4 μg P) until completely dry. Add conc. H_2SO_4 (specific gravity 1.84 = 36 N, 0.3 ml). It is advantagous to leave the tubes, stoppered, overnight or longer to partially digest larger samples since this makes wet oxidation easier. Heat the tubes on a microburner flame with slight shaking, confining the sample to the bottom of the tube until it is black and starts to fume. Add one drop of H_2O_2 (100 vol, 30% w/v) from a drawn glass tube so as to touch the wall of the digestion tube just above the acid, and shake vigorously. Resume heating until the acid is colorless, or add another drop of H_2O_2 if the acid does not clear in 30 s. Normally, one to two drops H_2O_2 should suffice and not more than four drops should be required. Blanks should contain H_2SO_4 (0.3 ml) and the same amount of H_2O_2 heated for the same time. Boil the tubes gently for 1 min to destroy most of the excess H_2O_2, then cool. Add water (3.8 – 4 ml), washing down the walls of the tubes, then add sulfite solution (33% w/v $Na_2SO_3 \cdot 7H_2O$; 0.1 ml) to neutralize residual H_2O_2, and vortex mix well. Add molybdate solution (2% w/v ($NH_{46}Mo_7O_{24} \cdot 4H_2O$; 1 ml) *directly into the acid*, mix immediately, add ascorbic acid (10 mg; or 0.1 ml of fresh solution, 1 g/10 ml), place the tubes in boiling water for 10–15 min, then cool and make up to 5 ml or 10 ml with water. Measure absorbance at 822 nm, correct for blanks, and calculate phosphorus content (1 µg P/5 ml gives 0.176 absorbance). Note that final acidity should be 1.6–2.5 N to give stoichiometric formation of heteropoly blue from phosphorus; low acidity gives a nonspecific molybdenum blue, while excess acidity or H_2O_2 inhibits formation of heteropoly blue.

3.5.2 Other Methods

Glycerophosphate (from LPL), hexose phosphate, and inorganic phosphate can be determined, after acid hydrolysis of starch, by ion exchange chromatography (Tabata et al. 1975; Hizukuri and Hisatsuka 1976). Glucose-6-phosphate can be determined enzymatically and glucose-3-phosphate by difference from total polysaccharide phosphate (Hizukuri et al. 1970)

4 Physical Analysis

4.1 Gelatinization Properties

When starch granules are placed in water at temperatures well below GT they undergo limited hydration and there is slight swelling which is reversible on drying. Further heating causes progressive disordering and irreversible swelling, with leaching of soluble polysaccharides. The processes are quite complex, and various definitions of gelatinization have been proposed (Atwell et al. 1980). In the author's laboratory gelatinization is regarded as the disordering of AP crystallites which occurs over a limited temperature range, while swelling is measured over the gelatinization range and at higher temperatures (Tester and Morrison 1990a, b; Tester et al. 1991).

4.1.1 Differential Scanning Calorimetry (DSC)

DSC is the method now used most widely to study gelatinization. The endotherm for disordering of AP crystallites is characterized by onset, peak, and conclusion temperatures (T_o, T_p, and T_c, respectively) and by negative enthalpy (ΔH). Gelatinization temperature (GT) and T_p are generally synonymous. An endotherm in the region 94–120 °C is usually due to dissociation of AM-lipid complexes, and an endotherm at 150–165 °C due to retrograded AM (the latter does not occur in native starches). Common conditions require 2–4 mg samples in 15–20 µl water, heated at 10 °C/min. When there is a limited amount of water, and when there are some types of anomalous AP, the thermogram shows additional features. For reviews, see Donovan (1979), Burt and Russell (1983), Blanshard (1987), Morrison (1988a, b), and Gidley and Robinson (1990).

4.1.2 Birefringence End-Point Temperature (BEPT)

Under the polarizing microscope most cereal starch granules show strong bire-fringence (a dark "Maltese Cross") due to radially oriented structures. If a gypsum plate is inserted, the black/white image changes to blue and yellow sectors against a red background, which can provide additional information (Nesse 1986). With a Koefler hot stage on a microscope the starch can be heated, and the temperature at which 98% of the granules have lost their birefringence (BEPT) can be determined. BEPT is close to T_p (Sect. 4.1.1) and synonymous with GT in older literature.

4.1.3 Other Methods

Noninvasive methods which can be used to study aspects of polysaccharide ordering in native starch granules include wide-angle and narrow-angle X-ray diffraction, small-angle scattering of polarized light, small-angle scattering of neutrons, and cross-polarization magic-angle nuclear magnetic resonance (^{1}H, ^{13}C, and ^{31}P). For reviews, see Zobel (1964, 1988a,b), and Gidley and Robinson (1990).

4.2 Granule Size Analysis

Accurate size analysis of populations is particularly useful for studying starch granules during grain (endosperm) development (Morrison and Gadan 1987; McDonald et al. 1991) and for characterizing starches from mature grain (Soulaka and Morrison 1985; Morrison and Scott 1986; Morrison et al. 1986; Morrison 1989). Granules suspended in saline or in other electrolyte solutions can be measured accurately with instruments such as the Coulter Multisizer with 256 size channels. Granules can also be measured in the dry state using laser light-scattering instruments but their resolution is generally poorer, although the particle size range may be much greater.

4.3 Granule Swelling Properties

The swelling curves of starches, and their rheological properties, are often important considerations for their industrial utilization. Swelling power can be determined by a simple colorimetric method that uses blue dextran (Tester and Morrison 1990a). Swelling, a property of the AP in the starch granule, becomes significant (and irreversible) at T_o, and above T_p and T_c swelling is markedly inhibited by amylose-complexing lipids (Tester and Morrison 1990a,b; Tester et al. 1991). Damaged starch granules (Sect. 3.2.4) swell in cold water similar to undamaged granules heated in the range T_o to T_c.

References

AACC (1984) Damaged Starch Method 76−30A. In: Approved methods of the American Association of Cereal Chemists, 8th edn. AACC, St. Paul, MN, 2pp

Adkins GK, Greenwood CT (1966) The isolation of cereal starches in the laboratory. Starch/Stärke 18:213−218

Atwell WA, Hood LF, Lineback DR, Varriano-Marston E, Zobel HF (1988) The terminology and methodology associated with basic starch phenomena. Cereal Foods World 33:306, 308, 310, 311

Baba T, Uemura R, Hiroto M, Arai Y (1987) Structural features of amylomaize starch. Denpun Kagaku 34:196−202, 213−217

Banks W, Greenwood CT (1975) Starch and its components. Edinburgh Univ Press, Edinburgh

Blanshard JMV (1979) Physiochemical aspects of starch gelatinisation. In: Blanshard JMV, Mitchell JR (eds) Polysaccharides in food. Butterworths, London, pp 139−152

Blanshard JMV (1986) The significance of the structure and function of the starch granule in baked products. In: Blanshard JMV, Frazier PJ, Galliard T (eds) Chemistry and physics of baking. R Soc Chem, London, pp 1−13

Blanshard JMV (1987) Starch granule structure and function: a physicochemical approach. In: Galliard T (ed) Starch properties and potential. John Wiley and Sons, Chichester, pp 16−54

Burt DJ, Russell PL (1983) Gelatinisation of low water content wheat starch-water mixtures. A combined study by differential scanning calorimetry and light microscopy. Starch/Stärke 35:354−360

Decker P, Höller H (1962) A time gradient method for fractionating granular materials particularly ion exchange materials through sedimentation. J Chromatogr 7:392−399

Donovan JW (1979) Phase transitions of the starch-water system. Biopolymers 18:263−275

Dubois M, Gilles KA, Hamilton JK, Rebers PA, Smith F (1956) Colorimetric method for determination of sugars and related substances. Anal Chem 28:350−356

Duprat F, Gallant D, Guilbot A, Mercier C, Robin JP (1980) L'Amidon (starch). In: Monties B (ed) Les polymères végétaux. Gauthier-Villars, Paris, pp 176−231

Farrand EA (1964) Flour properties in relation to the modern bread processes in the United Kingdom with special reference to α-amylase and starch damage. Cereal Chem 41:98−111

French AD, Murphy VG (1977) Computer modelling in the study of starch. Cereal Foods World 22:61−70

Gidley MJ, Robinson G (1990) Techniques for studying interactions between polysaccharides. In: Dey PM, Holborne JB (eds) Methods in plant biochemistry, vol 2. Acad Press, London, pp 607−642

Gidley MJ, Bulpin PV, Kay S (1986) Effect of chain length on amylose retrogradation. In: Philips GO, Wedlock DJ, Williams PA (eds) Gums and stabilisers for the food industry, vol 3. Elsevier Applied Science, London, pp 167−176

Goldner WR, Boyer CD (1989) Starch granule bound proteins and polypeptides: the influence of the waxy mutations. Starch/Stärke 41:250−254

Greenwell P, Schofield JD (1986) A starch granule protein associated with endosperm softness. Cereal Chem 63:379−380

Hizukuri S (1986) Polymodal distribution of the chain lengths of amylopectin and its significance. Carbohydr Res 147:342−347

Hizukuri S, Hisatsuka T (1976) Studies on the determination of wheat starch phosphorus as glycerol phosphates. J Agric Chem Soc Jpn 50:489−494

Hizukuri S, Takagi T (1984) Estimation of the distribution of the molecular weight for amylose by low-angle laser-light scattering technique combined with high-performance gel chromatography. Carbohydr Res 134:1−10

Hizukuri S, Tabata S, Nikuni Z (1970) Starch phosphate. 1. Estimation of glucose 6-phosphate residues in starch and the presence of other bound phosphate(s). Starch/Stärke 22:338−343

Karkalas J (1985) An improved enzymic method for the determination of native and modified starch. J Sci Food Agric 36:1019−1027

Karkalas J, Tester RF (1992) Continuous enzymic determination of α-glucans in eluates from gel-chromatographic columns. J Cereal Sci 15:175−180

Karkalas J, Tester RF, Morrison WR (1992) Properties of damaged starch granules I. A new micromethod for the enzymatic determination of damaged and gelatinized starch. J Cereal Sci (in press)

Koehler LH (1952) Differentiation of carbohydrates by anthrone reaction rate and color intensity. Anal Chem 24:1578−1579

Kodama M, Noda H, Kamata T (1978) Conformation of amylose in water. I. Light scattering and sedimentation equilibrium measurements. Biopolymers 17:985−1002

Manners DJ (1989a) Recent developments in our understanding of amylopectin structure. Carbohydr Polym 11:87−112

Manners DJ (1989b) Some aspects of the structure of starch and glycogen. Denpun Kagaku 36:311−323

McDonald AML, Stark JR (1988) A critical examination of procedures for the isolation of barley starch. J Inst Brew 94:125−132

McDonald AML, Stark JR, Morrison WR, Ellis RP (1991) Composition of starch granules from developing barley genotypes. J Cereal Sci 13:93−112

Mercier C, Kainuma K (1975) Enzymic debranching of starches from maize of various genotypes in high concentrations of dimethyl sulfoxide. Starch/Stärke 27:289−292

Morrison WR (1964) A fast simple and reliable method for the determination of phosphorus in biological materials. Anal Biochem 7:218−224

Morrison WR (1981) Starch lipids: a reappraisal. Starch/Stärke 33:408−410

Morrison WR (1988a) Lipids in cereal starches: a review. J Cereal Sci 8:1−15

Morrison WR (1988b) Lipids. In: Pomeranz Y (ed) Wheat: chemistry and technology, 3rd edn, vol 1. Am Assoc Cereal Chem, St Paul MN, pp373−439

Morrison WR (1989) Uniqueness of wheat starch. In: Pomeranz Y (ed) Wheat is unique. Am Assoc Cereal Chem, St Paul, MN, pp193−214

Morrison WR, Coventry AM (1985) Extraction of lipids from cereal starches with hot aqueous alcohols. Starch/Stärke 37:83−87

Morrison WR, Gadan H (1987) The amylose and lipid contents of starch granules in developing wheat endosperm. J Cereal Sci 5:263−275

Morrison WR, Karkalas J (1990) Starch. In: Dey PM (ed) Methods in plant biochemistry, vol 2. Acad Press, London, pp323−352

Morrison WR, Laignelet B (1983) An improved colorimetric procedure for determining apparent and total amylose in cereal and other starches. J Cereal Sci 1:9−20

Morrison WR, Scott DC (1986) Measurement of the dimensions of wheat starch granule populations using a Coulter Counter with 100-channel analyzer. J Cereal Sci 4:13−21

Morrison WR, Mann DL, Wong S, Coventry AM (1975) Selective extraction and quantitative analysis of non-starch and starch lipids from wheat flour. J Sci Food Agric 26:507−521

Morrison WR, Tan SL, Hargin KD (1980) Methods for the quantitative analysis of lipids in cereal grains and similar tissues. J Sci Food Agric 31:329−340

Morrison WR, Milligan TP, Azudin MN (1984) A relationship between the amylose and lipid contents of starches from diploid cereals. J Cereal Sci 2:257−271

Morrison WR, Scott DC, Karkalas J (1986) Variation in the composition and physical properties of barley starches. Starch/Stärke 38:374−379

Morrison WR, Greenwell P, Law CN, Sulaiman BD (1992) Occurrence of friabilin, a low molecular weight protein associated with grain softness, on starch granules isolated from some wheats and related species. J Cereal Sci 15:143−149

Nesse WD (1986) Introduction to optical mineralogy. Oxford Univ Press, Oxford

Osborne DR, Voogt P (eds) (1978) The analysis of nutrients in foods. Academic Press, London, pp175−178

Ough LD (1964) Chromatographic determination of saccharides in starch hydrolyzate. In: Whistler RL (ed) Methods in carbohydrate chemistry, vol 4. Acad Press, New York, pp91−98

Pfannemüller B (1986) Models for the structure and properties of starch. Starch/Stärke 38:401−407

Pfannemüller B, Mayerhofer H, Schulz RC (1971) Conformation of amylose in aqueous solution: optical rotary dispersion and circular dichroism of amylose-iodine complexes and dependance on chain-length of retrogradation of amylose. Biopolymers 10:243–261

Praznik W (1985) Application of gel permeation chromatography for the analytical estimation of starches. Ernährung/Nutrition 9:834–849

Praznik W (1986) GPC analysis of starch polysaccharides. Starch/Stärke 38:292–296

Radley JA (1968) Starch and its derivatives, 4th edn. Chapman and Hall, London

Radley JA (1976) Examination and analysis of starch and starch products. Applied Science, London

Robyt JF, Bemis S (1967) Use of the autoanalyzer for determining the blue value of the amylose-iodine complex and total carbohydrate by phenol sulfuric acid. Anal Biochem 19:56–60

Sargeant JG (1982) Determination of amylose: amylopectin ratios of starches. Starch/Stärke 34:89–92

Soulaka AB, Morrison WR (1985) The amylose and lipid contents dimensions and gelatinisation characteristics of some wheat starches and their A- and B-granule fractions. J Sci Food Agric 36:709–718

South JB, Morrison WR (1990) Isolation and analysis of starch from single kernels of wheat and barley. J Cereal Sci 12:43–51

South JB, Morrison WR, Nelson OE (1991) A relationship between the amylose and lipid contents of starches from various mutants of maize. J Cereal Sci 14:267–278

Sulaiman BD, Morrison WR (1990) Proteins associated with the surface of wheat starch granules purified by centrifuging through caesium chloride. J Cereal Sci 12:53–61

Tabata S, Nagata K, Hizukuri S (1975) Studies on starch phosphates. 3. Esterified phosphates of some cereal starches. Starch/Stärke 27:333–335

Takagi T, Hizukuri S (1984) Molecular weight and related properties of amylose determined by monitoring of elution from TSK gel PW high performance gel chromatography columns by the low angle laser light scattering technique and precision differential refractometry. J Biochem 95:1459–1467

Takeda Y, Hizukuri S, Juliano BO (1987) Structure of rice amylopectins with low and high affinities for iodine. Carbohydr Res 168:79–88

Takeda Y, Suzuki A, Hizukuri S (1988) Influence of steeping conditions for kernels on some properties of corn starch. Starch/Stärke 40:132–135

Tester RF, Morrison WR (1990a) Swelling and gelatinization of cereal starches. I. Effects of amylopectin amylose and lipids. Cereal Chem 67:551–557

Tester RF, Morrison WR (1990b) Swelling and gelatinization of cereal starches. II. Waxy rice starches. Cereal Chem 67:551–557

Tester RF, South JB, Morrison WR, Ellis RP (1991) The effect of ambient temperature during the grain filling period on the composition and properties of starch from four barley genotypes. J Cereal Sci 13:113–127

Ullmann M (1973) Handbuch der Stärke in Einzeldarstellungen, vol VII, part 2. Analytische Kennzeichnung von Amylose und Amylopectin. Paul Parey, Berlin

Whistler RL (1964) Methods in carbohydrate chemistry, vol 4. Acad Press, New York

Whistler RL, BeMiller JN, Paschall EF (1984) Starch: chemistry and technology, 2nd edn. Acad Press, Orlando

Wild D, Blanshard JMV (1986) The relationship of the crystal structure of amylose polymorphs to the structure of the starch granule. Carbohydr Polym 6:121–143

Yu L-P, Rollings JE (1987) Low angle laser light scattering aqueous size exclusion chromatography of polysaccharides. Molecular weight distribution and polymer branching determination. J Appl Polym Sci 33:1909–1921

Zobel HF (1964) In: Whistler RL (ed) Methods in carbohydrate chemistry, vol 4, Starch. Acad Press, New York, pp 109–113

Zobel HF (1988a) Starch crystal transformations and their industrial importance. Starch/Stärke 40:1–7

Zobel HF (1988b) Molecules to granules: a comprehensive starch review. Starch/Stärke 40:44–50

Food Properties of Amaranth Seeds and Methods for Starch Isolation and Characterization

O. PAREDES-LÓPEZ and D. HERNÁNDEZ-LÓPEZ

1 Introduction

1.1 Brief History

Grain amaranth is native to Mexico, Central America, and the Andes where it was once a staple food of early native American civilizations. The earliest archaeological record of pale-seeded grain amaranth is that of *Amaranthus cruentus,* found in Tehuacán, Mexico, about 4000 B. C., making it one of the oldest food crops (Pal and Khoshoo 1974; Grubben and van Sloten 1981; Stiebritz et al. 1985). Amaranth was a major grain crop in pre-Columbian times. Ancient Mexicans gave the name "huatle" to the seeds which have been identified as graín amaranth and "tzoalli" or "zoale" to the dough prepared for consumption (Sauer 1950a, b; Santin et al. 1986). Various amaranth species have been cultivated in many countries of the world since ancient times as grain crops, vegetable, ornamentals, and dye plants. The potential of both grain and vegetable amaranth as a food resource has been reviewed by Sánchez-Marroquin (1980), NRC (1984), Saunders and Becker (1984), Teutonico and Knorr (1985), and Paredes-López et al. (1990b). The red dye from amaranth leaves is used to color beverages and foods in Argentina, Bolivia, Ecuador, Mexico, and United States (Sauer 1950a; Teutonico and Knorr 1985).

1.2 Agronomical Traits

Three species of the genus *Amaranthus* produce large seed heads loaded with edible seeds. *A. hypochondriacus* and *A. cruentus* are native to Mexico and Guatemala; *A. caudatus* is native to Peru and other Andean countries. All three are still cultivated on a small scale in isolated mountain valleys of Mexico, Central America, and South America, where generations of farmers have continued to cultivate the crops of their forebears (NRC 1984). It appears that the Spanish introduced the seeds to Europe, where it soon became established (Oke 1983). About 60 members of this genus are now widely distributed throughout the world in tropical, subtropical, and temperate regions. Their growth habits vary from horizontal to erect, and branched to unbranched; their leaves and stems are of many shades of green, and seeds vary from black

Table 1. Agronomical traits of some species of amaranth[a]

Material	Plant density $\times 10^3$/ha	Flowering[a]		Harvest[b]		Yield (metric t/ha)
		Time (days)	Height (cm)	Time (days)	Height (cm)	
A. cruentus	320–360	44	60	102	125	4.1
A. hypochondriacus	230–260	43	43	100	130	3.8
A. hybridus	230–360	57	65	129	180	3.0
A. caudatus	323	50	80	109	150	2.0

[a] Adapted from Wagoner-Hass (1983); Stiebritz et al. (1985); Bressani et al. (1987); Espitia (1990, pers. comm.).
[b] After seeding.

to white (NRC 1984; Singhal and Kulkarni 1988). Amaranths can photosynthesize by the C_4 pathway and utilize CO_2 more efficiently to produce sugars than those plants possessing the classical C_3 pathway (Saunders and Becker 1984). They also perform better under adverse conditions such as drought. Amaranth has also been reported to be less susceptible to the toxigenic strains of *Aspergillus* than corn, rice, or winged bean (Tudor and George 1985).

Amaranth is a crop with brilliantly colored leaves, stems, and flowers of purple, orange, red, and gold. Some of its outstanding agronomical aspects appear in Table 1 (Wagoner-Hass 1983; Stiebritz et al. 1985; Bressani et al. 1987; E. Espitia, unpubl. results). The density of cultivated species may range from 230 to 360×10^3 plants/ha. At flowering time plants may be 43 to 80 cm in height, but at harvest time they may reach 180 cm with a seed yield of up to 4.1 t/ha, which is comparable to that of most cereals. The seeds, although barely bigger than a mustard seed (1–1.7 mm in diameter), occur in massive numbers, sometimes more than 50000 to a plant, and are of many colors (e.g., black, cream, white).

1.3 Nutritional Properties and Food Uses

Amaranth's potential rests in its valuable nutritive characteristics. It is one of the rare plants in which the leaves are edible and treated as a delicacy, and the seeds are also used as cereals. Very few plants have this double advantage (Oke 1983). With a protein content of 13%–18% (Table 2), amaranth seed compares well with the conventional varieties of maize (12%), wheat (12%–14%), and rice (7%–10%) (NRC 1984; Saunders and Becker 1984; Pedersen et al. 1987). The seed also contains a protein of unusual quality; it has nearly twice the lysine content of wheat protein, three times that of maize, and similar to milk – the standard of nutritional excellence (Oke 1983; NRC 1984) and shows acceptable levels of sulfur amino acids (Paredes-López and Mora-Escobedo 1989; Mora-Escobedo et al. 1990). It is, therefore, a nutritional

Table 2. Proximate composition of amaranth seeds compared to that of maize (g/100 g, dry basis)[a]

Component	*A. caudatus*[b]			*A. cruentus* (Cream)	*A. hypochondriacus* (Cream)	Maize
	Brown	Cream	Black			
Protein[c]	12.9	13.4	15.8	17.8	15.0	11.9
Fat	9.6	10.9	9.5	7.9	7.3	5.2
Crude fiber	8.0	7.6	16.4	4.4	2.8	2.7
Ash	3.0	3.2	3.0	3.3	3.3	1.6
Nitrogen-free extract[d]	66.5	64.9	55.3	66.6	73.3	78.6

[a] Adapted from Saunders and Becker (1984); Pedersen et al. (1987).
[b] Seeds were separated according to their color.
[c] Nitrogen conversion factors were 5.85 for amaranth, and 6.25 for maize.
[d] Calculated by difference.

complement to conventional cereals and even to legumes. The fact that most of the seed volume is occupied by the embryo may account for the high lysine content. Raw flour shows protein efficiency ratios (PER) from 1.6 to 2.2, while cooked samples exhibit values comparable to casein (2.3 to 2.4) (Bressani et al. 1987).

Amaranth seeds are parched or cooked into gruel or milled to produce a sweet, light-colored fluor suitable for biscuits, breads, tortillas, and cakes. It can also be used in breakfast cereals or as an ingredient in confections (Paredes-López et al. 1990b).

2 Structure and Composition of Amaranth Seeds

2.1 Seed Structure

Amaranth seeds are lenticular in shape, generally about 1 – 1.7 mm in diameter, with 1000 seeds weighing 0.6 – 1 g, and an average test weight of 77 kg/hl. Seed color ranges from white, through beige, light brown, red, brown, and black. Pigments are located in the outer layer of seed coat (Saunders and Becker 1984). There have been selections for pale-colored seed types, but not for seed size, although tetraploid lines have been created with a 2.5-fold increase in seed weight by treating *A. caudatus* with colchicine (Pal and Khoshoo 1977). The ultrastructure of amaranth seed (*A. hypochondriacus*) is presented in Fig. 1. Figure 1a shows the seed coat, radicle, cotyledons, and perisperm. It should be noted that the aleurone layer, the characteristic outermost layer of the endosperm of cereal grains, is absent in amaranth (Paredes-López et al. 1988b). Perisperm cells (Fig. 1b – d) are well defined in the central part of the seed. Perisperm is a storage tissue and starch is its major component. Protein

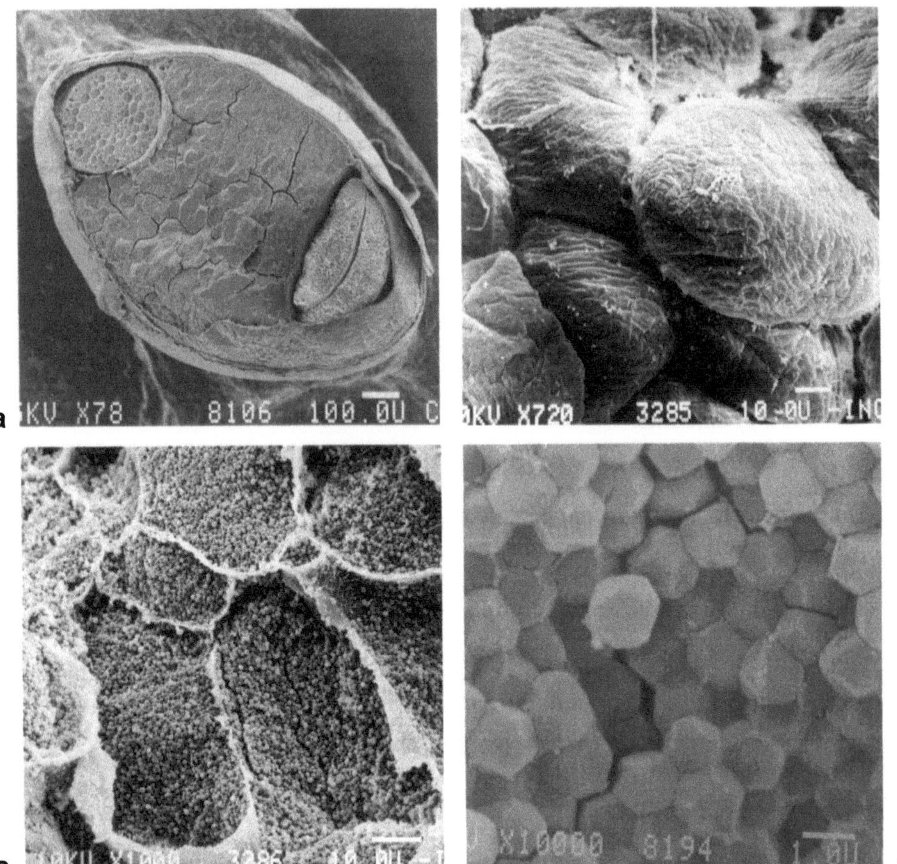

Fig. 1a–d. Scanning electron micrographs (SEM) of amaranth seed (*A. hypochondriacus*) and starch. **a** Amaranth seed showing seed coat, radicle, perisperm, and cotyledons; **b** perisperm cells; **c** fractured perisperm cells containing starch granules; **d** starch granules in perisperm. Techniques for preparation and observation of specimens have already been described by Paredes-López et al. (1988b)

in the perisperm occurs as very small deposits between starch granules (Saunders and Becker 1984). At a higher magnification (Fig. 1d) the uniform size of the granules becomes apparent. The polygonal configuration, dense packing, and smooth surface of these starch granules are evident.

2.2 Major Components of Seeds

Table 2 shows the proximate composition of seeds from *A. caudatus*, *A. cruentus*, and *A. hypochondriacus*. It ranges from 12.9% to 17.8% for protein, 7.3%

Table 3. Solubility fractionation of seed proteins from *A. hypochondriacus* (g/100 g protein)[a]

Protein fraction	Mercado cultivar[b]		Nonspecified cultivar[c]	
	Whole	Defatted	Waxy	Nonwaxy
Albumins + globulins + NPN[d]	45.7	48.6	56.9	59.4
Prolamins	2.8	1.4	1.6	1.2
Glutelins	33.3	29.9	22.4	29.4
Insoluble residue	15.9	22.3	8.5	12.6
N recovery (%)	97.7	102.2	89.4	102.6

[a] Extracted by an Osborne-modified procedure.
[b] Paredes-López et al. (1988c).
[c] Konishi et al. (1985).
[d] NPN = Nonprotein nitrogen.

Table 4. Characteristics and composition of amaranth starch granules compared to those from commercial sources[a]

Characteristic/ component (dry basis)	Starch source				
	Amaranth	Maize		Wheat	Potato
		Normal	Waxy		
Granule size (µm)	1 – 3	3 – 20	3 – 20	2 – 40	10 – 100
Granule shape[b]	P*, S	P, S	P, S	O, S	O
Protein (%)	0.49	0.60	0.15	0.40	0.06
Fat (%)	1.00	0.44	0.15	0.80	0.05
Ash (%)	0.70	0.10	1.10	0.20	0.40
Amylose (%)	0 – 22	25	0 – 3	25	20

[a] Wolf et al. (1950); Becker et al. (1981); Sugimoto et al. (1981); Tomita et al. (1981); Paredes-López et al. (1990b); Varriano-Marston (1983).
[b] O = Oval; P = polygonal; P* = mostly polygonal; S = spherical.

to 10.9% for fat, 2.8% to 16.4% for crude fiber, 3% to 3.3% for ash, and 55.3% to 73.3% for nitrogen-free extract.

Paredes-López et al. (1988c) reported that the protein fractionation of whole and defatted flour gives, by using an Osborne-modified procedure, 46% to 49% albumins + globulins, 1% to 3% prolamins, 30% to 33% glutelins, and 16% to 22% insoluble residue (Table 3); glutelins constitute the main single protein class. In relation to essential amino acids, albumins are rich in lysine, globulins in methionine and leucine, prolamins in threonine, whereas glutelins appear to have an even distribution (Duarte-Correa et al. 1986; Mora-Escobedo et al. 1990).

Edible oil can be extracted from coarsely milled amaranth flour (Lyon and Becker 1987; Becker 1989) and the defatted flour used as a source of protein and starch (Paredes-López et al. 1989). Procedures have been adapted to extract the fat from the seeds, which are first coarsely ground in a stone mill and refluxed with hexane. Then the residue is removed by filtration, and the solvent eliminated under vacuum from the fat. The crude oil is degummed and dewaxed by centrifugation and refined using sodium hydroxide and bleaching clay. The refined oil has a taste and appearance typical of vegetable oils.

The amaranth perisperm fraction is rich in starch, whereas in all cereals starch is stored in the endosperm. The starch granules, mostly polygonal in shape, are extremely small (Fig. 1, Table 4) compared to those from cereals, and may contain either waxy or nonwaxy starch (Wolf et al. 1950; Becker et al. 1981; Sugimoto et al. 1981; Tomita et al. 1981; Varriano-Marston 1983; Stone and Lorenz 1984; Paredes-López et al. 1990b).

3 Methods for Starch Isolation and Quantitation

3.1 Starch Isolation

Figure 2 shows the various steps involved in starch isolation. It is necessary that amaranth seeds be softened by steeping under specific conditions of temperature and time in buffer media; dry milling of the seed can lead to considerable physical damage of the starch granule. In addition, it is essential that starch-degrading enzymes be inactivated during the steeping process.

3.1.2 Mercuric Chloride Method

Steeping in acetate buffer (0.02 M sodium acetate) at pH 6.5, and rendered 0.01 M with respect to mercuric chloride, for 24 h at 5 °C has the desired effect of softening the seed and inhibiting amylase action (Adkins and Greenwood 1966; Stone and Lorenz 1984; Hernández-López 1991). The softened seeds are ground, and the resultant grits slurried in water before being sifted rapidly through sieves with meshes of different sizes (e.g., 150, 75 µm). The steps of grinding, slurrying, and sieving are repeated until the material left on the sieve is free of perisperm. Screening is performed more conveniently if screening materials are stretched over a frame attached to a shaking machine. In this case, nylon bolting fabric is most useful (Watson 1964a).

The starch suspension is contaminated with proteinaceous material and this contaminant must be removed without any concomitant degradation of the starch. For this reason, it is preferred to use the physical technique of shaking the aqueous suspension of starch with one-eighth of its volume of toluene overnight; the protein being denatured at the water-toluene interface. Mild centrifugation then separates the toluene layer, bearing denatured protein. Repeated shakings with fresh toluene are usually necessary to produce a starch of low protein content (Table 4; Adkins and Greenwood 1966; Guilbot and Mercier 1985).

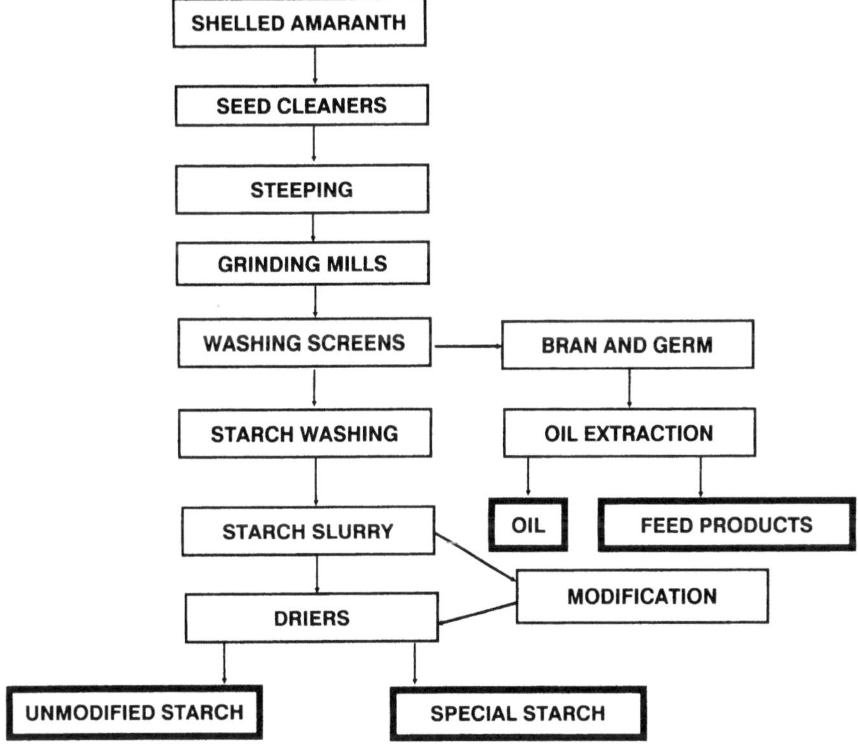

Fig. 2. Basic procedures for the isolation of amaranth starch (see text)

3.1.2 Sulfur Dioxide Method

At the laboratory scale, mercuric chloride is recommended (Banks and Greenwood 1975). However, due to pollution regulations and economic considerations with the previous reagent, sulfur dioxide is preferably used at the industry level to denature enzymes and soften the protein-starch matrix during steeping and wet milling (Watson 1964a). The remaining steps are similar to the method described above. Care must be taken to ensure that the loss of starch granules in the rejected toluene layer is not unacceptably high. Such losses may be avoided by adding several volumes of water to the toluene and aerating the mixture, so freeing the starch from its physical entrapment within the denatured protein.

3.1.3 Alkaline Method

Amaranth starch may also be isolated by the alkaline procedure described by MacMasters et al. (1955) and Paredes-López et al. (1989): cleaned seeds are soaked in deionized water containing 0.01% sodium metabisulfite for 12 h at room temperature. The swollen seeds are rinsed with water and blended with deionized water. The homogenate is stirred during 24 h at 4°C, and then centrifuged. The residue is extracted with 2% NaCl for 24 h at 4°C, centrifuged, and the supernatant discarded. The extract is mixed with 0.1 N sodium hydroxide, and shaken for 48 h at 4°C and centrifuged. The alkali extraction and centrifugation are repeated once. The resulting sediment, after removing the mucilagenous layer, is suspended in 80% aqueous ethanol, heated at 40°C for 1 h in a water bath, and centrifuged. The isolated starch is then freeze-dried. Yañez et al. (1986) reported an alkaline method which does not involve the steps of soaking in metabisulfite and sodium chloride extraction. The soaking and extraction are done solely with 0.25% sodium hydroxide. The advantage of the latter method is its rapidity, but the yielded starch contains more impurities than that obtained with the former technique.

Sometimes starch is defatted. Portions of starch are then slurried in methanol, and Soxhlet-extracted with hot methanol for a minimum of 150 siphonings (Adkins and Greenwood 1966).

3.2 Starch Quantitation

A number of methods for starch determination are reported in the literature. However, the procedures analyzed here are those that have given satisfactory results for the analysis of amaranth starch.

3.2.1 Chemical Method

This method (Hassid and Neufeld 1964; Hernández-López 1991) consists in the extraction of the soluble sugars from the dried material, which has been previously ground to pass a 50 to 80 mesh screen. The sample is immersed in a hot water bath, and hot 80% ethanol added; then the mixture is stirred several minutes and centrifuged. The extraction is repeated three more times in order to completely remove the soluble sugars. The sugar-free residue is then heated with water to gelatinize the starch. The suspension is cooled, 52% perchloric acid and water are added, and the mixture is centrifuged. The aqueous starch solution is extracted once more with perchloric acid. The starch extract is treated with 20% sodium chloride and iodine-potassium iodide reagent, then centrifuged and the supernatant removed. The precipitated starch-iodine complex is suspended in ethanolic sodium chloride wash solution, and centrifuged. The supernatant fluid is discarded, and the washing process is repeated. Ethanolic sodium hydroxide is added to the precipitate until all blue color is discharged, centrifuged, and washed several times with ethanolic sodium chloride as before. The liberated starch is then determined colorimetrically with an anthrone reagent: anthrone is added to an aqueous starch solution; at this point, the anthrone may precipitate but this will not affect the determination.

The sample is placed in a boiling water bath until maximum color develops, cooled, and absorbance measured at 620 nm. In calculating the amount of starch from the standard curve of D-glucose, the D-glucose reading must be multiplied by 0.925 (AOAC 1984, method 8.019). Alternatively, the starch may be hydrolyzed with acid (Barba de la Rosa 1989), and the reducing value of the hydrolysate determined as D-glucose by any of the standard procedures (Dubois et al. 1956; AACC 1984, method 80–60; AOAC 1984, method 31.038).

3.2.2 Polarimetric Method

Compounds possessing an asymmetric carbon atom have the ability to rotate the plane of polarization of polarized light; the optical rotation is the angle through which the light is rotated when the light passes through a solution of the optically active compound (Kearsley 1985). The polarimetric procedure (AACC 1984, method 76–20) is as follows: amaranth flour is mixed with mercuric chloride reagent (1 g mercuric chloride in 900 ml water plus 100 ml 95% ethanol), filtered, and washed twice with the same reagent. Acid calcium chloride (calcium chloride is acidified with acetic acid to pH 2.2–2.5 and adjusted to a density of 1.30 at 20 °C) is added to the residue; then it is boiled for 15 min, cooled, mixed with stannic chloride, and the volume adjusted with acid calcium chloride. The sample is filtered, and 30 to 50 ml collected for polarization which is done in a 2-dm tube, using sodium light, usually at 20 °C. Starch in percentage is equal to the polarimeter reading in degrees multiplied by 10. The use of stannic chloride in place of the original reagent (uranyl acetate) makes this procedure satisfactory for amaranth starch.

3.2.3 Enzymatic Method

Amaranth flour is mixed with 80% ethanol, placed in a water bath for 10 min at 60 °C, centrifuged, and the supernatant (containing free sugars) discarded. The extraction is repeated. The pellet is transferred into a culture tube with screw cap using calcium chloride solution (50% calcium chloride, pH 2), stirred, and heated for 15 min at 125 °C by immersion in a wax bath (Fleming and Reichert 1980; YSI 1989). Periodically, addition of water is required to maintain the original volume. After cooling, potassium hydroxide (0.05 M) and acetate buffer (0.1 M) are added to adjust the pH to 4.8. Glucoamylase and α-amylase (Diazyme L-100 and Tenase, respectively, Miles Laboratories Inc., Elkhart, IN) are mixed with the sample prior to incubation for 3 h at 48 °C. The hydrolysate is made up to a predetermined volume with water, filtered, and analyzed for glucose. Sample aliquots are mixed with a glucose oxidase/peroxidase reagent (Glox-Novum, Kabi-Diagnostica, Stockholm), incubated for 60 min at 37 °C, centrifuged, and absorbance of the supernatant measured at 450 nm. Standard solutions of anhydrous D-glucose are also incubated with the glucose oxidase/peroxidase reagent, and the glucose content computed by least squares linear regression. A blank containing water is incubated with the glucose oxidase/peroxidase reagent, and used for zero adjustment of the spectrophotometer.

The disodium salt of ethylenediamine tetraacetic acid (EDTA) may be included in the analyses (Fleming and Reichert 1980). EDTA is made to concentrations of 0.05, 0.1, and 0.2 M with distilled water, added to the enzymatic-hydrolyzed samples, and adjusted to a predetermined volume with acetate buffer before glucose analysis. It was found that minerals, such as the calcium used for starch dispersion or other minerals naturally present in plant products, are responsible for excessively high absorption values. The EDTA presumably acts as a chelating agent to render the dissolved ions unavailable for complexation with the glucose oxidase/peroxidase reagent.

Holm et al. (1986) have proposed using a thermostable α-amylase (Termamyl 120L, Novo A/G, Copenhagen) in place of Tenase. In this case, the amaranth flour (free of sugars) is suspended in distilled water, Termamyl added, and the suspension incubated in a boiling water bath for 15 min. After cooling, the sample is mixed with glucoamylase and acetate buffer (0.1 M, pH 4.8), incubated for 30 min at 60 °C, and the hydrolysate analyzed as described above. Nowadays enzymatic methods are widely used for starch analysis; in these methods, the complete availability of starch to enzymatic attack is essential (Barba de la Rosa and Paredes-López 1989; Paredes-López et al. 1990a). Attention must therefore be paid to some factors, such as the presence of retrograded starch, amylose-lipid complexes, and the protein structures encapsulating them.

Glucose determinations for starch analysis may also be done with the Industrial Glucose Analyzer (YSI model 27) equipped with a glucose oxidase membrane (YSI 1989). It gives high accuracy and precision, rapidity, and simple handling.

3.2.4 High-Performance Liquid Chromatography (HPLC) Method

Boley and Burn (1990) have developed an HPLC procedure to quantitate starch in various plant foodstuffs. Although amaranth starch was not included in these samples, this technique appears to offer various advantages compared to those described in Sections 3.2.1, 3.2.2, and 3.2.3. Free sugars may be extracted as described in Section 3.2.3. The remaining sample is heated under reflux for 3 h with deionized water containing concentrated hydrochloric acid to hydrolyze the starch. After cooling, the hydrolysate is made up to a fixed volume with deionized water, filtered, and diluted with acetonitrile. Further dilutions are carried out as necessary with acetonitrile/water (50:50 v/v) in order to bring the glucose concentration into the range 0.5 to 2.5 g/l, prior to analysis by HPLC. The HPLC apparatus is equipped with a Chrompack Chromsep (Middleburg, The Netherlands) cartridge system consisting of two 100×3 mm^2 glass columns packed with Hypersil-APS and an ODS guard column. The mobile phase may be acetonitrile/water (82 : 18 v/v) degassed using helium, pumped at a flow rate of 1 ml/min. Peak areas are determined using a computing integrator. Quantitation is achieved by chromatographic analysis, under identical conditions, of standard solutions of glucose, fructose, maltose, and sucrose at 0.1% in acetonitrile/water (50 : 50 v/v). The procedure described is simple and has been shown to be accurate. In addition, the procedure could be readily automated, which would make it considerably more cost-effective than any titrimetric, polarimetric, or enzymatic procedure. An initial concern regarding this technique was the concentration of acid which was being introduced onto the column, and the possible deleterious effects this could have on column performance and lifetime. Experience has demonstrated that no deterioration of column efficiency is observed after a larger number of analyses (>30).

4 Starch: Basic Characterization and Methods of Determination

4.1 Damaged Starch

The measurement of damaged starch determines the percentage of starch granules in flour starch preparations that are susceptible to hydrolysis by α-amylase (Paredes-López et al. 1988a). The percent starch damage is defined as the grams of starch subject to enzymatic hydrolysis per 100 g of sample on a 14% moisture basis (AACC 1984, method 76–30A). The AACC method consists basically in the following: a given amount of amaranth flour or starch is mixed with an α-amylase solution (from *Aspergillus oryzae*, Sigma Chemical Co., St. Louis, MO), and the mixture incubated 15 min at exactly 30°C. Sulfuric acid (3.68 N) and sodium tungstate solution are added, mixed, and filtered. The amount of reducing sugars in the filtrate is determined by any of the standard procedures (e.g., AACC 1984, method 80–60; AOAC 1984, method 31.038). A blank is determined by conducting previous steps, omitting flour or starch sample.

Sutton and Mouat (1990) have recently proposed the determination of starch damage by reversed-phase high performance liquid chromatography (RP-HPLC). This study describes the use of HPLC to determine the maltose concentration in flour samples incubated with α-amylase solutions and the relationship of these maltose levels to damaged-starch values determined by the AACC method. Although of greater capital cost and generally requiring more technical expertise, it appears that automated RP-HPLC has a much greater sample capacity than the AACC procedure and involves fewer reagents and steps, and less operator time. In view of these advantages, amaranth flour or isolated starch needs to be tested by this technique.

4.2 Starch Fractions

In 1941 Schoch demonstrated the fractionation of starch by adding butanol to a starch solution. Butanol formed an insoluble crystalline complex with the amylose fraction (Swinkels 1985). The percentages of amylose and amylopectin present in the granules may be the major granule compositional difference both between different species and within genetic strains of the same species. The starch granules containing mostly amylopectin, with an amylose composition of less than 1%, are called waxy, glutinous, or opaque types. When the proportion of the amylose composition equals or exceeds 1%, the granules are referred to as nonwaxy, nonglutinous, or translucent (Kintner 1987). These types of granules will also appear reddish-brown and blue respectively, when stained with iodine (Brimhall et al. 1945). Before 1975, all amaranth starch granules were thought to be of the amylopectin type based on the typical reddish-brown staining characteristic. Since this time, both waxy and nonwaxy types have been isolated from various genetic sources of amaranth (Sugimoto et al. 1981; Kintner 1987). Amylose concentrations of up to 22% (Table 4) have been reported in genetic strains of *A. hypochondriacus* (Tomita et al. 1981). It is now known that the proportion of starches found in the granules of seed crops, including amaranth, is determined by a single major gene, with the amylopectin allele recessive to amylose (Okuno and Sakaguchi 1982). This gene apparently controls the formation of two enzymes, one which couples via the (1→4) linkages (amylose), and the other which couples via the (1→6) linkages (amylopectin). The relative amounts of each starch is determined by the ratio of these two enzymes, which is an inherited characteristic.

4.2.1 Amylose

Amylose is essentially a linear polymer containing up to 6000 glucose units linked together by α-(1→4) bonds (Fig. 3a). Each macromolecule bears one reducing and one nonreducing end. However, the presence of anomalous linkages was first suggested by Peat et al. (1952). The main barrier was observed to be a minor degree of branching by α-(1→6) linkages as in amylopectins, because the combined action of β-amylase with yeast isoamylase or bacterial pullulanase completely or partly overcomes the barrier (Kjolberg and Manners 1963; Banks and Greenwood 1975). Subsequent studies have confirmed that the combined action of these enzymes invariably causes complete degradation of amyloses from a wide variety of starches (Guilbot and Mercier 1985). The branched amylose molecules may contain 3 to 20 chains, with an average chain length of about 500 glucose units.

Sargeant (1982) has developed an enzymatic method to quantitate amylose in starch. The isoamylase debranched starch is fractionated on a Sepharose CL-6B column in 0.25 N potassium hydroxide, separating chain populations of a degree of polymerization (DP) up to 550. The sample recovered in the first peak is attributed to amylose by its iodine absorption complex at 620 nm,

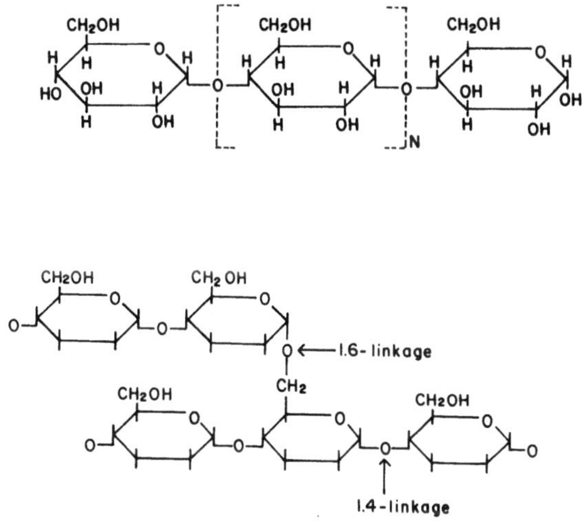

Fig. 3. **a** Linear chain structure of amylose molecules; **b** structure of amylopectin branching points

which corresponds to about the same proportion determined by potentiometric or amperometric measurements. In general, Sugimoto et al. (1981) have used the same methodology to characterize the starch from the seeds of two types of *A. hypochondriacus*. These workers used Sephadex G-75 and reported, on the basis of the elution profiles of isoamylase-debranched material, that one type was shown to be free of amylose and the other one to have 14% amylose.

Quantitation of amylose in amaranth starch has also been done by the blue-value method of Morrison and Laignelet (1983), which has been slightly modified by Paredes-López et al. (1989). Starch is mixed vigorously with urea (6 M) plus dimethylsulfoxide (DMSO) (1 : 9 v/v). The mixture is placed in a boiling water bath for 10 to 15 min until the solution is almost clear, then transferred to an oven at 100 °C for another 50 to 60 min, and cooled. The solution is diluted with deionized water, and solution iodine+potassium iodide (2 mg I_2+20 mg KI/ml) added. Absorbance (at 635 nm) is exactly read 15 min after the last reagent is added. The spectrophotometer is adjusted with a blank prepared following all the previous steps but without starch. Calculation for amylose is done with a regression equation derived using the amylose: amylopectin proportions 10:90, 20:80, 30:70, 40:60, 50:50, 60:40, 70:30, 80:20, and 90:10. In this method a distinction should be made between apparent amylose (measured in the presence of lipids which complex with amylose to reduce its iodine-binding capacity), and total amylose (measured on lipid-free starch, precipitated from urea-DMSO solution with ethanol). The coefficient of variations in replicate determinations (n = 6) may be 0.2% to 0.8% for nonwaxy starch, and 1.1% to 1.5% for waxy starches.

4.2.2 Amylopectin

Amylopectin is the branched fraction of starch (Fig. 3b); it is formed through chains of glucopyranose residues linked together mainly by α-(1→4) linkages but with about 5% of α-(1→6) bonds at the branch points. One of the models proposed for amylopectin consists of a cluster-type structure of highly ordered chains (Robin et al. 1974). In this model, the A- and B-chains are linear and have a DP of 15 and 45, respectively. The B-chains form the backbone of the amylopectin molecule and extend over two or more clusters. Each cluster contains two to four closely associated A-chains. The associated clusters of A-chains are primarily responsible for the crystalline regions within the granule. The intercrystalline (amorphous) areas occur at 60 to 70 Å intervals, contain the majority of the α-(1→6) linkages, and are relatively susceptible to hydrolytic (e.g., acids, enzymes) agents (Hood 1982). Within the granule, amylose may be located between amylopectin polymers and may associate with the linear regions of the amylopectin molecule. The molecular weight of amylopectin, one of the largest biological molecules, varies between the values of 10^6 and 10^9, the variation being dependent on the botanical origin of the starch, the conditions for the fractionation of amylopectin from amylose, and the method used to determine the molecular weight (Greenwood 1976; Guilbot and Mercier 1985). It must be emphasized that studies are lacking on the biochemical and physicochemical characterization of amylopectin from amaranth starch.

5 Physicochemical and Functional Properties of Amaranth Starch

5.1 Gelatinization and Methods of Determination

When the starch granule is heated in water, the weaker hydrogen bonds in the amorphous areas are ruptured and the granules swell with progressive hydration, leading to nonreversible changes. The process is called "gelatinization" and the temperature at which it occurs is called the "gelatinization temperature". With the swelling of amylose-containing granules, some of the amylose molecules are solubilized and leach out to form an extragranular network, which will precipitate if the starch concentration is low or will form a gel if the concentration is high. This is referred to as "retrogradation" or "setback" (Atwell et al. 1988). The summarized events during gelatinization are: disruption of molecular orders within the starch granule manifested in irreversible changes in properties such as granular swelling, native crystallite melting, loss of birefringence, and starch solubilization. The point of initial gelatinization and the range over which it occurs is governed by starch concentration, method of observation, granule type, and heterogenities within the granule population under observation (Smith 1982; Lund 1984; Atwell et al. 1988).

Table 5. Gelatinization temperature ranges of amaranth starch[a]

Sample	Range (°C)
A. caudatus	51 – 65
A. cruentus	63 – 78
A. hybridus	71 – 82
A. hypochondriacus	59 – 82

[a] Modi and Kulkarni (1976); Lorenz (1981); Okuno and Sakaguchi (1981); Sugimoto et al. (1981); Stone and Lorenz (1984); Konishi et al. (1985); Barba de la Rosa et al. (1989); Paredes-López et al. (1989, 1990b).

The degree of gelatinization of amaranth starch can be determined qualitatively and/or quantitatively by physical, chemical, and physicochemical methods. Table 5 shows ranges of gelatinization temperatures for this starch (isolated from a wide number of amaranth species) assessed by various techniques, such as those described in the following sections (Modi and Kulkarni 1976; Lorenz 1981; Okuno and Sakaguchi 1981; Sugimoto et al. 1981; Stone and Lorenz 1984; Konishi et al. 1985; Barba de la Rosa et al. 1989; Paredes-López et al. 1989, 1990b).

5.1.1 Birefringence End Point Method

Starch gelatinization is determined by estimating the percent loss of birefringence with a Kofler electrically heated microscopic hot stage and a polarizing microscope (Watson 1964b). An aqueous suspension of amaranth starch is prepared (0.1% to 0.2%), and a small drop of it is spotted on a microscope slide, and surrounded by a ring of high-viscosity mineral oil. The temperature is increased at a rate of 2°C/min, and the field is watched to record the temperature where 2%, 50%, and 98% of the granules lose their birefringence. These readings correspond to the initial, mid-point, and end-point gelatinization temperatures. Variations in starch gelatinization ranges by this technique may be ascribed to various factors, but especially to changes in the water/starch ratio of samples (Varriano-Marston 1983).

5.1.2 Differential Scanning Calorimetry (DSC) Method

Following the pioneering work of Zobel et al. (1965) and Stevens and Elton (1971), calorimetric methods have been frequently applied to study phase transitions which aqueous suspensions of granular starch undergo on heating. After this time, many groups have used DSC in the study of gelatinization

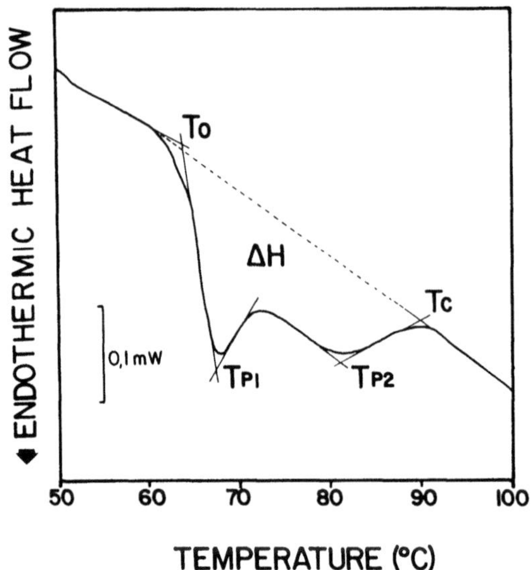

Fig. 4. Differential scanning calorimetry (DSC) thermogram of amaranth starch (*A. hypochondriacus*) with 0.53 g water/g total. The area of the endothermic curve (ΔH) was used to calculate gelatinization energy. *To* Onset temperature; *Tp1* low transition temperature; *Tp2* intermediate transition temperature or temperature at which additional melting of starch crystallites occurs; *Tc* completion temperature (see text; and also Paredes-López and Hernández-López 1991)

(Donovan 1979; Wootton and Bamunuarachchi 1980; Wright 1984; Robles et al. 1988). DSC detects the heat flow changes associated with both first-order (melting) and second-order (glass transition) thermal transitions, and it has proved to be of considerable value in studying order/disorder phenomena of granular starches. DSC measurements of amaranth starch gelatinization (Paredes-López and Hernández-López 1991) may show four endothermic transitions plus the onset (To) and completion (Tc) temperatures (Fig. 4). The low transition temperature, Tp1, called water-mediated gelatinization; the transition at intermediate temperature, Tp2, refers to the additional melting of starch crystallites; and two endotherms at high temperatures (110 to 140 °C), Tp3 and Tp4, reversible upon immediate reheating, have been attributed to the disorganization of the amylose-lipid complexes. Waxy and defatted or low lipid content starches thus show no evidence for the high temperature transitions. In addition to providing measurements of these transition temperatures, it is possible to determine gelatinization enthalpy to obtain a quantitative estimation of crystallinity and order in the amaranth starch granule (Barba de la Rosa et al. 1989; Knutson 1990).

The DSC method uses a small sample of starch (1 to 5 mg) at 10% to 20% total solids, which minimizes the thermal lag within the system. The instrument may be calibrated with indium; after sealing, the pan containing the sample is left to equilibrate (1 to 2 h), and scanned at a rate of about 10 °C/min. Figure 4 illustrates the DSC endotherm generated by isolated amaranth starch. Gelatinization energy (Δ H) in cal/g starch is calculated by measuring the peak area. The degree of gelatinization expressed as percent is determined from gelatinization energy. An amaranth starch suspension at 0.85 g water/g total, and above, has found to show no further increases in gelatinization energy

(Paredes-López and Hernández-López 1991). Therefore, this energy (5.9 cal/g starch) is equated with 100% gelatinization and the degree of gelatinization of remaining samples is calculated proportionally.

5.1.3 Other Methods

Other methods are available (Atwell et al. 1988) to measure amaranth starch gelatinization, in addition to those cited in Sections 5.1.1 and 5.1.2: pasting properties of starch suspensions (see Sect. 5.2); amylose/iodine blue value; X-ray diffraction; nuclear magnetic resonance; and enzymatic procedures. In each of these procedures, a specific event, related to the principle of the method used, is measured. However, details of these methods cannot be analyzed here due to space restrictions.

5.2 Rheological Properties and Methods of Determination

Swelling and the subsequent disruption of the starch granule are of great technological importance (Swinkels 1985). One of the best methods for following the rheological changes during cooking of an amaranth starch paste is with the Brabender viscoamylograph (Mora-Escobedo et al. 1991). This apparatus measures the viscosity of starch-water dispersions that are stirred and heated at a uniform rate, held at any desired temperature for a specific time, and then cooled at uniform rate. Slurries of various concentrations (Fig. 5) are prepared by dispersing amaranth starch in distilled water, transferring to the viscoamylograph bowl, and heating at a rate of 1.5 °C/min while stirring continuously. The sample is maintained at 95 °C for a fixed time, and cooled to 50 °C. Consistency (viscosity) in Brabender units at the three heating stages is recorded.

Initially, no viscosity effect is noted as the suspension of starch is heated until the pasting temperature is reached (Fig. 5). The temperature at which the viscosity begins to rise is termed the Brabender pasting temperature (Rasper 1982). This pasting temperature is usually higher than the gelatinization temperature measured by DSC (Hernández-López 1991). Consistency increases as the temperature increased, reaching a peak at 78 to 80 °C. Thereafter, between 78 to 95 °C and during the temperature holding period, consistency declines with the constant shear stress of the rotating viscoamylograph bowl, thus exhibiting a thixotropic behavior (Mora-Escobedo et al. 1991); viscosity increases again (setback) during the cooling period, which is a measure of the retrogradation process, due to the association of starch chains. These viscosity patterns are characteristic for this specific amaranth starch and will be different for other starches of different sources (Swinkels 1985).

The viscosity of amaranth starch pastes may also be measured by other procedures (Barba de la Rosa et al. 1989; Mora-Escobedo et al. 1991). The resistance under mechanical shear, which tears the swollen granules apart, can be

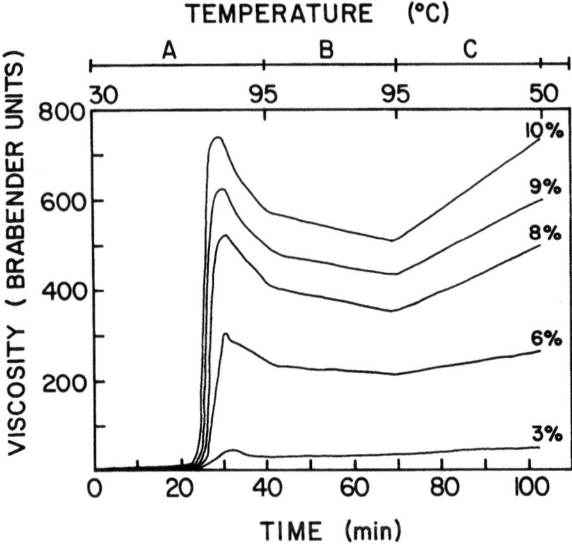

Fig. 5. Viscograms of amaranth starch (*A. hypochondriacus*) slurries in distilled water (3, 6, 8, 9, and 10%, w/w) at pH 7, produced with the heating (*A*), holding at 95 °C (*B*), and cooling (*C*) phases. A viscograph E-apparatus was used with the 700-cmg cartridge, at a speed of 75 rpm, and heating and cooling rate of 1.5 °C/min (see text; and also Mora-Escobedo et al. 1991)

tested by stirring starch pastes for a given time (e.g., 20 min), and then determining the apparent viscosity.

5.3 Starch-Water Interactions

Native starches are practically insoluble in water below their gelatinization temperature. The insolubility is mainly due to hydrogen bonds formed either directly via neighboring alcoholic OH groups of the individual starch chains or indirectly via water bridges. The hydrogen bonding forces are weak, but there are so many of them in a starch granule that insolubility in cold water prevails. Starch granules swell slightly in cold water, but this swelling is reversible (Christianson 1982). When starch granules are heated in water to progressively higher temperatures (see Sect. 5.1), the swollen starch granules begin to rupture and collapse, yielding a viscous colloidal dispersion of swollen granule fragments, hydrated starch aggregates, and dissolved molecules (Swinkels 1985).

5.3.1 Solubility and Swelling Assessment

Solubility and swelling of amaranth starch may be assessed by the procedure reported by Schoch (1964), slightly modified by Paredes-López et al. (1989).

Starch in dispersed is distilled water (1% w/v). This slurry is heated at the desired temperature (e.g., 60, 70, 90 °C) for 30 min in a water bath, cooled to room temperature, and centrifuged. Aliquots of supernatant are dried to constant weight to determine starch solubility (%). The residue with the water it retains is weighed (W2). Test tubes containing starch alone are weighed prior to adding distilled water (W1). Swelling of starch (dry weight basis) is calculated as follows: swelling (%) = ⟨(W2−W1)/starch weight⟩ 100.

The effect of pH on swelling and solubility is also assessed as another functional property of starch. Slurries at the same concentration in distilled water are prepared, and the pH adjusted to the desired value with 0.1 N chlorhydric acid and 0.1 N sodium hydroxide. The slurries are allowed to stand at room temperature for 30 min, centrifuged, and swelling and solubility determined as described above.

Information on these functional properties for amaranth starch is still very scarce. Swelling of amaranth starch tends to be lower than that reported for maize and wheat starches, but its solubility is higher (Becker et al. 1981; Lorenz 1981; Paredes-López et al. 1989).

5.3.2 Determination of Water Absorption

Water absorption of amaranth starch is measured by weighing a small amount of sample mixed with distilled water at room temperature. The suspension is agitated in a shaker for 30 min. After centrifugation, the sediment is weighed and its water content determined by drying to constant weight. The water-binding capacity of amaranth starch has been reported to be higher than that exhibited by maize and wheat starches (Lorenz 1981; Paredes-López et al. 1989). Water absorption, as well as swelling and solubility, is largely affected by the level of damaged starch. Therefore, this type of starch needs to be taken into account when the previous measurement is done.

6 Concluding Remarks

The role of amaranth as an underexploited plant with promising economic value has gained recent prominence. Amaranth is one of those few plants whose leaves are eaten as a vegetable while the seeds are used as cereals. The potential of amaranth seeds as a food resource has been briefly reviewed in this chapter. The nutritional quality of amaranth protein is so high that it approaches that of animal food proteins; this is an exceptional property restricted to only two or three vegetable proteins available in nature. One of the main challenges is to incorporate amaranth into existing food formulations to modify their functional and nutritional quality, as well as to create entirely new products, based mostly on this seed. Vargas-López et al. (1990) have developed methods for the transformation of amaranth seeds into tortillas and snacks of high nutritional and functional quality.

Some of the most important procedures for isolation and characterization of amaranth starch have also been reviewed here. There is a lack of experimental data on the biochemical, physicochemical, and functional properties of this particular polymer. At the same time, attention should be directed on using the basic knowledge of what amaranth starch is, what can be done to modify its basic properties, and how to choose and utilize their properties in the most effective way.

Acknowledgments. We acknowledge the Consejo Nacional de Ciencia y Tecnología-México for supporting our research work on amaranth starch. We are also thankful to our colleagues working on amaranth at Lab Biotecnología de Alimentos, Unidad Irapuato, CINVESTAV-IPN, for their cooperation, and to A. Cárabez, Inst Fisiología Celular, UNAM, for his assistance with SEM studies.

References

AACC (1984) Approved methods. American Association of Cereal Chemists, St Paul, MN

Adkins GK, Greenwood CT (1966) The isolation of cereal starches in the laboratory. Starch/Stärke 18:213–218

AOAC (1984) Official methods of analysis. Association of Official Analytical Chemists, Washington

Atwell WA, Hood LF, Lineback DR, Varriano-Marston E, Zobel HF (1988) The terminology and methodology associated with basic starch phenomena. Cereal Foods World 33:306–311

Banks W, Greenwood CT (1975) Starch and its components. Edinburgh University Press, Edinburgh

Barba de la Rosa AP (1989) Procedimiento enzimático para producir harinas de amaranto de alto contenido de proteína y jarabes con potencial edulcorante. Thesis, CINESTAV-IPN, Irapuato, México

Barba de la Rosa AP, Paredes-López O (1989) Development of an enzymatic procedure to produce high-protein amaranth flour. Biotechnol Lett 11:417–422

Barba de la Rosa AP, Paredes-López O, Cárabez-Trejo A, Ordorica-Falomir C (1989) Enzymatic hydrolysis of amaranth flour − differential scanning calorimetry and scanning electron microscopy studies. Starch/Stärke 41:424–428

Becker R (1989) Preparation, composition, and nutritional implications of amaranth seed oil. Cereal Foods World 34:950–953

Becker R, Wheeler EL, Lorenz K, Stafford AE, Grosjean OK, Betschart AA, Saunders RM (1981) A compositional study of amaranth grain. J Food Sci 46:1175–1180

Boley NP, Burn MJS (1990) The determination of starch in composite foodstuffs by high-performance liquid chromatography. Food Chem 36:45–51

Bressani R, Elías LG, Gonález JM, Gómez-Brenes R (1987) The chemical composition and protein quality of amaranth grain germplasm in Guatemala. Arch Latinoam Nutr 37:363–377

Brimhall B, Sprague CF, Sass JE (1945) A new waxy allele in corn and its effect on the properties of the endosperm starch. J Am Soc Agron 37:937–945

Christianson DD (1982) Hydrocolloid interaction with starches. In: Lineback DR, Inglett GE (eds) Food carbohydrates. AVI Publ, Westport, CT, p 399

Donovan JW (1979) Phase-transitions of the starch-water system. Biopolymers 18:263–275

Duarte-Correa A, Jokl L, Carlsson R (1986) Amino acid composition of some *Amaranthus* sp. grain proteins and of its fractions. Arch Latinoam Nutr 36:466–476

Dubois M, Gilles KA, Hamilton JK, Rebers PA, Smith F (1956) Colorimetric method for determination of sugars and related substances. Anal Chem 28:350–355

Fleming SE, Reichert RD (1980) Note on a modified method for the quantitative determination of starch. Cereal Chem 57:153–154

Greenwood CT (1976) Starch. In: Pomeranz Y (ed) Advances in cereal science and technology, vol 1. American Association of Cereal Chemists, St Paul, MN, p 119

Grubben GJH, van Sloten DH (1981) Genetic resources of amaranths. Int Board for Plant Genetic Resources, FAO, Rome

Guilbot A, Mercier C (1985) Starch. In: Aspinal GO (ed) The polysacharides, vol 3. Academic Press, New York, p 209

Hassid WZ, Neufeld EF (1964) Quantitative determination of starch in plant tissues. In: Whistler RL (ed) Methods in carbohydrate chemistry, vol 4. Academic Press, New York, p 33

Hernández-López D (1991) Aislamiento y caracterizatión fisicoquímica del almidón de amaranto – efectos hidrotérmicos y de solutos. Thesis, CINVESTAV-IPN, Irapuato, México

Holm J, Bjorck I, Drews A, Asp NG (1986) A rapid method for the analysis of starch. Starch/Stärke 38:224–226

Hood LF (1982) Current concepts of starch structure. In: Lineback DR, Inglett GE (eds) Food carbohydrates. AVI Publ, Westport, CT, p 217

Kearsley MW (1985) Physical, chemical and biochemical methods of analysis of carbohydrates. In: Birch GG (ed) Analysis of food carbohydrate. Elsevier, London, p 15

Kintner PK (1987) Starch. Rodale Research Center, Kutztown, PA

Kjolberg O, Manners DJ (1963) The action of isoamylase on amylose. Biochem J 86:258–262

Knutson CA (1990) Annealing of maize starches at elevated temperatures. Cereal Chem 67:376–384

Konishi Y, Fumita Y, Ikeda K (1985) Isolation and characterization of globulin from seeds of *Amaranthus hypochondriacus* L. Agric Biol Chem 49:1453–1459

Lorenz K (1981) *Amaranthus hypochondriacus* – characteristics of the starch and baking potential of the flour. Starch/Stärke 33:149–153

Lund D (1984) Influence of time, temperature, moisture, ingredients, and processing conditions on starch gelatinization. Crit Rev Food Sci Nutr 20:249–273

Lyon CK, Becker R (1987) Extraction and refining of oil from amaranth seed. J Am Oil Chem Soc 64:233–237

MacMasters MM, Baird PD, Holzapfel MM, Rist CE (1955) Preparation of starch from *Amaranthus cruentus* seed. Econ Bot 9:300–302

Modi JD, Kulkarni PR (1976) New starches: the properties of the starch from *Amaranthus paniculatus* Linn. Acta Aliment 5:399–402

Mora-Escobedo R, Paredes-López O, Ordorica-Falomir C (1990) Characterization of albumins and globulins from amaranth. Food Sci Technol (Lebensm Wiss Technol) 23:484–487

Mora-Escobedo R, Paredes-López O, Gutiérrez GF (1991) Effect of germination on the rheological and functional properties of amaranth seeds. Food Sci Technol (Lebensm Wiss Technol) 24:287–292

Morrison WR, Laignelet B (1983) An improved colorimetric procedure for determining apparent and total amylose in cereal and other starches. J Cereal Sci 1:9–20

NRC (1984) Amaranth. Modern prospects for an ancient crop. National Research Council, National Academy Press, Washington, DC

Oke OL (1983) Amaranth. In: Chan HT (ed) Handbook of tropical foods. Marcel Dekker, New York, p 1

Okuno K, Sakaguchi S (1981) Glutinous and non-glutinous starches in perisperm in grain amaranths. Cereal Res Commun 9:305–310

Pal M, Khoshoo TN (1974) Grain amaranths. In: Hutchinson J (ed) Evolutionary studies in world crops. Cambridge University Press, Cambridge, p 129

Pal M, Khoshoo TN (1977) Evolution and improvement of cultivated amaranths. VIII. Induced autotetraploidy in grain types. Z Pflanzenzücht 78:135–148

Paredes-López O, Hernández-López D (1991) Application of differential scanning calorimetry to amaranth starch gelatinization – influence of water, solutes and annealing. Starch/Stärke 43:57–61

Paredes-López O, Mora-Escobedo R (1989) Germination of amaranth seeds: effects on nutrient composition and color. J Food Sci 54:761–762

Paredes-López O, Maza-Calviño EC, Montes-Rivera R (1988a) Effect of the hard-to-cook phenomenon on some physicochemical properties of bean starch. Starch/Stärke 40:205–210

Paredes-López O, Cárabez-Trejo A, Pérez-Herrera S, González-Castañeda J (1988b) Influence of germination on physico-chemical properties of amaranth flour and starch microscopic structure. Starch/Stärke 40:290–294

Paredes-López O, Mora-Escobedo R, Ordorica-Falomir C (1988c) Isolation of amaranth proteins. Food Sci Technol (Lebensm Wiss Technol) 21:59–61

Paredes-López O, Schevenin ML, Hernández-López D, Cárabez-Trejo A (1989) Amaranth starch – isolation and partial characterization. Starch/Stärke 41:205–207

Paredes-López O, Barba de la Rosa AP, Cárabez-Trejo A (1990a) Enzymatic production of high-protein amaranth flour and carbohydrate rich fraction. J Food Sci 55:1157–1161

Paredes-López O, Barba de la Rosa AP, Hernández D, Cárabez A (1990b) Amaranto – características alimentarias y aprovechamiento agroindustrial. Monografía. Organización de los Estados Americanos, Washington, DC

Peat S, Whelan WJ, Thomas GJ (1952) Evidence of multiple branching in waxy maize starch. J Chem Soc 35:4546–4548

Pedersen B, Kalinowski LS, Eggum BO (1987) The nutritive value of amaranth grain (*Amaranthus caudatus*). 1. Protein and minerals of raw and processed grain. Plant Foods Hum Nutr 36:309–324

Rasper V (1982) Theoretical aspects of amylographology. In: Shuey W, Tipples KH (eds) The amylograph handbook. American Association of Cereal Chemists, St Paul, MN, p 1

Robin JP, Mercier C, Charbonnière R, Guilbot A (1974) Lintnerized starches. Gel filtration and enzymatic studies of insoluble residue from prolonged acid treatment of potato starch. Cereal Chem 51:389–406

Robles RR, Murray ED, Paredes-López O (1988) Physicochemical changes of maize starch during the lime-heat treatment for tortilla making. Int J Food Sci Technol 23:91–98

Sánchez-Marroquín A (1980) Potencialidad agroindustrial del amaranto. Centro de Estudios Económicos y Sociales del Tercer Mundo, México DF

Santín C, Lazcano M, Morales J (1986) Pasado, presente y futuro del amaranto. Cuad Nutr (Méx) 9:16–32

Sargeant JG (1982) A simple method for the physicochemical characterization of amylose. Starch/Stärke 34:89–92

Sauer JD (1950a) Amaranths as dye plants among the Pueblo peoples. Southwest J Anthropol 6:412–417

Sauer JD (1950b) The grain amaranths. A survey of their history and classification. Ann Mo Bot Gard 37:561–632

Saunders RM, Becker R (1984) *Amaranthus*: a potential food and feed resource. In: Pomeranz Y (ed) Advances in cereal science and technology, vol 6. American Association of Cereal Chemists, St Paul, MN, p 357

Schoch TJ (1964) Swelling power and solubility of granular starches. In: Whistler RL (ed) Methods in carbohydrate chemistry, vol 4. Academic Press, New York, p 106

Singhal RS, Kulkarni PR (1988) Review: amaranths – an underutilized resource. Int J Food Sci Technol 23:125–139

Smith PS (1982) Starch derivatives and their use in foods. In: Lineback DR, Inglett GE (eds) Food carbohydrates. AVI Publ, Westport, CT, p 237

Stevens DJ, Elton GAH (1971) Thermal properties of the starch/water system. I. Measurement of heat of gelatinization by differential scanning calorimetry. Starch/Stärke 23:8–12

Stiebritz S, Harrington M, Gilbert LC (1985) Amaranth: food research and development. Summary report. Rodale Research Center, Kutztown, PA

Stone LA, Lorenz K (1984) The starch of amaranthus – physicochemical properties and functional characteristics. Starch/Stärke 36:232–237

Sugimoto Y, Yamada S, Sakamoto S, Fuwa H (1981) Some properties of normal- and waxy-type starches of *Amaranthus hypochondriacus* L. Starch/Stärke 33:112–116

Sutton KH, Mouat CL (1990) Determination of damaged starch in wheat flours by reversed-phase high-performance liquid chromatography. J Cereal Sci 11:235–242

Swinkels JJM (1985) Sources of starch, its chemistry and physics. In: van Beynum GMA, Roels JA (eds) Starch conversion technology. Marcel Dekker, New York, p 15

Teutonico RA, Knorr D (1985) Amaranth: composition, properties, and applications of a rediscovered food crop. Food Technol 39(4):49–61

Tomita Y, Sugimoto Y, Sakamoto S, Fuwa H (1981) Some properties of starches of grain amaranths and several millets. J Nutr Sci Vitaminol 27:471–483

Tudor F, George B (1985) Production of aflatoxins on amaranth seeds by *Aspergillus* species. Food Chem 17:33–39

Vargas-López JM, Paredes-López O, Espitia E (1990) Evaluation of lime heat treatment on some physicochemical properties of amaranth flour by response surface methodology. Cereal Chem 67:417–421

Varriano-Marston E (1983) Polarization microscopy: application in cereal science. In: Bechtel DB (ed) New frontiers in food microstructure. American Association of Cereal Chemists, St Paul, MN, p 71

Wagoner-Hass P (1983) Amaranth density report. Rodale Research Center, Kutztown PA

Watson SA (1964a) Corn starch. In: Whistler RL (ed) Methods in carbohydrate chemistry, vol 4. Academic Press, New York, p 3

Watson SA (1964b) Determination of starch gelatinization temperature. In: Whistler RL (ed) Methods in carbohydrate chemistry, vol 4. Academic Press, New York, p 240

Wolf MJ, MacMasters MM, Rist CE (1950) Some characteristics of the starches of three South American seeds used for food. Cereal Chem 27:219–222

Wootton M, Bamunuarachchi A (1980) Application of differential scanning calorimetry to starch gelatinization. III. Effect of sucrose and sodium chloride. Starch/Stärke 32:126–131

Wright DJ (1984) Thermoanalytical methods in food research. In: Chan HWS (ed) Biophysical methods in food research. Blackwell Scientific Publ, Oxford, p 1

Yañez GA, Messinger JK, Walker CE, Rupnow JH (1986) *Amaranthus hypochondriacus*: starch isolation and partial characterization. Cereal Chem 63:273–276

YSI (1989) YSI model 27 industrial analyzer instruction manual. Yellow Springs Instrument Co., Yellow Springs, OH

Zobel HF, Senti FR, Brown DS (1965) Studies of starch gelatinization by differential thermal analysis. Cereal Sci Today 10:154

Glycolipid Analysis in Wheat Grains

N.G. LARSEN

Introduction

Cereal lipids are broadly classified according to the grain tissues (germ, aleurone, endosperm) from which they are extracted and, like proteins, their extraction behaviour in solvents. For example, endosperm glycolipids can be extracted from outside ("non-starch" glycolipids) or inside ("starch" glycolipids) the starch granules (Chung 1986).

Wheat glycolipids have been studied by cereal chemists and plant breeders over the last 30 years because these lipids improve bread and cookie quality. A role in pasta has also been implicated. Detailed studies have appeared from cereal laboratories in many countries, relating wheat or flour glycolipid content to bread quality. The main influence has been the prospect that flours with more glycolipids will bake better. If ways to genetically control flour glycolipid levels can be found, then plant breeders will be able to improve the value of their wheats.

Much of the information on wheat glycolipids has been published in studies of wheat lipids in general. Hence, a number of modern reviews and books contain information on glycolipids. See for example, Morrison (1978, 1988a, 1989), Barnes (1987), Pomeranz (1988) and Chung (1989). Christie (1982) has published an extensive guide to the analysis of lipids in both plant and animal tissues. This review attempts to collate and discuss methods specific to wheat glycolipid analysis. It also briefly looks at defining glycolipids, where they are found in the grain, and their importance in wheat products.

1.1 Defining Glycolipids

"Glycolipid" describes all compounds comprising a saccharide (sugar) joined through a glycosyl linkage to a lipid. The nomenclature is described by Lundberg (1984), and is based on the parent lipid. The parent lipid name is generic (glycerol, ceramide, etc.), and complex carbohydrates associated with them have been given trivial names for simplicity. Based on the *parent*, the wheat glycolipids are glycosyl*glycerides*, glycosyl*sterols*, glycosyl*ceramides*, glycosyl-*phosphoceramides* and acylglycosyl*diols* (Morrison 1983). Glycosyglycerides predominate. The main sugar is galactose and the major glycosylglycerides are monogalactosyldiacylglycerol and digalactosyldiacylglycerol (Morrison 1978).

Petroleum solvents (e.g. hexane) extract "free" non-starch lipids, including glycolipids. Following petroleum extraction, solvents such as cold water-saturated 1-butanol extract "bound" non-starch lipids. Free and bound lipids are further subclassified as "neutral" if they are eluted by chloroform from silicic acid, or "polar" when eluted by methanol after chloroform.

Glycolipids comprise 65% – 70% of the free non-starch polar wheat lipids, and 45% – 50% of the bound non-starch polar lipids (Chung 1986). They are a minor component (see Sect. 1.2) of starch lipids.

1.2 Glycolipid Distribution in Wheat Grain Tissues

At a subcellular level, wheat glycolipids are found in plastid membranes of amyloplasts, chloroplasts and etioplasts (Morrison 1988a). They are structural lipids (Morrison 1989) with aqueous phase behaviour that is relevant to their biological and technological activity (Barnes 1987).

Hargin and Morrison (1980) have analyzed whole wheat and wheat grain tissues in four varieties. Whole wheat lipids comprised 8% – 14.4% glycolipids, compared to 30.6% – 41.5% phospholipids and 44.4% – 56.9% neutral lipids. Most glycolipids were in the non-starch endosperm lipid extracts (20.4% – 38.3% glycolipids), followed by aleurone (2.2% – 9.8%), starch (1.2% – 6.7%), and germ (0% – 3.5%) lipid extracts.

1.3 Technological Importance of Wheat Glycolipids

Fractionation studies show that free wheat glycolipids improve the loaf volumes from defatted flours (Daftary et al. 1968; Lin et al. 1974b; MacRitchie 1977). Synthetic glycolipids act similarly (Pomeranz 1980). This is supported by positive correlations between loaf volume and free lipid carbohydrate (Chung et al. 1982) or free glycolipid concentration (Bekes et al. 1986; Panozzo et al. 1990). In the United Kingdom, there was no correlation between loaf volume and free glycolipids in 1983 and 1984 (Bell et al. 1987), and in New Zealand negative correlations were reported (Larsen et al. 1986, 1989b). Indeed, flour reconstitution confirmed the beneficial effect of glycolipids on loaf volume, despite these negative correlations (Larsen et al. 1989b). This emphasized the danger of interpreting correlations (Larsen et al. 1990), as sometimes the lipids may only be acting as markers for other effects (Morrison 1989).

A number of authors have reviewed possible mechanisms for the bread-improving effects of free glycolipids (Larsson 1983; MacRitchie 1983; Chung 1986; Barnes 1987). The two current schools of thought put forward binding to proteins (Chung 1986) or the interfacial and phase behaviour of glycolipids (Larsson 1983; Barnes 1987) as explanations.

Cookie quality of hexane defatted flours was restored by adding lipid fractions rich in digalactosyldiacylglycerols (DGDG). In contrast, monogalactosyldiacylglycerols (MGDG) were ineffective (Clements and Donelson 1981).

Lin et al. (1974b) showed that MGDG increased spaghetti firmness while DGDG decreased it.

2 Glycolipid Extraction from Wheat Tissues

Glycolipid analyses are governed by the aim of the study and should be carefully planned. To study the relationship between flour glycolipids and loaf volume, the analyst should not measure the total glycolipid content without first extracting the free non-starch glycolipids. At present there is no in situ measure of glycolipids, although near infrared reflectance has shown promise (Simmons and McDonald 1985; Simmons 1986). Hence, they are solvent-extracted with other lipids from tissues.

Soxhlet extraction is convenient for free non-starch lipids. A single solvent (e.g. light petroleum, hexane) is used (Chung et al. 1982; Larsen et al. 1989a, b; Panozzo et al. 1990). Slurry extractions work well with binary solvents (e.g. water-saturated 1-butanol) for bound non-starch lipids (Bekes et al. 1986) or starch lipids (Morrison 1988b), and large-scale (>0.5 kg) extractions. However, Clements (1977) has extracted flour lipids form 6-kg samples using a Soxhlet apparatus.

Assessing the completeness of solvent extraction, by acid hydrolysis, has been discussed by Morrison (1978). For many studies complete lipid extraction is not vitally important. For example, I have found that with light petroleum (40−60°C boiling point, bp) Soxhlet extractions of 10 g flour samples, over 90% of free lipids are extracted within 15 min. As found by Panozzo et al. (1990), 1-h extraction was sufficient for repeatable results (Larsen et al. 1989a, b) compared to 16 h used by others (Chung et al. 1982; Bell et al. 1987). Longer extraction times will be needed for larger samples.

An alternative to organic solvents is supercritical fluid CO_2 extraction (Hamman et al. 1987). An important future application might be extracting residue-free glycolipids for re-use in food processing.

2.1 Effect of Solvent on Glycolipid Extraction

2.1.1 Choice of Solvent

This, as well as extraction conditions, is important and governs whether free non-starch, bound non-starch or starch glycolipids are extracted. Without prior extraction, hot water-saturated 1-butanol will remove all three classes of glycolipid together. If glycolipids are going to be analyzed colorimetrically, polar solvents such as water-saturated 1-butanol should be avoided due to the risk of color interference from soluble, non-lipid carbohydrate (Larsen et al. 1990).

Chung et al. (1980) studied a range of solvent polarities and found that free lipids extracted using non-polar solvents like hexane best differentiated wheats varying in bread-making quality. A second study confirmed this (Chung et al. 1982). Despite varying results in other countries (Bekes et al. 1986; Bell et al. 1987; Larsen et al. 1989b; Panozzo et al. 1990), for studies of wheat quality the free glycolipids should be analyzed. Genetic studies by Morrison et al. (1984, 1989) support this view.

2.1.2 Solvent Purity

It is good analytical practice to use clean solvents. For someone new to lipid analysis, there are pitfalls to avoid.

Lipid isolation usually involves evaporating the solvent and drying the residues. This can be done under vacuum (Larsen et al. 1989a) or under a stream of dry nitrogen (Bekes et al. 1986). My experience has been that up to 20 mg of the apparent lipids may be non-volatiles from 500 ml of analytical grade light petroleum (Larsen, unpubl.). Therefore, all solvents should be carefully redistilled and minimum solvent volumes used, particularly for gravimetric glycolipid analysis.

2.2 Other Effects on Glycolipid Extraction

In general, more DGDGs are extracted, with polar and non-polar solvents, as flour moisture increases (Chung et al. 1984). At flour moistures of 1.2%, 7.2% and 13.8%, hexane extracted 8.5, 8.5 and 11.0 mg DGDG respectively. However, Chung et al. (1984) concluded that solvent affected extractability more than flour moisture. While Larsen et al. (1989b) found no moisture effect, this was probably due to the narrow range of moistures (13.2% – 14.6%; 12.3% – 13.7%) and wide range of free glycolipids (0.38 – 0.63 g; 0.39 – 0.57 g galactose per kg flour) in the two sample sets.

Sample preparation also affects glycolipid analyses. Morrison et al. (1980) found that bran and aleurone-rich samples need fine grinding. Larsen et al. (1989a) compared Brabender Junior Quadrumat and Buhler MLU-202 flours milled from the same wheat. The Brabender mill flour had 0.64 g/kg lipid galactose, compared to 0.54 g/kg in the Buhler flour. Commercial flour from the same wheat had 0.58 g/kg lipid galactose. Wheat feedrate on to the first break rolls of each laboratory mill had no effect on flour glycolipid concentration (Larsen et al. 1989a). Whole meal is expected to have less glycolipid than flour (Larsen et al. 1989b) since glycolipid concentrations are highest in endosperm (Hargin and Morrison 1980). Morrison et al. (1982) found that the proportions of glycolipids were almost constant in the 23 mill streams of a commercial mill.

3 Methods for Glycolipid Analysis

In the history of wheat lipid studies, there appears to have been three phases to the development and application of glycolipid analyses:

Glycolipid Characterization. This coincided with early attempts to identify wheat lipids by paper, thin layer or column chromatography. Carter et al. (1956) first isolated crude glycolipid mixtures and identified MGDG and DGDG. Later, they established the structural nature of these two lipids (two fatty acids per molecule) and fatty acid compositions (Carter et al. 1961a, b). As thin-layer chromatography (TLC) became established, identification of glycolipids in wheat lipid extracts turned routine (Pomeranz 1965). Gas chromatography (GC) was applied to analyzing glycolipid fatty acid composition (Carter et al. 1961a; McKillican 1964; McKillican and Sims 1964).

Glycolipid Functionality. This phase was the establishment of the functional involvement of glycolipids in baked products, particularly bread. Glycolipids needed to be purified in sufficient quantities for test baking. Hence, Daftary et al. (1968) showed the bread loaf volume restoration properties of glycolipids. Lin et al. (1974b) concluded that DGDG was the most effective glycolipid for restoring the bread loaf volume of defatted flours.

Glycolipid Quantification. This followed from knowledge that glycolipids were functionally important. Chung et al. (1980) showed that free lipids best differentiated varying wheat quality. Hence, a number of studies ensued. These looked at the glycolipid contents of wheats in several countries. The glycolipids were analyzed by column liquid chromatography (Bekes et al. 1986; Panozzo et al. 1990) and colorimetry of lipid sugars (Chung et al. 1982; Bell et al. 1987; Larsen et al. 1989b). During this time, HPLC was first applied to wheat glycolipid analysis (Tweeten et al. 1981; Christie and Morrison 1988; Walker 1988).

3.1 Thin-Layer Chromatography

Cereal chemists have been using thin-layer chromatography (TLC) for about 30 years. It is still the best way to quickly identify lipids in their native state (Morrison 1978). Morrison et al. (1980) have written a concise and clear description of methods used in Morrison's laboratory. However, despite these advances, unequivocal identification can only be made using multiple analytical and physical tests (Morrison 1978), of which TLC is one component.

3.1.1 Lipid Sample Preparation

Some workers fractionated lipids by column liquid chromatography (McKillican 1964; McKillican and Sims 1964; Pomeranz et al. 1966; Lin et al.

1974a; Chung et al. 1977; Berger 1983) or solvent extractions (Lin et al. 1974b) prior to TLC analysis. For quantitative work the complexity of these prefractionations is a big disadvantage. Hence, total lipids are more easily separated into neutral and polar fractions by preparative TLC using added esterified steryl glucoside as an internal standard (Lin et al. 1974a). The plates are eluted with 100:1.5 v/v diethyl ether-acetic acid. Bands above the standard are neutral, and remaining bands including the standard are the polar glyco- and phospholipids. The glycolipids can be removed for further analysis.

With the methods used by Morrison et al. (1980), prior fractionation of lipid samples is not necessary. Neither is it necessary for qualitative TLC.

3.1.2 TLC Plates and Coatings

Most laboratories use 20×20 cm glass plates coated with $0.25 - 0.3$ mm silica gel G (McKillican 1964; Daniels et al. 1966; Pomeranz et al. 1966; Lin et al. 1974b; Chung et al. 1977; Morrison et al. 1980). Silica gel H is an alternative (Morrison et al. 1980; Christie 1982). For preparative TLC, coatings of $0.75 - 1$ mm are successful (Lin et al. 1974a; Clements and Donelson 1981; Berger 1983). In all cases the plates are activated before use by heating them in an oven. Morrison et al. (1980) recommend $110 \, ^\circ C$ for at least 1 h.

I have found 20×20 cm aluminium TLC plates (silica gel 60) convenient (Larsen, unpub.) because each can be cut into 18 small plates (about 3×6.5 cm) for monitoring glycolipid fractions coming off columns. These plates corrode if treated with acidic eluents or sprays.

3.1.3 Eluents for Glycolipid TLC

Combinations of chloroform, methanol and water (Daniels et al. 1966; Pomeranz et al. 1966; Graveland 1968; Youngs et al. 1970; Lin et al. 1974a, b; Chung et al. 1977) have been used, usually in the ratio 65:25:4 v/v/v. With this combination MGDG is found near the top of the plate (Pomeranz et al. 1966; Chung et al. 1977). If neutral lipids are present, there may be inadequate separation. Daniels et al. (1966) used 80:25:2 chloroform-methanol-water. This has the advantage of decreasing the mobility of MGDG relative to the neutral lipids, and keeps phospholipids nearer the origin. Morrison et al. (1980) caution that while more chloroform gives better glycolipid separation, contamination with phospholipids occurs. Instead, Morrison et al. (1980) favour initial development to 15 cm with 10:90:2:3 chloroform-acetone-acetic acid-water. Acetone retards the migration of phospholipids relative to glycolipids (Christie 1982). The plate is then developed to 18 cm with 99:1 diethyl ether-acetic acid to take the neutral lipids clear of MGDG (Morrison et al. 1980). The neutral lipids are found in two bands at the 18-cm solvent front, and phospholipids near the origin. There is some overlap between phosphatidylethanolamine (if present) and digalactosylmonoacylglycerol (Morrison et al. 1980).

For routine TLC monitoring of column separations, I have found that 90:20 chloroform-methanol is convenient and effective (Larsen, unpubl.).

3.1.4 Visualization of Glycolipids

After TLC separations, the cereal chemist has to find the colorless lipids on the plates and identify the glycolipids.

A common visualization method is the absorption of iodine vapour onto the plate (Christie 1982). This can be done prior to further lipid manipulation (Pomeranz 1965; Graveland 1968; Lin et al. 1974a). Iodine-developed plates can be photocopied for permanent record (Larsen, unpubl.). A light spray with water will reveal hydrophobic zones which can be scraped from the plate for analysis (Lin et al. 1974a; Clements and Donelson 1981).

Methanol or ethanol sprays of fluorescein (Williams et al. 1975) or 2,7-dichlorofluorescein (McKillican and Sims 1964; Daniels et al. 1966; Morrison et al. 1980) reveal lipids. Daniels et al. (1966) viewed the sprayed plates under ultraviolet light, but Morrison et al. (1980) did not favour this because impurities fluoresce. Instead, after drying the sprayed plates, a second spray with 8 M ammonia enhanced the pink lipid bands.

Finally, the lipid bands can be revealed by charring, that is, heating the TLC plates after a spray with acid-based reagents. Reagents include 3% cupric acetate in 8% phosphoric acid (Lin et al. 1974b; Clements and Donelson 1981), 50% sulphuric acid (Daniels et al. 1966; Graveland 1968; Youngs et al. 1970; Lin et al. 1974a) and potassium dichromate in sulphuric acid (Hoseney et al. 1969; Chung et al. 1977, 1984).

The next step is to identify the bands that are glycolipids. For this, a glycolipid-specific spray reagent is best. If two samples are spotted in parallel and then separated by TLC, simple masking can be used to reveal all the lipids in one lane and only the glycolipids in the parallel lane (McKillican 1964). Filter paper works quite well as a mask (Larsen, unpubl.). Various spray reagents have been used to identify glycolipids. Some of these reagents are not very specific. For example, Dragendorff reagent (Pomeranz 1965; Graveland 1968), a mixture of bismuth subnitrate, potassium iodide, acetic acid and water (Gordon and Ford 1972), also identifies choline-containing phospholipids (McKillican 1964; Pomeranz 1965; Pomeranz et al. 1966; Youngs et al. 1970). Glycolipid-specific reagents include 20% perchloric acid (Lin et al. 1974a, b), diphenylamine (Youngs et al. 1970; Christie 1982); α-naphthol (Chung et al. 1977; Christie 1982), and orcinol-sulphuric acid (Clements and Donelson 1981, Christie 1982). Orcinol-sulphuric acid is easily made (200 mg orcinol in 100 ml 75% sulphuric acid) and has the advantage over α-naphthol of being a single-spray process. However, orcinol-sulphuric acid is not stable and will keep for about 1 week when refrigerated in the dark (Christie 1982). Christie (1982) recommends heating the orcinol-sprayed plates at 100°C for 10–15 min. The glycolipids turn blue-violet. When using aluminium-backed TLC plates, 3-min heating gave a better result (Larsen, unpubl.).

3.1.5 Quantifying Glycolipids Separated by TLC

The two basic procedures have been densitometry after charring (Graveland 1968; Youngs et al. 1970; Chung et al. 1984) and measurement of fatty acid methyl esters by gas chromatography.

Densitometry requires calibration standards and considerable skill (Morrison 1978). How much each lipid chars relates to its %-carbon composition (Graveland 1968), so calibration standards, which are not always available (Morrison 1978; Chung et al. 1984), should be chosen carefully. Chung et al. (1984) used a least square regression from standard lipids for converting band density to lipid weight.

Gravimetric analysis, by extracting the lipids from the TLC silica gel, is not recommended due to the risk of low lipid recovery (Morrison 1978). For example, Lin et al. (1974a) calculated recoveries of 85% to 97%.

The most used and reliable method is measurement of fatty acid methyl esters by gas chromatography (see Sect. 3.2).

3.2 Gas Chromatography

All aspects of gas chromatography (GC) of lipids, from theory to practical, are specifically addressed by Christie (1982, 1989) in two books. GC for quantifying wheat glycolipids is best done after TLC separations (McKillican 1964; Daniels et al. 1966; Graveland 1968; Lin et al. 1974a; Morrison et al. 1980). Glycolipids can be measured as their trimethylsilyl ether derivatives or trimethylsilyl derivatives after full or partial deacylation (Williams et al. 1975). Morrison (1978) cautions that after full deacylation, mono- and diacylated glycosyl-glycerides cannot be differentiated. However, to date, most GC analyses have been on fatty acid methyl esters (FAME) obtained by acidic methanolysis of the glycolipids. The main fatty acids found are 16:0, 16:1, 18:0, 18:1, 18:2 and 18:3 (Morrison 1978).

3.2.1 GC of FAME

Methanolysis in the presence of TLC silica gel is preferable if oxidation losses are to be avoided (Morrison et al. 1975). This is done by transferring the glycolipid band, visualized with 2,7-dichlorofluorescein, from the TLC plate to the methanolysis tubes (Morrison et al. 1980). Oxidation of unsaturated fatty acids is also minimized by manipulating the lipids under nitrogen. Daniels et al. (1966) added traces of quinol to prevent oxidation.

Methanolysis can be done by reacting fatty acids with diazomethane (Carter et al. 1961a). However, much safer reagents are 14% (w/v) boron trifluoride in methanol (Graveland 1968; Morrison et al. 1980; Tweeten et al. 1981), methanol with 10% sulphuric acid (Daniels et al. 1966; Williams et al. 1975) or methanol saturated with hydrochloric acid (McKillian 1964; Kates 1964). In

Christie's opinion (Christie 1982), methanol-hydrochloric acid is generally the best esterifying reagent. Boron trifluoride-methanol needs to be refrigerated, and its effectiveness is age- and concentration-dependent. However, the 20-min reaction times for glycolipids in boron trifluoride-methanol at 100 °C (Morrison et al. 1980) are short compared with $1-3$ h for methanol-sulphuric and methanol-hydrochloric acid (Kates 1964; Daniels et al. 1966; Williams et al. 1975).

After esterification, the FAME are extracted into light petroleum (Kates 1964; Daniels et al. 1966), heptane (Graveland 1968) or pentane (Morrison et al. 1975, 1980) and an aliquot is injected into the gas chromatograph.

Heptadecanoic acid (17 : 0) is the preferred internal standard as it is not usually found in cereal lipids (Daniels et al. 1966; Morrison 1978). Daniels et al. (1966) applied a known weight of heptadecanoic acid to each of the lipid bands separated by TLC, before removing them for methanolysis and GC. Williams et al. (1975) added methyl heptadecanoate just prior to injection in their study of plant galactolipids, whereas Morrison et al. (1980) added this ester to the methanolysis tubes before esterifying the cereal glycolipids.

For GC of glycolipid FAME a conventional instrument, with a flame ionization detector, is used (Morrison et al. 1975, 1980). Columns are about 2 m long and have a 3 mm diameter (Graveland 1968; Lin et al. 1974a; Morrison et al. 1975, 1980; Tweeten et al. 1981), though shorter columns have been used (McKillican 1964; Daniels et al. 1966). Columns are operated at $185-195$ °C using nitrogen or argon as the carrier gas (McKillican 1964; Daniels et al. 1966; Graveland 1968; Lin et al. 1974a; Morrison et al. 1975, 1980; Tweeten et al. 1981), and will last several months before resolution deteriorates.

Polar polyester liquid phases on GC columns packed with inert silanized support material ($100-120$ mesh) are suited to separating wheat glycolipid fatty acids (Morrison et al. 1975, 1980; Christie 1982). Common liquid phases include methylsilicone copolymers of polymeric ethylene glycol succinate. Retention times vary with liquid phase concentration, which is usually 15% (Christie 1982).

Early workers relied on measuring fatty acid peak areas by hand. There is no need to do this now with electronic integrators, but overlapping peaks still present problems. If this arises, Christie (1982) suggests altering column conditions or trying a different liquid phase. Equipment is calibrated with standard methyl ester mixtures (Graveland 1968; Morrison et al. 1980) from which correction factors can be derived for each of the fatty acids analyzed (Morrison et al. 1980; Christie 1982). Then a conversion factor for each type of glycolipid is multiplied by the weight of the FAME to give the weight of original glycolipid (Morrison 1978; Morrison et al. 1975, 1980). If digalactosylmonoacyl-glycerol and phosphatidylethanolamine have not been resolved by TLC, the phospholipid has to be accounted for by a phosphorus analysis (Morrison et al. 1980).

3.2.2 GC of Glycolipid Derivatives

Little has been done to measure derivatized wheat glycolipids directly by GC. Williams et al. (1975), using linear temperature programming, resolved trimethylsilyl ether derivatives of broad bean galactolipids. The method works for wheat glycolipids (Larsen and Winter, unpub.) Wheat DGDG were reacted with hexamethyldisilazine-trimethylchlorosilane-pyridine (Williams et al. 1975). The derivatives were not stable and had to be prepared just before GC analysis. There were eight peaks in a typical run (Larsen and Winter unpubl.). However, none was identified and in view of Tweeten et al. (1981) reporting three DGDG isomers, more work needs to be done.

3.3 Column Liquid Chromatography

In this section, column liquid chromatography refers to low resolution preparative scale separations. High resolution analytical scale high performance liquid chromatography is discussed later (see Sect. 3.5). Nevertheless, column liquid chromatography has been used for gravimetric analysis of wheat glycolipids (Bekes et al. 1986; Bell et al. 1987; Panozo et al. 1990).

Column liquid chromatography has the advantage of preparative scale (Christie 1982). As such, the method was first used for isolating wheat flour MGDG and DGDG for structural characterization (Carter et al. 1961a, b). Later, Fujino and Ohnishi (1983) isolated pure glycosylceramides for identification. But nobody has used column liquid chromatography to purify wheat glycolipids for functional studies; batch separations (Daftary et al. 1968; Lin et al. 1974b; MacRitchie 1977) or preparative TLC (Clements and Donelson 1981) being preferred. Indeed, liquid chromatography columns need careful preparation, and separations are time-consuming, use a lot of solvent and risk lipid degradation (Morrison 1978; Christie 1982).

3.3.1 Chromatography Media

Pure silicic acid, also called silica or silica gel, is the most used separation medium for wheat lipids (Carter et al. 1961a; McKillican 1964; Pomeranz et al. 1966; Daftary et al. 1968; Hoseney et al. 1969; Chung et al. 1977, 1980, 1982, 1984; Bekes et al. 1983, Bell et al. 1987; Larsen et al. 1989a, Panozzo et al. 1990). Other media include "Florisil" (Bekes et al. 1986) and diethylaminoethyl (DEAE) cellulose (Lin et al. 1974a; Berger 1983). Florisil is a tradename for silicic acid co-precipitated with magnesia (Christie 1982). It has a higher solvent flow rate than silicic acid, but magnesium silicate may elute with polar solvents (Christie 1982).

3.3.2 Column Preparation

Preparation governs the success of glycolipid separations. Important factors include the column size, adsorption medium particle size, and cleaning and activating the adsorption medium.

Column size is governed by the internal diameter of the glass support, and the amount of adsorption medium. This in turn may depend on what column performance is required. With silicic acid, approximately 20-mm internal diameter glass supports (McKillican 1964; Pomeranz et al. 1966) are quite adequate (Larsen, unpubl.). The column is loaded with about 20 g of silicic acid to separate up to 200 mg of lipids (McKillican 1964; Pomeranz et al. 1966). Bekes et al. (1986) used 4-mm diameter columns, loaded with Florisil to a depth of 120 mm, to separate glycolipids from about 50 mg wheat-flour lipid extracts.

Particle size determines solvent flow rates and resolution and therefore a compromise is sought (Christie 1982). Precipitated silicic acid is a very fine powder. Hence, the solvent needs to be forced through the column under nitrogen (Larsen et al. 1989a) to achieve an adequate flow rate. Cereal chemists appear to agree on media particle sizes of about 100 mesh (Carter et al. 1961a; Chung et al. 1984; Bekes et al. 1986; Panozzo et al. 1990) for wheat lipid, including glycolipid, separations. For plant lipid analyses, in general, a finer particle size (200 mesh) has been commonly used (Christie 1982).

Before use, batches of the separation medium are activated (Carter et al. 1961a; Pomeranz et al. 1966; Christie 1982), then cleaned with solvents that represent the range of polarity to be put through the column (Carter et al. 1961a; Pomeranz et al. 1966). Activation controls lipid binding and, therefore, resolution. Batches of silicic acid are activated by drying them in an oven. Pomeranz et al. (1966) dried silicic acid (20 g) at 120°C for 4 h, whereas the larger batches (1 kg) used by Carter et al. (1961a) were dried at 110°C for 12 h. The silicic acid was then washed with 7:1 and 15:1 chloroform-methanol followed by chloroform (Pomeranz et al. 1966), or methanol and chloroform (Carter et al. 1961a). Christie (1982) suggests drying at 110°C or using acetone to dehydrate silicic acid, and adding water to get about 5% hydration. Either way, the separation medium is then loaded onto the column as a slurry in the solvent that will be used first for eluting the lipids (Carter et al. 1961a; Pomeranz et al. 1966; Christie 1982).

In my experience with separating glycolipids, columns should be prepared the day before they are run (Larsen, unpubl.).

3.3.3 Running the Column

The lipid is carefully loaded, in an absolute minimum of solvent, onto the top of the prepared column (Christie 1982). The elution sequence is started only when all the lipid has been adsorbed onto the separation medium.

Chloroform and then methanol are used to separate lipids into neutral and polar (glyco- and phospholipids) fractions (Pomeranz et al. 1966). However, if acetone is used after the chloroform, the glycolipids can be preferentially isolated as a class and their residues weighed (Bekes et al. 1983; Bell et al. 1987; Panozzo et al. 1990). Progress at all stages should be monitored by TLC (Pomeranz et al. 1966; Panozzo et al. 1990). Where gravimetric analysis has not been a concern, others have run solvent gradients to get several lipid fractions that progress from neutral to polar (Carter et al. 1961a; McKillican 1964). Each fraction is anaylzed by TLC to identify the lipids present. Fractions of similar purity can be combined. Some fractions will be rich in glycolipids, or even be pure glycolipid species. For example, I have used a modified method of Carter et al. (1961a) to obtain over 100 mg of nearly pure DGDG from a one-column silica gel (60–120 mesh) separation of 630 mg of polar wheat lipids (Larsen, unpubl.). However, the separations took 2–3 days to complete. The modifcation comprised eluting the silica gel-hexane column first with n-hexane (300 ml), then 1:1 (200 ml) and 80:20 (1000 ml) chloroform-hexane. Abrupt changes in solvent can cause columns to crack. The glycolipids were then eluted with the chloroform-methanol sequences of Carter et al. (1961a). MGDG, seldom obtained as pure fractions, were eluted by a chloroform-hexane (90:10 v/v), chloroform and 98:2 chloroform-methanol sequence. DGDG came off the column in the 98:2, 96:4, 90:10 and 90:20 chloroform-methanol sequence. The latter fractions were mostly pure DGDG. The results varied slightly from column to column (Larsen, unpubl.) but usually agreed with those to Carter et al. (1961a).

3.4 Colorimetric Analysis

Colorimetric analysis is used to quantify wheat glycolipids by reacting the lipid sugar with a sugar-specific reagent. The absorbance of the colored solution is measured and is proportional to the amount of glycolipid. Calibration is against standard sugar solutions. The results are reported as the weight of sugar equivalents (Lin et al. 1974a; Chung et al. 1980, 1982; Larsen et al. 1989a, b) or multiplied by a conversion factor to an equivalent glycolipid weight (Bell et al. 1987). Three studies have used colorimetric analyses of wheat glycolipids and related them to flour quality (Chung et al. 1980, 1982; Bell et al. 1987; Larsen et al. 1989b).

Two sugar-specific reagents have been used. These are anthrone-sulphuric acid (Fujino and Ohnishi 1983; Bell et al. 1987; Larsen et al. 1989a, b) and phenol-sulphuric acid (Lin et al. 1974a; Chung et al. 1980, 1982). Both work in the presence of TLC adsorbents (Yamamoto and Rouser 1970; Morrison 1978). However, lipids extracted with polar solvents such as water-saturated butanol should not be analyzed colorimetrically without further purification (e.g. TLC). Water-soluble carbohydrates are likely to give false results (Larsen et al. 1990).

3.4.1 Phenol-Sulphuric Acid

A phenol-sulphuric acid method was developed by Dubois et al. (1956) for measuring simple sugars. Later, it was adapted to measure the sugars liberated by hydrolysis of glycolipids (Kates 1964; Kushwaha and Kates 1981). However, its major disadvantage is interference by free sterols (Lin et al. 1974a).

Wheat lipids containing glycolipids are hydrolyzed with methanol-hydrochloric acid (Lin et al. 1974a; Chung et al. 1982) or sulphuric acid (Chung et al. 1980). With methanol-hydrochloric acid, the released sugars are extracted into an aqueous phase to which phenol and sulphuric acid are added. The absorbance is read at 490 nm (Chung et al. 1982). After hydrolysis with sulphuric acid, aqueous-phase sugar extraction is not needed (Chung et al. 1980).

In Kates' (1964) early method (Chung et al. 1982), methanol-hydrochloric acid hydrolysis takes 2 h. The phase extraction method needs to be followed carefully to avoid emulsions forming. In contrast, sulphuric acid hydrolysis (Chung et al. 1980) takes 15 h and risks sugar decomposition.

Kushwaha and Kates (1981) developed a more precise and simple phenol-sulphuric acid method for bacterial glycolipid sugars. Hydrolysis and color development take 5 min. After 30-min cooling, the solution absorbance is read. Nobody has yet applied this method to wheat glycolipids.

3.4.2 Anthrone-Sulphuric Acid

Anthrone sugar determinations (Scott and Melvin 1953) predate phenol-sulphuric acid. The disadvantages of anthrone were cited by Dubois et al. (1956). These included its instability, expense and limited use for pentoses, and interference by phenol from paper chromatography of sugars. Nevertheless, for measuring wheat glycolipid sugars, anthrone-sulphuric acid is successful (Fujino and Ohnishi 1983; Bell et al. 1987; Larsen et al. 1989a, b).

Yamamoto and Rouser's (1970) method for glycosphingolipids is quick (about 30 min) and direct. Glycolipid sugars were quantified for individual lipids after TLC separation (Yamamoto and Rouser 1970). Larsen et al. (1989a, b) applied this method to measure total glycolipid sugars, as galactose equivalents, in wheat lipids extracted by petroleum ether (bp 40–60°C). The procedure gives direct color development without extracting the hydrolyzed sugars. In contrast, Bell et al. (1987) applied on anthrone method to wheat glycolipid sugars extracted after lipid hydrolysis.

Yamamoto and Rouser (1970) give a maximum refrigerated shelf-life of 3 weeks for stock anthrone reagent (2% anthrone in 98% sulphuric acid). The anthrone method's standard deviation for the samples analyzed by Larsen et al. (1989b) was 0.03 g lipid galactose per kg flour, compared to 0.01 g/kg for phenol-sulphuric acid (Chung et al. 1982). In agreement with Scott and Melvin (1953), anthrone reagent is very sensitive to extraneous carbohydrate such as from dust and flour particles, and glassware residues (Larsen, unpubl.). This was accounted for by running blanks, several of which can be conveniently tested at the same time. The method's long-term reliability was confirmed by repeated analyses of standard solutions (Larsen and Humphrey-Taylor, unpubl.).

3.4.3 Standard Solutions

To analyze the major wheat glycolipids, galactose standard solutions are prepared (Chung et al. 1980, 1982; Bell et al. 1987; Larsen et al. 1989a, b). The presence of sugars other than galactose would not affect the results for two reasons. Firstly, galactose is the major wheat glycolipid sugar (Morrison 1978). Secondly, with anthrone the common lipid hexose sugars have similar extinction coefficients (Christie 1982). For wheat sphingolipids, Fujino and Ohnishi (1983) used glucose standards.

Standard solutions of sugars should be preserved against microbial fermentation. Scott and Melvin (1953) and Larsen et al. (1989a, b) used phenyl mercuric acetate (2 mg per 100 ml of standard solution) as a preservative.

3.5 High Performance Liquid Chromatography

The application of high performance liquid chromatography (HPLC) to the analysis of wheat glycolipids is comparatively recent. Indeed, only three studies have been reported (Tweeten et al. 1981; Christie and Morrison 1988; Walker 1988).

The development of lipid HPLC has been hampered by the lack of a universal lipid detection system (Christie 1985). Hence, the potential advantages of HPLC have yet to be fully utilized. These include easy quantification, less risk of decomposition and ease of collecting separated lipids for further analysis (Hamilton et al. 1987), and automated throughput.

Ten potential detection systems for the HPLC of lipids have been discussed by Christie (1987). In this review, I will only discuss the detection systems applied to wheat glycolipid analyses so far, that is, refractive index (Tweeten et al. 1981) ultraviolet (Walker 1988) and mass (Christie and Morrison 1988) detection.

3.5.1 Refractive Index Detection

The best separations of glycolipids by TLC (see Sect. 3.1) or column liquid chromatography (see Sect. 3.3) are achieved with solvent gradients. On the other hand, HPLC refractive index detection can only be used where the solvent composition stays constant, i.e. isocratic (Christie 1987; Hamilton et al. 1987). This also means that volatile components of isocratic solvents should be avoided (Christie 1987). The potential application of refractive index detection for wheat glycolipids is, therefore, limited.

Tweeten et al. (1981) quantified MGDG and DGDG, but no other glycolipids, from wheat and flour samples. Their method required prefractionation, by silicic acid colum liquid chromatography, of the free lipids into neutral and polar fractions. The polar lipids were then analyzed by HPLC, eluting with 88:12 methanol-water from an octadecylsilane derivatized column (10 µm particle size). Overall though, the method was not easy. They identified two DGDG and two MGDG peaks by HPLC separation of crude DGDG and MGDG preparations respectively. However, when applied to the polar lipid fraction the method did not resolve these four glycolipid peaks adequately. Hence, Tweeten et al. (1981) had to calculate empirical coefficients for quantifying the MGDG and DGDG in their samples. Standard MGDG and DGDG for calibrations were isolated from wheat flour by a combination of silicic acid column liquid chromatography and semi-preparative HPLC. Mass spectroscopy identified three MGDG and three DGDG isomers (see Sect. 3.6.3).

During this study, Tweeten et al. (1981) compared deflection and interferometric refractive index detectors. They concluded the latter type has better sensitivity and lower detection limits.

3.5.2 Ultraviolet Detection

Chromophores that strongly absorb ultraviolet (UV) light are rare in lipids (Christie 1987; Hamilton et al. 1987). Therefore, if UV detection is used on native lipids, reliance must be made on the weak absorption between 200 and 210 nm (Heemskerk et al. 1986; Christie 1987). For proper quantification the response factors and number of double bonds must be known for each lipid (Heemskerk et al. 1986). Also, solvents such as chloroform that absorb in this range cannot be used (Christie 1987) and oxidized lipids interfere (Christie and Morrison 1988).

Fortunately, for glycolipids these problems can be overcome by derivatizing them with benzoyl chloride in pyridine. Both N-acylation and O-acylation occur (Christie 1982). The benzolyated glycolipids strongly absorb UV and are readily separated. Walker (1988) appears to have been the only person to apply benzoylation to identifying and quantifying wheat glycolipids. Benzoylated MGDG and DGDG each eluted as one peak from a neutral silica gel (10 µm particle size) column held at 30 °C. The derivatives were formed by adding benzoyl chloride (0.1 ml) and pyridine (0.2 ml) to a 1/25th aliquot of the hexane-extracted lipids from 2.5 g of flour. Work-up comprised removing excess reagents and heptane extractions. The HPLC solvent was 200:1 heptane-2-propanol, and detection was at 254 nm. There was no overlap between the two principal glycolipid derivatives or resolution within them. Walker (1988) did not identify any other glycolipids.

Walker's method is simple and direct, and readily applicable to studies of the relationships between free glycolipid levels and the bread-making quality of flours. No prior separation of polar and neutral lipids is needed. Lipid recoveries, determined from standards, averaged 99% (90% – 109%).

3.5.3 Mass Detection

"Mass detection", for quantifying samples separated by HPLC, does not apply mass spectroscopy principles (Christie 1985). It is a method where light scattered by the solute passing through a light beam is related to solute concentration (Christie 1985, 1987). The detectors, also referred to as "evaporative analyzers" and "light-scattering detectors", have had little application to lipid analysis (Christie 1987). An early example was for measuring the main lipid classes in animal tissues (Christie 1985).

More recently, Christie and Morrison (1988) applied "mass detection" to cereal lipid analysis. They devised a stepwise elution scheme to separate glycolipids and phospholipids in a glycolipid-rich fraction from wheat flour. This fraction comprised MGDG, DGDG and their monoacyl analogues, phosphatidylethanolamine and lysophosphatidylethanolamine. The column was packed with 3 μm silica. It was first eluted with hexane-butan-2-one-acetic acid (35:65:0.4 v/v/v) for 8 min, then hexane-chloroform-isopropanol-aqueous buffer (42:5:45:3 v/v/v) for 15 min to resolve the glycolipids, and particularly digalactosylmonoacylglycerol from phosphatidylethanolamine. Recall that overlap between these two lipids is a problem for TLC (see Sect. 3.1.3) and GC (see Sect. 3.2.1). One further solvent gradient (32:5:50:8 v/v/v/v hexane-chloroform-isopropanol-aqueous buffer) separated the phospholipids. Using this method, Christie and Morrison (1988) separated and quantified all six lipids in their test samples. Reference lipids were isolated by TLC for HPLC peak identification.

The mass detection method for cereal glycolipids looks very promising. The lipids do not need derivatizing and, using a stream splitter (Christie 1987), they can be collected for further analysis. There is little baseline drift with changes in solvent (Christie 1987). This allows improved resolution using solvent gradients, which is not achievable with refractive index detectors (see Sect. 3.5.1). There is the bonus of resolving digalactosylmonoacylglycerol from phosphatidylethanolamine. However, further research is needed to look at problems with calibrating mass detectors for HPLC. Detector response is non-linear for glycolipids and separate calibrations are needed for each component (Christie and Morrison 1988). Examples of non-linear detector response are shown by Christie (1985).

3.6 Other Methods

Three methods that have had little application to wheat glycolipid analysis are near-infrared, nuclear magnetic resonance and mass spectroscopy. The latter two in particular have tremendous potential to increase our understanding of wheat glycolipid composition, structure and technical function.

3.6.1 Near-Infrared Spectroscopy

Already in wide use for routine analyses, near-infrared (NIR) spectroscopy would be useful for screening wheats where glycolipids have proven to be a useful predictor of baking potential, for example, to screen American hard red winter (Chung et al. 1982) and hard red spring wheats (Simmons 1986) or Canadian spring wheats (Bekes et al. 1986). In contrast, NIR could not be used for screening English (Bell et al. 1987), New Zealand (Larsen et al. 1986, 1989 b) or Australian (Panozzo et al. 1990) wheats.

Only one application of NIR to wheat glycolipid analysis has appeared (Simmons and McDonald 1985; Simmons 1986). Simmons first measured NIR absorption at 3375 cm^{-1} (glycolipid O-H stretch) but replicate results were inconsistent. Consequently, multiple-filter calibrations were used. Glycolipids in calibration samples were analyzed by the phenol-sulphuric acid method (Simmons and McDonald 1985).

3.6.2 Nuclear Magnetic Resonance

Nuclear magnetic resonance (NMR) has proven to be a useful probe for phospholipids in wheat gluten (Marion et al. 1987), but no studies are reported for wheat glycolipids. There is potential to use NMR in two ways. Firstly, it can elucidate glycolipid structure in non-crystalline states. For example, structural studies of glycosphingolipids (Yu et al. 1986). NMR might be used to look at the role of cereal glycolipids in bread-making. Secondly, carbon-13 NMR can analyze mixtures of sugars (Blunt and Munro 1976). The advantages, compared to GC, included no response variations or column and derivatization losses. However, a large sample size is required compared to GC. Long data accumulation times would severely limit applications of quantitative NMR in cereal glycolipid studies.

3.6.3 Mass Spectroscopy

Tweeten et al. (1981) applied electron impact mass spectroscopy to wheat glycolipid structural analysis. They were able to show at which glycerol carbon the linoleic (18:2), palmitic (16:0) and oleic (18:1) fatty acids were found in MGDG and DGDG isolated by HPLC. MGDG and DGDG each had three "positional isomers", but only two MGDG and two DGDG peaks were resolvable by HPLC (see Sect. 3.5.1)

A preliminary look at trimethylsilyl-derivatized wheat DGDG (see Sect. 3.2.2; Larsen and Winter, unpubl.) revealed problems in calibrating electron impact mass spectroscopy, due to the high molecular weights (>1000) of the derivatives. Tris(perfluorononyl)-s-triazine was found suitable for calibrations up to molecular weights of about 1500.

4 Conclusions

Wheat glycolipid analysis has evolved considerably in the number and sophistication of available methods. For example, HPLC has tremendous potential to make analytical life easy. Even so, to calibrate these new methods cereal chemists have to resort to more traditional and basic analytical techniques such as TLC. This will continue until chemical suppliers sell a full range of wheat glycolipid standards.

A major goal for cereal chemists is to resolve, identify and quantify the "positional isomers" of glycolipids, such as MGDG and DGDG, that result from combinations of different fatty acids. This may be possible with GC or HPLC, in tandem with mass spectroscopy. This sort of information is needed for mechanistic studies of phenomena like glycolipid binding, phase behaviour contributions to bread improvement, and staling.

References

Barnes PJ (1987) Wheat grain lipids and their role in the bread-making process. In: Hamilton RJ, Bhati A (eds) Recent advances in chemistry and technology of fats and oils. Elsevier, London, pp 79–107

Bekes F, Zawistowska U, Bushuk W (1983) Protein – lipid complexes in the gliadin fraction. Cereal Chem 60:371–378

Bekes F, Zawistowska U, Zillman RR, Bushuk W (1986) Relationship between lipid content and composition and loaf volume of twenty six common spring wheats. Cereal Chem 63:327–331

Bell BM, Daniels DGH, Stewart BA (1987) Lipid compositions, baking qualities and other characteristics of wheat varieties grown in the UK. J Cereal Sci 5:277–286

Berger M (1983) Composition lipidique de huit farines de blé tendre. Ind des Céréales 22:5–26

Blunt JW, Munro MHG (1976) An automated procedure for qualitative and quantitative analysis of mixtures by means of carbon magnetic resonance spectroscopy: applications to carbohydrate analysis. Aust J Chem 29:975–986

Carter HE, McCluer RH, Slifer ED (1956) Lipids of wheat flour. I. Characterization of galactosylglycerol components. J Am Chem Soc 78:3735–3738

Carter HE, Ohno K, Nojima S, Tipton CL, Stanacev NZ (1961a) Wheat flour lipids. II. Isolation and characterization of glycolipids of wheat flour and other plant sources. J Lipid Res 2:215–222

Carter HE, Hendry RA, Stanacev NZ (1961b) Wheat flour lipids. III. Structure of the mono- and digalactosylglycerol lipids. J Lipid Res 2:223–227

Christie WW (1982) Lipid analysis, 2nd edn. Pergamon Press, Oxford

Christie WW (1985) Rapid separation and quantification of lipid classes by high performance liquid chromatography and mass (light-scattering) detection. J Lipid Res 26:507–512

Christie WW (1987) High performance liquid chromatography and lipids. A practical guide. Pergamon Press, Oxford

Christie WW (1989) Gas chromatography and lipids: a practical guide. Oily Press, Ayr, Scotland

Christie WW, Morrison WR (1988) Separation of complex lipids of cereals by high-performance liquid chromatography with mass detection. J Chromatogr 436:510–513

Chung OK (1986) Lipid-protein interactions in wheat flour, dough, gluten, and protein fractions. Cereal Foods World 31:242–256

Chung OK (1989) Functional significance of wheat lipids. In: Pomeranz Y (ed) Wheat is unique. AACC, St Paul, MN, pp 341–368

Chung OK, Pomeranz Y, Finney KF, Hubbard JD, Shogren MD (1977) Defatted and reconstituted wheat flours. I. Effects of solvent and Soxhlet types on functional (breadmaking) properties. Cereal Chem 54:454–465

Chung OK, Pomeranz Y, Jacobs RM, Howard BG (1980) Lipid extraction conditions to differentiate among hard red winter wheats that vary in breadmaking. J Food Sci 45:1168–1174

Chung OK, Pomeranz Y, Finney KF (1982) Relation of polar lipid content to mixing requirement and loaf volume potential of hard red winter wheat flour. Cereal Chem 59:14–20

Chung OK, Pomeranz Y, Jacobs RM (1984) Solvent solubility parameter and flour moisture effects on lipid extractibility. J Am Oil Chem Soc 61:793–797

Clements RL (1977) Large scale laboratory Soxhlet extraction of wheat flours, and of intact and cracked grains. Cereal Chem 54:865–874

Clements RL, Donelson JR (1981) Functionality of specific flour lipids in cookies. Cereal Chem 58:204–206

Daftary RD, Pomeranz Y, Shogren M, Finney KF (1968) Functional breadmaking properties of wheat flour lipids. 2. The role of flour lipid fractions in breadmaking. Food Technol 22:327–330

Daniels NWR, Richmond JW, Eggitt PWR, Coppock JBM (1966) Studies on the lipids of flour. III. Lipid binding in breadmaking. J Sci Food Agric 17:20–29

Dubois M, Gilles KA, Hamilton JK, Rebers PA, Smith F (1956) Colorimetric method for determination of sugars and related substances. Anal Chem 28:350–356

Fujino Y, Ohnishi M (1983) Sphingolipids in wheat grain. J Cereal Sci 1:159–168

Gordon AJ, Ford RA (1972) The chemists companion. A handbook of practical data, techniques, and, references. John Wiley & Sons, New York, p 379

Graveland A (1968) Combination of thin layer chromatography and gas chromatography in the analysis on a microgram scale of lipids from wheat flour and wheat flour doughs. J Am Oil Chem Soc 45:834–840

Hamilton RJ, Mitchell SF, Sewell PA (1987) Techniques for the detection of lipids in high performance liquid chromatography. J Chromatogr 395:33–46

Hammam H, Sivik B, Schwengers D (1987) Baking qualities of CO_2- and ethanol-extracted gluten and gluten lipids. Acta Agric Scand 37:130–136

Hargin KD, Morrison WR (1980) The distribution of acyl lipids in the germ aleurone, starch and non-starch endosperm of four wheat varities. J Sci Food Agric 31:877–888

Heemskerk JWM, Bögemann G, Scheijen MAM, Wintermans JFGM (1986) Separation of chloroplast polar lipids and measurement of galactolipid metabolism by high-performance liquid chromatography. Anal Biochem 154:85–91

Hoseney RC, Finney KF, Pomeranz Y, Shogren MD (1969) Functional (breadmaking) and biochemical properties of wheat flour components. V. Role of total extractable lipids. Cereal Chem 46:606–613

Kates M (1964) Simplified procedures for hydrolysis or methanolysis of lipids. J Lipid Res 5:132–135

Kushwaha SC, Kates M (1981) Modification of phenol-sulphuric acid method for estimation of sugars in lipids. Lipids 16:372–373

Larsen NG, Baruch DW, Humphrey-Taylor VJ (1986) Lipid quality factors in breeding New Zealand wheats. In: Williams TA, Wratt GS (eds) Plant Breeding Symp DSIR, Agronomy Society of New Zealand Spec Publ 5. Agron Soc of NZ, Christchurch, pp 278–281

Larsen NG, Baruch DW, Humphrey-Taylor VJ (1989a) The effect of laboratory and commercial milling on lipids and other physicochemical factors affecting the breadmaking quality of wheat flour. J Cereal Sci 9:139–148

Larsen NG, Humphrey-Taylor VJ, Baruch DW (1989b) Glycolipid content as a breadmaking quality determinant in flours from New Zealand wheat blends. J Cereal Sci 9:149–157

Larsen NG, Levick SM, Mouat CH, Sutton KH (1990) The effect of ball-milling on phospholipid extractability and the breadmaking quality of flour. J Cereal Sci 12:155–164

Larsson K (1983) Physical state of lipids and their technical effects in baking. In: Barnes PJ (ed) Lipids in cereal technology. Academic Press, London, pp 237–251

Lin MJY, Youngs VL, D'Appolonia BL (1974a) Hard red spring and durum wheat polar lipids. I. Isolation and quantitative determinations. Cereal Chem 51:17–33

Lin MJY, D'Appolonia BL, Youngs VL (1974b) Hard red spring and durum wheat polar lipids. II. Effect on quality of bread and pasta products. Cereal Chem 51:34–45

Lundberg WO (1984) Lipidology. In: Mangold HK (ed) CRC handbook of chromatography. Lipids, vol I. CRC Press, Boca Raton, p 23

MacRitchie F (1977) Flour lipids and their effects in baking. J Sci Food Agric 28:53–58

MacRitchie F (1983) Role of lipids in baking. In: Barnes PJ (ed) Lipids in cereal technology. Academic Press, London, pp 165–188

Marion D, Le Roux C, Akoka S, Tellier C, Gallant D (1987) Lipid-protein interactions in wheat gluten: a phosphorus nuclear magnetic resonance spectroscopy and freeze-fracture electron microscopy study. J Cereal Sci 5:101–115

McKillican ME (1964) Studies on the phospholipids, glycolipids and sterols of wheat endosperm. J Am Oil Chem Soc 41:554–557

McKillican ME, Sims RPA (1964) The endosperm lipids of three Canadian wheats. J Am Oil Chem Soc 41:340–344

Morrison WR (1978) Cereal lipids. In: Pomeranz Y (ed) Advances in cereal science and technology, vol II. AACC, St Paul, MN, pp 221–348

Morrison WR (1983) Acyl lipids in cereals. In: Barnes PJ (ed) Lipids in cereal technology. Academic Press, London, pp 11–32

Morrison WR (1988a) Lipids. In: Pomeranz Y (ed) Wheat chemistry and technology, vol I, 3rd edn. AACC, St Paul, MN, pp 373–439

Morrison WR (1988b) Lipids in cereal starches. A review. J Cereal Sci 8:1–15

Morrison WR (1989) Wheat lipids are unique. In: Pomeranz Y (ed) Wheat is unique. AACC, St Paul, MN, pp 319–339

Morrison WR, Mann DL, Soon W, Coventry AM (1975) Selective extraction and quantitative analysis of non-starch and starch lipids from wheat flour. J Sci Food Agric 26:507–521

Morrison WR, Tan SW, Hargin KD (1980) Methods for the quantitative analysis of lipids in cereal grains and similar tissues. J Sci Food Agric 31:329–340

Morrison WR, Coventry AM, Barnes PJ (1982) The distribution of acyl lipids and tocopherols in flours millstreams. J Sci Food Agric 33:925–933

Morrison WR, Wylie LJ, Law CN (1984) The effect of Group 5 chromosomes on the free and total galactosyldiglycerides in wheat endosperm. J Cereal Sci 2:145–152

Morrison WR, Law CN, Wylie LJ, Coventry AM, Seekings J (1989) The effect of Group 5 chromosomes on the free polar lipids and breadmaking quality of wheat. J Cereal Sci 9:41–51

Panozzo JF, O'Brien L, MacRitchie F, Bekes F (1990) Baking quality of Australian wheat cultivars varying in their free lipid composition. J Cereal Sci 11:51–57

Pomeranz Y (1965) Thin layer chromatography in studies of cereal lipids. Qual Plant Mater Veg XII:321–341

Pomeranz Y (1980) What? How much? Where? What function? in bread making. Cereal Foods World 25:656–662

Pomeranz Y (1988) Composition and functionality of wheat flour components. In: Pomeranz Y (ed) Wheat chemistry and technology, vol II, 3rd edn. AACC, St Paul, MN, pp 285–328

Pomeranz Y, Chung O, Robinson RJ (1966) The lipid composition of wheat flours varying widely in bread-making potentialities. J Am Oil Chem Soc 43:45–48

Scott TA, Melvin EH (1953) Determination of dextran with anthrone. Anal Chem 25:1656–1661

Simmons CT (1986) The development and application of new infrared methods for determining the glycolipid fraction in hard red spring wheats. Diss Abstr Int 47(1):20

Simmons CT, McDonald CE (1985) A near infrared reflectance method for the glycolipid fraction in wheat flour. Cereal Foods World 30:566

Tweeten TN, Wetzel DL, Chung OK (1981) Physicochemical characterization of galactosyldiglycerides and their quantitation in wheat flour lipids by high performance liquid chromatography. J Am Oil Chem Soc 58:664–672

Walker GC (1988) Determination of flour glycolipids as their benzoyl derivatives by high-performance liquid chromatography with ultraviolet detection. Cereal Chem 65:433–435

Williams JP, Watson GR, Khan M, Leung S, Kukis A, Stachnyk O, Myhler JJ (1975) Gas-liquid chromatography of plant galactolipids and their deacylation and methanolysis products. Anal Biochem 66:110–122

Yamamoto A, Rouser G (1970) Spectrophotometric determination of molar amounts of glycosphingolipids and ceramide by hydrolysis and reaction with trinitrobenzenesulfonic acid. Lipids 5:442–444

Youngs VL, Medcalf DG, Gilles KA (1970) The distribution of lipids in fractionated wheat flour. Cereal Chem 47:640–649

Yu RK, Koerner TAW, Scarsdale JN, Prestegard JH (1986) Elucidation of glycolipid structure by protein nuclear magnetic resonance spectroscopy. Chem Phys Lipids 42:27–48

Proteinaceous Inhibitors of Lipase Activities in Soybean and Other Oil Seeds

A. H. C. HUANG and S. M. WANG

1 Introduction

There are at least three major mechanisms whereby an inhibitor of an enzyme may exert its effect. These three mechanisms are all exemplified by lipase inhibitors. In the first mechanism, the inhibitor acts directly on the enzyme molecule, either binding to or altering it. Examples are the apolipoproteins of the lipoproteins, which modulate (activate as well as inhibit) epithelium lipase activities (Cheng et al. 1990). In the second mechanism, the inhibitor acts as a substrate analog, which exerts its effect on the active site reversibly or irreversibly. Lipase inhibitors employing this mechanism include the reversible inhibitors of alkyl and aryl boronic acid derivatives (Garner 1980), and the irreversible inhibitors of synthetic 1,6-di(O-(carbamyl)cyclohexanone oxime)hexane (Sutherland and Amin 1982) and *Streptomyces toxytricini* lipstatin (Borgstrom 1988; Hadvary et al. 1988). In the third mechanism, the inhibitor acts on the substrate, thereby reducing the availability of the substrate to the enzyme. It is exemplified by many artificial surfactants and amphiphatic proteins which bind to the surface of the substrate emulsion of lipase (Ransac et al. 1990).

The above-mentioned lipase inhibitors can be natural or synthetic molecules. The physiological roles of the naturally occurring inhibitors are unknown, with the exception of the apolipoproteins of lipoproteins in mammals. Lipase inhibitors in mammalian systems have been studied to great extents partly because of their possible use in eliminating hyperlipidemic problems in humans. In plants, lipase inhibitors have been found in soybean and several other oil seeds. They inhibit in vitro lipase activity by acting as surfactants on the substrate emulsions. These inhibitors are the subject of the current chapter.

2 Lipase and Its Substrate Emulsions

A true lipase (EC 3.1.1.3) hydrolyzes triacylglycerols at an oil-water interface and does not hydrolyze monomolecular triacylglycerols in solution (Galliard 1980; Huang 1987). This property distinguishes the enzyme from an esterase which hydrolyzes triacylglycerols and other esters that are water-soluble. The distinction can be best illustrated in Fig. 1. The figure is a hypothetical plot

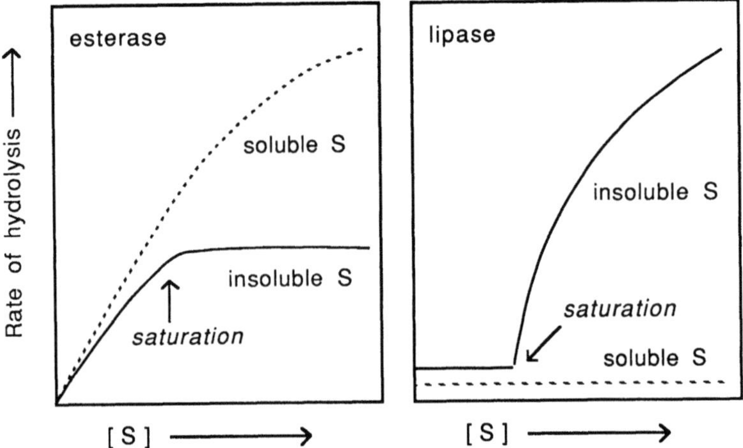

Fig. 1. Hypothetical plots of activities of esterase and lipase versus substrate concentrations. The activities on a water-soluble and a moderately water-soluble substrate are shown. In the latter case, the concentration of substrate increases to a point of saturation beyond which additional substrate monomers start to aggregate

of activities of the two enzymes versus increasing concentrations of a water-soluble substrate and a moderately water-soluble substrate.

Esterase activity increases with increasing concentrations of the water-soluble substrate. The enzyme also acts on the moderately water-soluble substrate, but only when the substrate is present as monomer molecules. When the substrate concentration has increased beyond the saturation point, there is no further increase in the esterase activity. This highest esterase activity represents action on the monomer molecules at the saturated concentration. On the contrary, lipase does not act on the water-soluble substrate at any concentration. On the moderately water-soluble substrate, lipase has no activity below the saturation point. After the saturation point beyond which the additional substrate monomers start to aggregate into emulsions, lipase activity appears and increases with increasing substrate concentrations.

In a typical assay of lipase activity, the substrate is water-insoluble triacylglycerol which is emulsified with a stabilizing agent to form stable emulsions. The release of fatty acid is monitored with a pH stat continuously or with a colorimetric reagent at time intervals. A radioactive assay can also be performed using commercially available radioactive triolein, and monitoring the release of radioactive fatty acid. In our assay with the colorimetric method, the production of fatty acid is linear until about 40% of the substrate triacylglycerols has been consumed.

3 Lipases in Seeds

Most seeds contain triacylglycerols as food reserves which are synthesized during seed maturation and utilized during germination and postgerminative growth. In most seed, lipase activity is absent in ungerminated seeds and increases during germination. Lipases from castor bean and maize have been studied most extensively (Galliard 1980; Huang 1987).

In the study of seed lipase, as in most cases of studying enzymes in relation to metabolism, the detected in vitro activity of the enzyme should be sufficient to account for the rate of in vivo lipolysis. This is indeed valid in castor bean, maize, rapeseed, and mustard (Wang and Huang 1984). However, in soybean, sunflower, peanut, and cucumber, the detected lipase activity, as measured at diverse pH's, is too low to account for the expected rate of lipolysis during germination and postgerminative growth. The deficiency is due to the presence of lipase inhibitors in the seed extract.

4 Occurrence of Lipase Inhibitor in Seeds of Diverse Species

Lipase inhibitors have been detected in the crude extracts of germinated seeds from diverse plant species, including soybean, sunflower, peanut, and cucumber (Wang and Huang 1984). They are present in the soluble fractions obtained after centrifugation of the crude seed extracts for 150000 g for 2 h. They exhibit inhibitory effects on the activities of partially purified maize lipase and commercially prepared pancreatic lipases (Fig. 2). The activities of these two lipases are reduced when increasing amounts of seed extracts are added to the assay medium. In the presence of roughly equal activities of the two lipases in the assays, the pancreatic lipase is more sensitive to the inhibitors than the maize lipase. Of the several oil seed species examined, the soybean possesses the highest inhibitory activity in its extract. Because of this highest inhibitory activity and because of the extensive use of soybean meal in animal feed, the soybean inhibitors have been studied to a greater extent (Satouchi et al. 1974; Satouchi and Matsushita 1976; Widmer and Kaplan 1977; Gargouri et al. 1984; Wang and Huang 1984).

5 Properties of the Lipase Inhibitors from Soybean

In general, a crude extract of mature soybean, or the supernatant obtained after centrifugation of the crude extract for 100000 g, is used as a source of lipase inhibitor. The soybean inhibitors inhibit the activities of lipases from plants (maize, cotton, mustard, castor bean, and rapeseed), animals (porcine

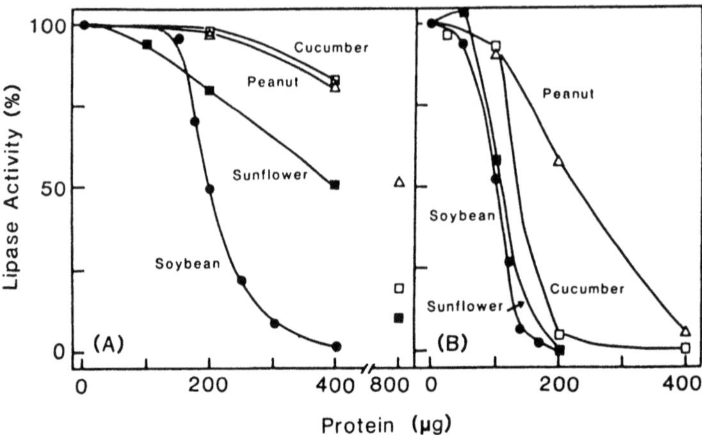

Fig. 2. Inhibitory effects of extracts of increasing amounts from various oil seeds on **A** maize lipase and **B** pancreatic lipase. The 150000 g supernatants from seedlings of various species were used as the sources of inhibitors (Wang and Huang 1984)

pancreas), and fungi (*Rhizopus* and *Aspergillus*). The degree of inhibition exhibited by the soybean inhibitors is independent of the amount of lipase used within the range suitable for the enzyme assay (a sixfold difference).

Incubation of the soybean inhibitors at pH 3–10 at room temperature for 18 h gives no reduction in their inhibitory effect. Treatments of the soybean inhibitors with ribonuclease, amylase, and β-galactosidase also result in no loss of their inhibitory effect. However, treatments of the inhibitors with proteases (pronase, protease V8), or heating the inhibitors at 50 °C for 5 min, destroy their inhibitory effect. The inhibitors appear to be proteins.

The inhibitory effect of the inhibitors is expressed irrespective of the sequence of addition of the three components, lipase, substrate, and inhibitors, into the reaction mixture. In addition, the inhibitory effect is immediately and equally expressed when the inhibitors are added either before or after the lipase reaction is in progress (Fig. 3). Furthermore, the inhibitors exert their effect on various oil substrate emulsions, including those of triolein, trilinolein, soybean oil, linseed oil, and Ediol.

Although the inhibitory effect of the inhibitors is independent of the amount of lipase present in the assay, it is dependent on the amount of substrate added. In our routine lipase assay, 5 mM of sonicated substrate is used, which is always sufficient to give maximal enzyme activity (Fig. 4). At this amount of substrate, the inhibitors drastically reduce the lipase activity, and the reduction is roughly proportional to the amount of inhibitors added. However, the reduction is gradually lessened as the amount of substrate increases. At higher amounts of substrate, the inhibitors exert no inhibition; they actually enhance the lipase activity. This phenomenon is observed using maize, pancreatic, and *Rhizopus* lipases. The enhancement is not due to the presence of lipase activity in the inhibitor preparation, which by itself gives no detectable

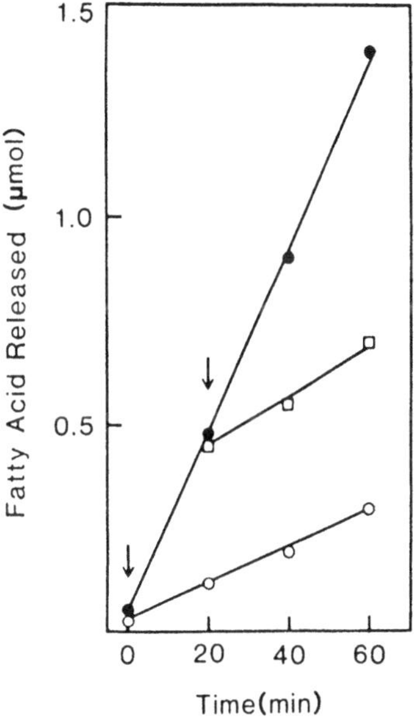

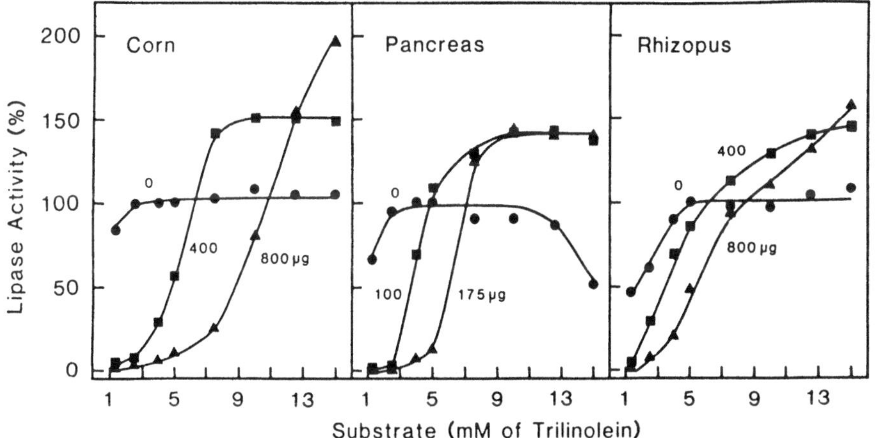

Fig. 3. A time course of lipolysis by pancreatic lipase without or with the addition (*arrows*) of 600 µg soybean proteins at 0 or 20 min (Wang and Huang 1984)

Fig. 4. Effect of increasing substrate concentrations on the activities of lipases in the absence or presence of soybean proteins of the indicated amounts. A suspension of 25 mM trilinolein in a 5% gum acacia was prepared by sonication and added to the assay medium to make the desired concentration in a final volume of 1 ml (Wang and Huang 1984)

lipase activity. There is no apparent alteration of the substrate emulsions by the inhibitors, as judged from the turbidity of the assay suspension or the number of emulsions per volume observed under a light microscope.

During the course of soybean seedling growth, there is a gradual decrease in the content of the inhibitors in the cotyledons, concomitant with the decrease in the total proteins. In differential centrifugations of the cotyledon extract of 6-day-old seedlings, essentially all the inhibitors are present in the supernatant after centrifugation at 150000 g for 2 h. This recovery of the inhibitors in the soluble fraction is similar to that from the mature, ungerminated soybean.

The total proteins in ungerminated soybean seeds can be extracted with 1 M NaCl, and separated into three fractions of 11S, 7S, 2–4S proteins according to their sedimentation coefficients by sucrose gradient centrifugation. The lipase inhibitors are present only in the 2–4S protein fraction. The 11S and 7S proteins are well-known storage proteins, whereas the 2–4S proteins are a mixture of storage proteins and other proteins such as protease inhibitors. The soybean trypsin inhibitors (purchased from Sigma Corp., St. Louis, MD, USA) have no appreciable inhibitor activity on maize lipase.

The soybean inhibitors belong to a heterogeneous population of proteins, as judged from the profiles of proteins and inhibitor activity separated from the soluble soybean extract by ammonium sulfate fractionation, ion exchange chromatography, and gel filtration. An inhibitor protein of about 70–80 kD has been isolated from the soybean extract (Gargouri et al. 1984). The protein preparation is apparently homogeneous as shown by the appearance of a single protein band in SDS-polyacrylamide gel after electrophoresis, and by the presence of a single N-terminal residue of alanine. Its amino acid profile reveals a high content of nonpolar amino acid residues.

6 Assay

The most important step is to set up a reliable assay of lipase activity (Huang 1985). Commercially available, completely or partially purified, lipase from pancreas, fungi, or bacteria can be used. Triolein is used as the lipase substrate. It should be free of contaminating monoacylglycerols and diacylglycerols which are also substrates for nonlipase acyl hydrolases. The substrate preparation obtained commercially should be checked by thin layer chromatography for purity, irrespective of the claims of purity made by the manufacturers. The preparation is dissolved in chloroform (5–10 µg/10 µl for each spot) and spotted onto a TLC plate coated with 250 µm of Silica Gel G (Brinkman Instruments, Inc., Westbury, NY, USA). The plate isdeveloped in 80:20:2 (v/v/v) hexane: diethyl ether: acetic acid, and allowed to react with iodine vapor (crystals of iodine are placed in a TLC tank). Standards of tri-, di-, and mono-olein and oleic acid (5–10 µg/10 µl chloroform for each spot) are run on the same TLC plate. The mobilities of the components in descending order are triacylglycerol, fatty acid, 1,3-diacylglycerol, 1,2-diacylglycerol, and monoacylglycerols. A small percent of free fatty acid in the triacylglycerol preparation will not interfere with the assay. If monoacylglycerol or diacylglycerol is present, the triacylglycerol is purified on a similar TLC plate. The triacylglycerol preparation is applied as a line across the whole plate (200 μg per one 20×20 cm plate). A similar but smaller plate (20×5 cm) is run in the same tank as a marker. After development, the small plate is allowed to react with iodine vapor. By making a comparison between the two plates, the position of triacylglycerol in the large plate can be identified. The silica gel containing the triacylglycerol line is scraped and extracted with chloroform. After centrifugation to remove the silica gel, the chloroform supernatant is obtained, and the chloroform is evaporated.

The activity of true lipase is measured by a colorimetric method. Although the released fatty acids can be measured more rapidly using an automatic titrator, the colorimetric method has the advantage of requiring no specific equipment or setup. In addition, if many enzyme samples are to be assayed, the colorimetric method probably consumes about the same amount of time per assay, and all the assays can be performed simultaneously for a more uniform quantitation. In the colorimetric assay, the fatty acids produced are converted to copper soaps and measured using 2,2′-diphenylcarbazide. The reaction is performed at room temperature in a 5-ml tube. The 1-ml reaction mixture contains 0.1 Tris HCl, pH 7.5 (for pancreatic lipase), 5 mM dithiothreitol, 5 mM substrate, and enzyme preparation. Triolein (50 mM) is first emulsified in 2 ml of 5% gum acacia for 1 min at low speed with a Bronwill Biosonic IV ultrasonic generator fitted with a microprobe. The reaction is stopped at time intervals (ranging from 5 min to 2 h, depending on the amount of enzyme used) to ensure that proper kinetics are observed. Each 0.1-ml aliquot of the reaction mixture is put in a 7-ml tube and boiled in a boiling water bath for 5 min. After cooling to room temperature, 4 ml of chloroform heptane methanol (4:3:2, v/v/v) is added. The tube is closed with a Teflon screw cap, and shaken horizontally for 15 min. Two milliliters of 0.1 M sodium phosphate, pH 2.5, is added, and the tube is shaken horizontally for 3 min. After centrifugation in a table-top centrifuge, the upper layer of methanol water is pipetted out and discarded. One milliliter of 0.01 M HCl is added, and the tube is shaken horizontally for 3 min. After centrifugation, the upper layer of HCl solution is pipetted out and discarded. One and a half milliliters of copper reagent [0.1 M $Cu(NO_3)_2$, 0.2 M triethanolamine, 0.06 N NaOH, and 6 M NaCl] is added. The tube is closed with a Teflon screw cap, and shaken horizontally for 5 min. After centrifugation in a table-top centrifuge, 2 ml of the chloroform layer is transferred to a tube, and 0.1 ml of color reagent (10 ml freshly prepared 0.4% 2,2′-diphenylcarbazide solution in 100% ethanol, plus 0.1 ml 1 M triethanolamine added immediately before use) is added. After 10 min, the absorbance is read at 550 nm. Oleic acid is used to produce a standard curve that is linear up to a concentration of 0.05 μmol per 1.5 ml chloroform.

A crude extract of the soybean lipase inhibitors is prepared (Wang and Huang 1984). All operations are performed at 4 °C. Dehulled, mature soybean is chopped into small pieces with a razor blade in a Petri dish and ground gently with a mortar and pestle. The grinding medium contains 1 mM EDTA, 10 mM KCl, 1 mM $MgCl_2$, 2 mM DTT, 0.5 M sucrose, and 0.15 M Tricine adjusted to pH 7.5 with KOH. The homogenate is filtered through a Nitex cloth of pore size of 20×20 µm. The filtrate can be used directly as a crude preparation of lipase inhibitors or subjected to further fractionation. In the latter case, the filtrate is centrifuged at 150000 g for 2 h. The resulting supernatant below the oil layer is used directly or dialyzed against 0.1 M Tris-HCl buffer (pH 7.5). Fifty to 800 µg proteins of the inhibitors (in crude extract, undialyzed, or dialyzed 150000 g supernatant) are added to a lipase assay mixture as described above, but without lipase, making a volume of 0.9 ml. The mixture is incubated at 24 °C for 5 min. Lipase preparation in 100 µl of grinding medium (pH 7.5) is added to initiate the lipase reaction. The lipase activity is monitored as described above. Alternatively, the inhibitors can be added to the assay medium together with, or after the addition of, lipase (Fig. 3). For the convenience of assaying for lipase inhibitors, the amount of lipase used should enable the production of free fatty acid linearly for 60 min.

7 Discussion

The soybean lipase inhibitors are amphipathic proteins which can bind to the surface of the substrate emulsions. The binding is immediate. The inhibitors modify the emulsion surface properties, or displace the lipase from the emulsion surface, or hinder the normal catalysis of the interfacial lipase. Addition of extra substrate increases the total emulsion surface and therefore releases the inhibitory effects. It is unlikely that the inhibitors exert their effect by binding to the lipase molecules. Such a direct binding would not be released by the increase in substrate, and would be released by the increase in lipase molecules; these are contrary to the experimental results. Thus, the inhibitor proteins are unlike some enzyme inhibitors or effectors which bind to the enzyme directly and affect the function of the enzyme. Amphiphiles or surfactants including proteins (e.g., bovine serum albumin, ovalbumin, α-lactoglobulin) and nonprotein molecules (e.g., many detergents) are known to bind to the substrate emulsions and thus affect lipase activities (Gargouri et al. 1986, Ransac et al. 1990). Whether the inhibitors in the oil seed exert some physiological functions or merely represent proteins of amphipathic nature that fortuitously can bind to the substrate emulsions in lipase assay remains to be seen.

Acknowledgment. The work was supported by a grant from the National Science Foundation.

References

Borgstrom B (1988) Mode of action of tetrahydrolipstatin: a derivative of the naturally occurring lipase inhibitor lipstatin. Biochim Biophys Acta 962:308–316

Cheng Q, Blackett P, Jackson KW, McConathy WI, Wang CS (1990) C-terminal domain of apolipoprotein CII as both activator and competitive inhibitor of lipoprotein lipase. Biochem J 269:403–407

Galliard T (1980) Degradation of acyl lipids: hydrolytic and oxidative enzymes. In: Stumpf PK, Conn EE (eds) The biochemistry of plants, vol 4. Acad Press, New York, pp 85–116

Gargouri Y, Julien R, Pieroni G, Verger R, Sarda L (1984) Studies on the inhibition of pancreatic and microbial lipases by soybean proteins. J Lipid Res 25:1214–1221

Gargouri Y, Piéroni G, Riviére C, Sarda L, Verger R (1986) Inhibition of lipases by proteins: a binding study using dicaprin monolayers. Biochemistry 25:1733–1738

Garner CW (1980) Boronic acid inhibitors of porcine pancreatic lipase. J Biol Chem 255:5064–5068

Hadvary P, Lengsfeld H, Wolfer H (1988) Inhibition of pancreatic lipase in vitro by the covalent inhibitor tetrahydrolipstatin. Biochem J 256:357–361

Huang AHC (1985) Lipid bodies. In: Linskens HF, Jackson JF (eds) Modern methods of plant analysis. New Series, vol 1. Cell components. Springer, Berlin Heidelberg New York, pp 145–151

Huang AHC (1987) Lipases. In: Stumpf PK, Conn EE (eds) The biochemistry of plants, vol 4. Acad Press, New York, pp 91–119

Ransac S, Riviére C, Soulie JM, Gancet C, Verger R, de Haas GH (1990) Competitive inhibition of lipolytic enzymes. I. A kinetic model applicable to water insoluble competitive inhibitors. Biochim Biophys Acta 1043:57–66

Satouchi K, Matsushita S (1976) Purification and properties of a lipase-inhibiting protein from soybean cotyledons. Agric Biol Chem 40:889–897

Satouchi K, Mori T, Matsushita S (1974) Characterization of inhibitor protein for lipase in soybean seeds. Agric Biol Chem 38:97–101

Sutherland CA, Amin D (1982) Relative activities of rat and dog platelet phospholipase A_2 and diglyceride lipase. J Biol Chem 257:14006–14010

Wang SM, Huang AHC (1984) Inhibitors of lipase activities in soybean and other oil seeds. Plant Physiol 76:929–9343

Widmer F, Kaplan NO (1977) Pancreatic lipase effectors extracted from soybean meal. Agric Food Chem 25:1142–1145

Conductivity Testing of Seeds

D. K. PANDEY

1 Introduction

Electro-analytical approaches to seed testing can be envisaged in two ways: the direct measurement of bioelectrical potential or electrical resistance or current on partially hydrated seed, and conductivity measurement of seed steep water which determines the extent of electrolyte efflux out of seed into imbibition medium. While the first approach, involving measurement of an electrical current passed through seed, remains confined to an experimental stage; the second approach, measuring conductivity of seed steep water, has been extensively explored for seed quality assessment. In a broader perspective, seed quality in the context of the present chapter refers to the overall structural and functional state of seed which includes and determines readily measurable parameters of vigour and viability.

Conductivity testing of seeds as applied in the context of this chapter includes electromotive measurements regarding the ability to conduct electric current directly on seed or seed steep water, whether it be in terms of current or conductance, since the term conductivity is widely used in seed science and technology. The conductivity testing usually involves electromotive measurements on imbibed seeds or on seed steep water in a series of seeds or seed lots with varying vigour and viability under standard conditions. The electromotive measurement values obtained with test seed lot are compared with those of the standard seed lots, and the vigour and viability status of the seed lot in question is predicted, presuming that similar values of electromotive measurements in a standard and a test seed lot are associated with similar vigour and viability status. Thus, the method requires vigour and viability measurements on standard seed lots and electromotive measurements on seeds or seed steep water of both the standard and test seed lots. Other uses of conductivity testing of seeds may include the detection of membrane integrity and, to a lesser extent, other associated characteristics such as solute leakage, imbibition injury, permeability characteristics and mechanical damage/injury etc.

Vigour and viability are economically important manifestations of seed quality. The viability of a seed is the ability to germinate or to produce a normal seedling in a standard germination test performed usually under optimal growth conditions (ISTA 1985; AOSA 1986). Seed vigour is the sum of all those qualities which determine the level of activity and performance potential, also under suboptimal conditions (Perry 1981, 1984; Halmer and Bewley 1984; Hampton and Coolbear 1990). Besides the traditional quality criteria of germination, purity and health, vigour is an important component of seed quality revealed not only in practice but also in specific situations. Especially under suboptimal conditions, it may determine the rapidity and uniformity of emergence, stand establishment and final yield, or the storage potential of a seed lot. Among the important objectives of seed testing for vigour and viability are (1) to predict and select seeds which perform well when sown in the field; (2) to monitor and predict seed quality during processing and storage; and (3) to provide discriminating, physiological seed treatments. The basic quality assessment through the standard germination test (ISTA 1985; AOSA 1986), is inevitably time-consuming, requiring more than 1 week to complete and, furthermore, may not always correlate well with emergence under field conditions over a range of environmental conditions. Therefore, it is desirable to have a simple, reliable, accurate, rapid to perform, physiologically informative and relatively inexpensive test of seed quality involving vigour and viability.

2 Basis of Conductivity Testing of Seeds

Conductivity testing of seeds usually involves cellular membrane integrity or its manifestation determined by potentiometric measurements on hydrated seeds or on seed steep water (Parrish and Leopold 1978; Fensom 1985) and the association of such measurements with seed quality involving vigour and viability.

An introduction to various aspects of seeds and membrane integrity of seed tissues at varying hydration levels and their manifestation may be relevent in the conductivity testing of seeds.

2.1 Classification of Seeds in Regard to Storage Behaviour

Seeds have been broadly placed into two broad categories, the desiccation-tolerant, i.e. "orthodox", and desiccation-sensitive, i.e. "recalcitrant", seeds (Roberts 1973; Roberts and King 1980; Berjak et al. 1984; Farrant et al. 1986, 1988, 1989; Berjak et al. 1990).

Orthodox seeds usually terminate their development by maturation drying, during which there is a metabolic switch from the developmental phase to that of germination (Kermode and Bewley 1986, 1988). After they are shed from the parent plant, the moisture content of these seeds remains in equilibrium with the relative humidity of their surroundings. Such seeds can usually be further dried even to a ultra-low moisture content ($<5\%$) without irreparable damage (Roberts and Ellis 1989) and can retain viability for remarkably long periods provided relative humidity and temperature are kept constantly at low levels (Ellis and Roberts 1980). While under natural conditions, desiccation tolerance may help in resisting the vicissitudes of the environment, this advantage is widely utilized for storage of seeds for germplasm conservation in gene banks (Roberts and Ellis 1984) and for various other agricultural purposes. Such seeds would, unless dormant or deteriorated through ageing or otherwise, switch to germination processes and resume metabolic activities, growth and development under a conducive environment.

Recalcitrant seeds, by contrast, do not undergo maturation drying. The water content does not diminish at all during their development (Berjak et al. 1990). They germinate shortly after they are shed under conductive conditions or continue germination processes with the available water upon shedding until the water becomes a limiting factor (Farrant et al. 1988). This characteristic makes them highly susceptible to desiccation injury. Thus, they are short-lived and can not be stored under conditions suitable for orthodox seeds (Farrant et al. 1988). These seeds are most sensitive to chilling injury at lower temperatures; drying to a low moisture content ($<20\%$) results in irreversible damage, resulting in death (King and Roberts 1979).

2.2 Seed Coat and Its Permeability

Typically, seeds have an outer coat or testa, an inner coat or tegma (both derived from ovular integuments), food reserves in the endosperm or cotyledons, and an embryo. The seed coats are permeable to water except in the seeds of members of some families including particularly species of Leguminosae, Malvaceae and Chenopodiaceae (Tran and Cavanagh 1984), which may be correlated with various histological features of the seed coat and with the presence of waterproofing substances (Rolston 1978; Bevilacqua 1987).

2.3 Cell Wall and Its Permeability

Once it has entered through the coat, water will move freely in the apoplast due to physical forces. The cellular membranes, especially the plasmalemma and tonoplast, act as the major barriers against water movement and dissolved solutes. Thus, over and above the cellular membranes, the permeability of the seed coat may influence the rate and quantity of water and solute movement to and from the imbibition medium (Simon 1984).

2.4 Membrane Integrity

The fluid mosaic model of the lipoprotein membrane (Singer and Nicolson 1972) envisages that, under aqueous conditions (moisture content $>20\%$), the lamellar configurations will have polar head groups of phospholipids facing the aqueous phase on either side of the bilayer with integrated proteins and their polar and ionic moieties in conformity with the lipid molecule bilayer (Simon 1984). The amphopathic nature of the proteins and phospholipids forms a bilayer configuration by constructing a central hydrophobic field lined by polar and charged regions facing the hydrophilic environment on either side of the bilayer. The aqueous environment stabilizes the bilayer configuration of the membranes (Simon 1984; Steponkus 1984). Membranes in their lamellar form, performing normal physiological functions, are considered to have complete integrity. However, under dry conditions when the water level falls below 20%, the lipid molecules change the configuration from lamellar to attain a hexagonal mode of arrangement with the lipid polar head groups lining a series of narrow, water-filled cylinders running at right angles to the plane (Simon 1984; Steponkus 1984). In this state the membranes will cease to function as semi-permeable barriers. The bilayer stability in the membranes could be maintained in some cases even at hydration levels down to 5% water level in some structures like pollen, and in seeds in some cases. This has been suggested to be a consequence of lipid sterols, as has been extrapolated from reports on the effects of cholesterol on the phase transition of artificial lipid mixtures. Thus, survival of most seeds and pollen suffering severe desiccation has been envisaged to be a consequence of (1) the retention of a bilayer arrangement rather than forming a hexagonal phase (Steponkus 1984; (2) the ability to conserve essential components of metabolism, like the protein-synthesizing complex in the dry state; (3) the ability to form membrane structures that can withstand a water-depleted microenvironment (Osborne 1980; Berjak et al. 1984; Priestley 1986; Roberts and Black 1989); and (4) the capacity to reconstruct membranes (Bewley and Black 1978, 1982; Priestley 1986; Simon 1984) upon rehydration, as is evident in orthodox seeds.

2.5 Significance of Membrane Integrity

Cellular membrane organization is essential for the maintenance of structure and function since, apart from serving as a permeability barrier, the membranes play a vital role in the compartmentalization of cellular components. Additionally, the cooperative enzymes of a metabolic pathway may be linked together in association with, or integrated into, membrane structures (Bewley and Black 1982; Halmer and Bewley 1984). Thus, the extent of membrane organization would have direct bearing on seed quality.

2.6 Factors Affecting Membrane Integrity

Membrane integrity is a function of damage and repair, which depend, among other factors, on the rate of the metabolism and ageing status. The factors influencing membrane integrity are likely to change the membrane permeability. The plasma membrane of a plant cell serves as the principal barrier to the free movement of molecules between the external environment and the cytoplasm.

The plasma membrane is relatively permeable to ions, nutrients and other solutes, but has a greatly reduced permeability to highly charged macromolecules such as nucleic acids and water-soluble proteins (Lindsey and Jones 1987). Several factors can change the permeability of the membrane, which also involves mechanical or physical injury, and constitutional changes and stresses (Levitt 1980) like temperature, moisture and chemical factors. They include physical breakdown, the saturation level of fatty acid components, the sterol: phospholipid ratio (McKersie et al. 1978; Borochov et al. 1982), lipid molecular species (Lynch and Steponkus 1987), polyamines (Roberts et al. 1986), and water content (Chapman and Wallach 1968); chemical agents including solvents such as dimethyl sulphoxide (DMSO; Delmer 1979) and toluene (Lerner et al. 1978); antibiotics (McDaniels 1983); pathological disorders (Benson 1990); osmotic shock (Tanaka et al. 1985); temperature treatment (Phodes and Stewart 1974; Yoshida and Niki 1979; Lockwood 1984); electrical treatments (Zimmerman et al. 1974; Linder et al. 1977; Weaver et al. 1984); growth regulators (Penny and Penny 1978); and deterioration through ageing (Pauls and Thompson 1980, 1981, 1984; Wilson and McDonald 1986). Membrane damage due to ageing, senescence, pathological disorders and environmental stresses may involve oxidative stress and free radical-mediated damage. Since conductivity testing of seeds is mostly confined to the assessment of the quality of ageing seed lots, we focus mainly on the membrane damage that is manifested by a detectable increase in permeability and its association with seed vigour and viability.

2.7 Membrane Integrity in Imbibed Orthodox and Recalcitrant Seeds

From the preceding section it is apparent that imbibed orthodox and recalcitrant seeds would have membranes in their lamellar configuration, thus having normal structure and function. In such a state, provided water is not a limiting factor, non-dormant orthodox and recalcitrant seeds would have normal germination. In the hydrated state, a consequence of peroxidation is the accumulation of gel-phase liquid domains within the membrane and thus the decreased motional freedom of fatty acyl chains (Pauls and Thompson 1980, 1981), resulting in a loss of membrane integrity and an increase in membrane permeability (Barber and Thompson 1980). Thus, any change in the membrane integrity in imbibed orthodox seeds will be expected to be due to ageing and microbial deterioration. On the other hand, in recalcitrant seeds that fail to sustain a water supply from an external source, the available water may become a limiting factor, resulting in desiccation injury; ageing and microbial deterioration may further aggravate the deterioration (King and Roberts 1979). Thus, membrane damage in seeds in such a situation would be due to desiccation, ageing and microbial deterioration.

2.8 Membrane Integrity in Orthodox and Recalcitrant Seeds in the Dry State

Drying of seeds upon maturity results in the loss of membrane integrity (Halmer and Bewley 1984), since the minimum level of hydration (>20%) necessary for the retention of the normal lamellar phase will not be available (Simon 1984; Steponkus 1984).

2.9 Membrane Integrity Reorganization in Imbibing Orthodox and Recalcitrant Seeds

Following hydration, in dry orthodox seeds, the constituent components of the membranes reorganize in a lamellar form to revive a normal, continuous configuration. Such restoration of membrane integrity is lacking in recalcitrant seeds (Berjack et al. 1984).

When hydrated seeds are placed in free water with a potential of zero, they will absorb a low amount of water at a relatively slow rate. The relatively low potential gradient thus developing would not be significant enough to cause an inrush of water. Membrane integrity will be the function of turnover and the repair mechanisms involved, ageing and microbial deterioration (Bewley and Black 1982). The rapid inrush of water due to a high potential gradient between very dry seeds and the imbibition medium may lead to the displacement of the membrane components, phospholipids and proteins, far from their original positions, thus disrupting membrane reorganization into a continuous bilayer configuration. In the absence of intact membranes, deleterious mixing of cellular constituents and their loss through leakage into the aqueous medium by diffusion into, and dispersal by, the turbulent flow of water may follow (Simon 1984). An excessively rapid inrush of water into the seeds may result in imbibition injury due to cellular membrane rupture (Powell et al. 1986). However, this situation is reversed with time, with the membranes either physically reverting to their most stable configuration, and/or else being repaired by some unidentified mechanism, probably metabolic repair (Simon 1984; Tilden and West 1985; Pandey 1988a, 1989a). In low or non-viable seeds such a repair mechanism might be inefficient or absent, or excessive membrane disruption might make repair impossible (Halmer and Bewley 1984; Simon 1984; Pandey 1989a). Thus, the extent of leakage of cellular constituents would depend on membrane reorganization and repair processes.

2.10 Manifestations of Membrane Integrity

The increase in permeability due to loss of membrane integrity can be easily shown by measuring the leakage of electrolytes and other cytoplasmic components to an external medium (Parrish and Leopold 1978).

In stored, dry seeds, oxidative stress and free radical-mediated damage, involving lipid peroxidation damage to membranes, probably disrupts membrane reconstitution and repair during imbibition, which results in more leakage of substances from deteriorated seeds (Abdul Baki and Anderson 1972; Halmer and Bewley 1984; Wilson and McDonald 1986). Thus, the extent of membrane disorganization can usually be estimated by the magnitude of solute leakage into seed steep water. However, the unaged (most viable) seeds also leak a variety of constituents having small molecules, e.g. salts and sugars, into the imbibition medium. The movement of small molecules across the membrane from a highly concentrated cellular medium to the very low concentration outside does not implicate changes in membrane integrity (Abdul Baki and Anderson 1972). On the other hand, relatively more leakage and leakage of larger molecules, such as polypeptides or polynucleotides, whose movement across the intact membrane is usually restricted, would imply a disruption of membrane integrity or a rupture of membranes.

2.11 Seed Composition at the Cellular Level

Seeds contain a variety of cellular constitutents and metabolites, including storage reserves, to sustain and support the nutritional supply to growing embryonic axes until the autotrophic seedling establishes. Basically, in addition to normal metabolites, the seeds have various other organic and inorganic compounds in different storage tissues and in the axes. Constituents like sugars, amino acids, organic acids and a variety of other substances may be components of the metabolic pool (Bewley and Black 1978, 1982).

2.12 Quality of Solutes Leaking Out of the Seeds

Solute loss from seeds through leakage has been reviewed by Simon (1984). A variety of solutes leak out of normal seeds when they are placed in water; this is even more obvious when dry seeds are placed in water (Simon 1984). Almost all cellular constituents, including inorganic salts, sugars, organic acids, amino acids, phenolics, potassium, phosphate, growth substances, polysaccharides, nucleotides and polypeptides, may leak out of the seeds into the imbibition medium. The extent and quality of substances that leak out may depend, among other factors, on the species and seed quality and the factors affecting seed directly or indirectly.

2.13 Electrolyte Leakage in Relation to Total Solute Leakage from Seeds and Seed Quality

Most cellular constituents are acids, bases or their salts, thus electrolytes. The rate of leakage of different substances may not be the same (Simon 1984). However, in some cases, electrolyte leakage may represent the total solute leakage or the extent of leakage of individual substances (Pandey 1989b). The amount of electrolytes escaping an imbibing seed will be a function of its ability to reorganize or repair cell membranes, its mineral ion content and damage to seed coat and tissues. While the first parameter is directly related to the performance potential, the other may not be directly related to it (Hampton and Coolbear 1990).

3 Conductivity

3.1 Basic Considerations

Electrochemical implications of conductivity and conductometry can be found elsewhere (Lingane 1958; Braunstein and Robbins 1971; Willard et al. 1974; Holler and Enke 1984). Conductivity is a measure of the ability of a conductor to carry an electric current. Electrical conductivity of the hydrated seeds or seed steep water will be due to the transport of positively and negatively charged ions (cations and anions) which are free to move through the medium to the electrode with opposite charge under the influence of an applied electric potential. It depends on the concentration, mobility and charge of the particles present in the medium.

The effect of temperature on the conductivity of an ionic solution is not simple, however, in practice, the conductivity of a solution increases at a rate of approximately 2%/°C with rising temperature (US Salinity Laboratory Staff 1954).

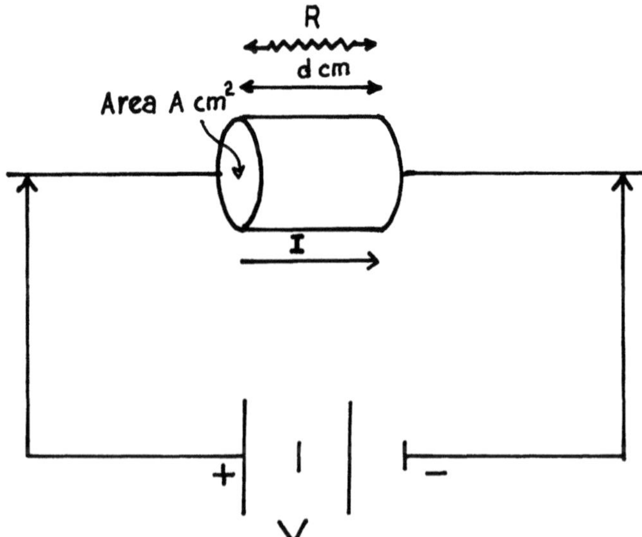

Fig. 1. Current flow through a conductor

When an electric potential is applied through a conductor (Fig. 1), the current-voltage relationship is given by Ohm's law. This states that the flow of current I (in amperes, A) depends on the magnitude of applied voltage V (in volts, V) and resistance R (in ohm, Ω) of the medium between the electrodes, as given by the equation

$$I = \frac{V}{R} = VC \tag{1}$$

where C is the conductance $\left(\dfrac{1}{R}\right)$.

This equation is the basis for all conductance or resistance measurements.

The conductance can be determined by measuring the 1/R ratio. The unit of R (V/A) is ohm (Ω). Since C is the reciprocal of R, the unit of C (A/V) is $\Omega^{-1}\,cm^{-1}$, smaller values are expressed as $m\Omega^{-1}$ or $\mu\Omega^{-1}$ and its SI equivalent S (siemens)/cm, mS or μS/cm. Conductivity testing of seeds may involve current measurement of hydrated seeds or seed steep water involving smaller currents expressed in milliamperes (mA) or microamperes (μA).

3.2 Electromotive Measurement Through Seeds

Since the current flow is inversely proportional to resistance (Eq. 1), when the voltage is held constant, any change in current flow will arise due to the change in resistance. The basic technique involves subjecting seeds to a constant low electrical potential and measuring the electric current which passes through the seed. The use of current measurement for seed testing is based on the reported association of differences in seed vigour and viability with varying resistance (Nelson and Burr 1946; Levengood et al. 1975; Cimino and Belcher 1986). The level of current conducted through the hydrated testas or seed tissues of partially imbibed seeds was related to the seed vigour and viability (Levengood et al. 1975; Cimino and Belcher 1986). Higher resistance, as reflected by the lower current passage through the seed, is considered to be due to dense tissue with strong cellular turgor pressure. Whereas lower resistance, resulting in higher current readings through the seed, has been attributed to weak and deteriorating cells allowing the free exchange of water, and to the failure of empty seeds to provide an impediment against the movement of water. Thus, the difference in current readings in the seeds differing in vigour and viability could at least, to some extent, be attributed to membrane integrity, as both apoplast and symplast are probably relatively good conductors. The main resistance tends to occur at membranes, particularly the plasmalemma (Fensom 1985).

There is no commercially available equipment to perform the current measurements on seeds described in this section. However, the components can be procured commercially and the simple instrument required to suit the experimental conditions can be easily constructed.

Problems inherent to current measurement in plants or their tissues have been highlighted by Fensom (1985). Several precautions are necessary in order to avoid the current drain of the instrument when very small amounts are measured through the medium. The instrument used for current measurements should have a high input impedance.

3.2.1 Preparation of Seeds

Presoaking of seeds by immersion in water (preferably distilled water) is necessary to measure current output (Levengood et al. 1975; Cimino and Belcher 1986). The primary criterion for soaking time was to allow seeds to imbibe water about 5% above the level in the quiscent state, this being a standard level for all experiments. The soaking period may vary with species and even with cultivar, ranging from about 5–90 min to 2 h. After the presoaking, the seeds were blotted dry with absorbent tissue to remove excess surface water and then placed under the electrodes to determine the current.

3.2.2 Current Conduction Through Hydrated Testas

The measurement of current conducted through the hydrated testas of partially imbibed seeds to select individual seeds according to vigour and viability, using a suitable apparatus, has been described by Levengood et al. (1975). The assembly used consisted of a simple circuit having a microammeter, a source of electrical potential and electrodes to connect the seed. The conductivity of the seed was monitored by measuring the current after applying a potential of 1.5 V across two stainless steel electrodes. The partially hydrated seeds were placed between stainless steel electrodes covered with wet paper towelling to ensure electrical contact. An exponential decay reaching a steady state level 2–5 s later followed the rapid rise in current when contact was made. The current is measured on the seeds and their vigour and viability determined under standard conditions. The seed materials tested by this method include varieties of French bean (*Phaseolus vulgaris*), groundnut (*Arachis hypogea*) and cotton (*Gossypium hirsutum*). Selection of individual seeds for growth and vigour prior to germination was based on the measurement of an electric current originating during the initial stage of germination.

3.2.3 Current Conduction Through the Seed Tissues

The device used by Cimino and Belcher (1986) consisted essentially of the same components as described for the measurement of current through the hydrated seed coat except that the electrodes were especially designed to penetrate the seed coat to ensure (1) contact with the imbibed tissue, (2) the application of a relatively higher constant potential of 10 V across the electrodes; and (3) the insertion of a 4800 Ω resistor placed in series with the seed to protect the monitoring meter in case of a short circuit. The value of the series resistor was selected so as to permit a maximum current of 2 mA on the monitoring meter when a seed showed zero resistance. The initial current readings (in mA) taken at the point of contact with the seed tissue were recorded. The method was designed for the tree seed of long-leaf pine (*Pinus palustris*), which has a thick seed coat. Based on the correlation of current passed through the seeds with viability, a formula was developed to estimate the viability of the seed.

3.3 Electromotive Measurement on Seed Steep Water

When seeds are steeped in water, a variety of cellular constituents, including electrolytes involving acids, bases and salts, and ions, leak into the medium. The magnitude of leakage depends on the extent of membrane integrity, its reconstruction and its repair. The ionic species and their concentration may vary with species and cellular constituents, the extent of leakage of electrolytes and their ionization. The conductivity of an electrolyte solution depends on the ion concentration, the characteristic mobility of the ions present and the temperature (Holler and Enke 1984).

3.3.1 Deionized Water

Pure water is a very poor conductor of electricity. Water prepared by distillation, using condensers of silver, tin or resistance glass, can be employed for the conductivity test. Distilled water is used as such or it is further deionized using special ion-exchange resins to decrease the conductivity to the lowest level (Adam 1962). This is referred to as equilibrium water having a very low specific conductance, which is due to a normal amount of dissolved CO_2 present in the air. In the present context the term deionized water is used to designate distilled water having a very low conductivity. For all practical purposes, the conductivity of the water is subtracted from the measurements.

3.3.2 Current Measurements on Seed Steep Water

This test is based on a basic association in individual seeds between high current levels and poor seed quality. This involves steeping of seeds in a definite quantity of deionized water for a standardized period of time at a constant temperature to cause the seeds to release electrolytes into the imbibition medium, the subsequent application of an electromotive potential across the electrodes and the measurement of the electric current passing through the solution. The results of current on seed lots of known vigour and viability under standard conditions (ISTA 1985; AOSA 1986) or in the field, involving suboptimal conditions, are used to ascertain the quality of a test sample using the thus obtained partition point.

Furthermore, information on membrane integrity for physiological purposes and on mechanical damage during processing and handling of seeds can be acquired by using this method.

3.3.2.1 Instrument

There is a commercially available instrument designed for measuring current through a large number of the seed exudates simultaneously. Steere et al. (1981) have described the instrument, so called electronic automatic seed analyzer (Model ASA-610, Agro Sciences, Inc., Ann Arbor, MI, USA), and how it functions. A modified version of this instrument (ASA-220 and ASAC-1000, Neogen Corp. Lansing, MI, USA) has been designed for small seeds and has additional facilities, e.g., a built-in computer system which prints out the germination rate as part of the statistical analysis. This equipment circumvents the mechanical difficulties of placing a large number of seeds in physical contact simultaneously for steep water current measurements necessary for seed quality assessment.

The instrument consists of (1) a plastic soaking tray containing 100 small individual compartments or cells each with about 4-ml capacity; (2) a multielectrode head containing 100 pairs of especially designed electrodes connected to an electric system. When the electric system is placed on top of the soaking tray, one pair of electrodes is submerged in each cell, thus the applied voltage exerts a uniform electric potential across each electrode pair in the cell. (3) Seed analyzer with the LED (light emitting diode) display of the predicted germination percent based on steep water current values obtained from standard seed lots with known vigour and viability; and (4) a printer to give the current values. The measurements can be made on a sample of 100 seeds at a time.

3.3.2.2 Methodology

To measure the individual seed steep water current level, using an automatic seed analyzer, one seed (preferably preweighed) is placed in each of 100 small cells in the soaking tray prewashed with deionized water to completely remove ionic impurities. After filling the cells with deionized water to a standard level above the seed to ensure the equal exposure of electrodes or after adding an equal quantity, e.g. 4 ml to each of the cells, the seeds are allowed to imbibe for a suitable period of time depending on the type of seed (usually $18-24$ h at a suitable temperature, e.g. 25 °C). Thereafter the electrode head, which should be completely clean, is placed over the seeds on the tray. A suitable potential difference (e.g. 0.25 V) is uniformly applied to each electrode pair and the current (mA) passing through each cell's exudate is automatically monitored and printed. Voltage can be changed to 0.5, 1 or 2 to enhance the sensitivity in detecting small amounts of electrolytes in small seed species.

3.3.3 Electrical Conductance Measurements on Seed Steep Water

Thus far, among all the methods involving electromotive measurements, this is the most versatile and frequently used test of seed quality. This involves seed steeping and conductance measurement of the steep water.

3.3.3.1 Bulk Steeping of Seeds in Cold Water

This involves the collective soaking of seeds in replications (Matthews and Bradnock 1967, 1968; Bradnock and Matthews 1970; Yaklick et al. 1979; Brouwer and Mulder 1982; Loeffler et al. 1988; Pandey 1988b). Preferably, the seeds are equilibrated to a moisture content of about 10% (fresh weight basis) to minimize variation arising from moisture differences among seed lots and to prevent the rapid influx of water into seed which would cause excessive imbibition injury (Powell 1986; Loeffler et al. 1988). The seeds are washed quickly with deionized water and dried under a fan to their original weight before being used for steeping. This step may be avoided if the seeds are not treated with chemicals in any way and also if dry seed washings yield no conductance in an initial trial test. The optimum sample size for a species may have to be standardized. Usually 1 to 10 g seeds in each of three to five replications will suffice. In some circumstances, smaller samples are used if steep water conductivity is in a sufficiently readable range. At times, for convenience, one may prefer to use seed number in some cases rather than weight. However, weighing of seeds before steeping is obligatory, since it may avoid variations in results that are likely to arise from differences in seed size. The quantity of water used for seed steeping may depend on the type and weight of the seeds, the extent of electrolyte leakage, statistical considerations and on the minimum quantity of steep water necessary for conductivity measurements with the instrument available. Fifty to 100 ml is sufficient and convenient in most cases. In special cases, an even lower quantity may be used for small samples of tiny seeds, provided the conductivity of the steep water is high enough to be detected satisfactorily. After weighing, the seed samples are placed in suitable containers such as beakers, flasks or high-quality plastic containers with a lid. A suitable quantity of water is added to the seeds. Steeping is carried out at a constant standard temperature, e.g. 25 °C, for a set period, depending on the seed material and the worker, normally ranging from 2 to 24 h or more (Duke and Kakefuda 1981; Givelberg et al. 1984; Schmidt and Tracy 1989). About 16–18 h is suitable for many crops. After steeping, the seeds are removed and the volume of steep water is made up to the original level; the samples are then ready for conductance measurement.

3.3.3.2 Single Seed Steeping in Cold Water

This method has been developed for large seeds exuding a sufficient amount of electrolytes into the steep water (Mullett 1978; Siddique and Goodwin

1985). This has an advantage over bulk steeping in that a few abnormal or bad seeds will not prejudice the seed quality evaluation of a bulk sample. A large number of individual seeds, e.g. 100, are preferably weighed and steeped in about 50 to 80 ml of deionized water with a sufficient number of replications at 25 °C for 16 – 18 h. The time and temperature of steeping and the quantity of water to be used for steeping are adjustable to suit the requirements. The abnormally high value of conductivity arising from defective seeds is eliminated and measurements on a certain number of seeds is averaged and treated as a replication.

3.3.3.3 Bulk Steeping in Hot Water

Steeping of seeds in hot water for 30 min to 1 h may substitute cold water steeping (Pandey 1988b), thus drastically reducing the duration of the test. A suitable number of preweighed seeds with the desired number of replications were placed into 20 ml deionized water in 75-ml test tubes maintained at 58 °C in a stirred water bath. The seeds were retained in hot water for 30 min to 1 h, depending on seed material (Pandey 1988b). At the end of the incubation period, the tubes were immediately removed and placed in another stirred water bath at 25 °C to minimize further efflux of electrolytes into the incubation medium. The incubation water was decanted and the seeds were washed twice with deionized water. The washings were pooled and the volume was made up to 50 ml for conductivity measurements.

3.3.3.4 Single Seed Steeping in Hot Water

As for single seed steeping in cold water, application of this method is restricted to large seeds only. The method followed is essentially the same as for bulk steeping of seeds in hot water except that single, preweighed seeds were steeped in large number to drastically reduce the duration of the test and to avoid the chance of prejudice by a few abnormal or bad seeds. Seeds were incubated in hot water for steeping and the final volume was attained exactly as for the bulk steeping method.

3.3.3.5 Conductance Measurements

Electrolytic conductance measurements usually involve the determination of the resistance of part of the solution between two parallel electrodes. This is converted into reciprocal values for conductance.

In order to facilitate the comparison of values obtained with various electrode assemblies, geometric factors are eliminated from the measurements by expressing the values as a standard unit of specific conductance (SC). The SC of a solution is the conductance of a cylinder of the solution in question 1 cm long and 1 cm^2 in cross-sectional area at a specified temperature:

$$SC = C\frac{d}{A}\,.\tag{2}$$

The factor $\dfrac{d}{A}$ is often referred to as the cell constant, k. The specific conductance is determined by multiplying the conductance value with the cell constant.

The cell constant is rarely determined by actual geometric measurements, but is generally determined empirically by measuring the conductance of a cell filled with a solution of exactly known conductance at a specified temperature. The most common method is the conductance measurement of a standard KCl solution, using the equation:

$$k = \frac{SC}{C}\,,\tag{3}$$

where SC is the known specific conductance of the standard solution (at the specified temperature), and C is the conductance of the standard solution measured with the given cell.

Various methods of conductometric measurements and their advantages and disadvantages have been elaborately discussed by Holler and Enke (1984).

Instrument

Currently, a variety of conductivity meters with direct reading or digital devices, using AC or DC current, are available commercially. The instrument used to measure the conductance of an electrolytic solution has a conductivity cell and an AC or DC circuit. Although equipment having a DC circuit may be simpler, the AC circuit is often preferred since it avoids interferences arising from polarization effects at the electrodes. Is also prevents any buildup of material on a given electrode. The Wheatstone bridge is most often used to determine the resistance or conductivity of an electrolytic solution.

Various kinds of conductance cells are available, their use depending upon the application. The most popular type uses platinum electrodes (1 cm^2 area) separated about 1 cm from each other and oriented in a vertical position to avoid the collection of solids at the electrode surface. The electrodes are coated with a colloidal deposit of platinum black. The electrodes thus facilitate quick and easy measurements as the coating considerably increases the interfacial area and allows faster equilibrium. The conductivity evaluation involves a comparative procedure.

Methodology

Thorough washing of the electrodes and of all items to be used with deionized water is essential. The instrument is prepared for measurements following the instructions of the manufacturer and according to the experimental conditions. The conductivity meter, whether digital or manual, is calibrated with a standard solution with a known conductance at a specified temperature. Most commonly, KCl solutions are used for this purpose, since their specific conductances have been determined with high precision. The specific conductance of 0.01 normal KCl solution at 25 °C is 0.001413 m Ω^{-1} cm^{-1}. The specific conductance of different concentrations of KCl solutions at different temperatures can be found elsewhere (Lange 1967). Preferably, the steep water samples are used for conductivity measurements as such. However, if required, they can be filtered through very low ionic filter paper discs (e.g. Whatman No. 1) or centrifuged to remove coarser particles in samples showing excessive leakage of constituents. If filteration is used, necessary corrections may be incorporated by filtering the standard calibration solution and distilled water and measuring any change in conductivity. Immediate measurement of conductivity is advisable, but in some cases the samples can be preserved at ambient temperature for a few days by adding one or two drops (up to 0.2 ml) of toulene or at 10 °C with traces of toluene, or they can even be frozen and stored for long periods without significant variations in the results. The sample solution is shaken well and the conductance cell is placed so that the electrode bulb is completely immersed inside but does not touch the base of the container; the conductance is then measured. The conductivity of the deionized water used is measured and the value obtained is subtracted from all measurements to eliminate errors. Since the conductance changes with temperature, either all standardization and measurements are carried out at a standard temperature, e.g. 25 °C, or temperature compensation is affected. All values can be standardized by converting them to equivalent values at a reference temperature, e.g. 25 °C, either by using a conversion table (US Salinity Laboratory Staff 1954) or by using a thermocompensator within the measuring unit itself. It may be desirable to express the results in specific conductance on a per gram seed fresh weight basis, so as to nullify the effect of seed size and to facilitate the comparison of results.

4 Factors Affecting Conductivity Measurements

Conductivity measurements may depend on large number of factors that influence the conductivity itself and solute leakage from seeds. These include genotype and pedigree (Schmidt and Tracy 1988), stage of seed maturation at time of harvest (Powell 1986), changes in seed structure and composition during development (Styer and Cantliffe 1983), temperature and drying rate during dehydration (Biddington et al. 1982; Seyedin et al. 1984), degree of seed ageing

(Powell 1986; Priestley 1986; Wilson and McDonald 1986; Senaratna et al. 1988), deterioration and damage to membranes (Givelberg et al. 1984; Powell 1986) including by microbial infection (Harman and Granett 1972) and the physiological effects of drying temperature (Seyedin et al. 1984). Further, conductivity is affected by seed size (Hoy and Gamble 1985; Tao 1978) and seed weight (Hepburn et al. 1984; Siddique and Goodwin 1985), imbibition temperature (Leopold 1980; Murphy and Noland 1982; Givelberg et al. 1984), duration of imbibition (Schmidt and Tracy 1989), steep water volume (Tao 1978), initial seed moisture (Tao 1978; McDonald and Wilson 1979; Loeffler et al. 1988), imbibition injury (Powell 1986; Powell et al. 1986; David 1989) including injury related to a specific temperature-dependent moisture level (David 1889), integrity of testa (Samad and Pearce 1978; Duke and Kakefuda 1981) and its permeability (Powell and Matthews 1979) or pericarp (Wann 1986), seed injury (Tao 1978), seed dehulling (Kuo 1986), cell intactness and rupture (Simon and Raja Harun 1972; Powell and Matthews 1979; McKersie and Stinson 1980; Duke and Kakefuda 1981), effect of temperature mediated by viscosity (Murphy and Noland 1982), various seed treatments including those meant to maintain or improve seed quality (Gorecki and Harman 1987), or to evaluate seed quality (Mugnisjah and Nakamura 1986), condition of steep water, i.e. whether microbial growth persists on stored samples, and finally it is effected by the instrument used for measurement (Hepburn et al. 1984) and the expression of the results, e.g. on the number or on a per gram seed basis (Mullett and Wilkinson 1979; Siddique and Goodwin 1985). Virtually all factors affecting membranes (as discussed in Sect. 2.6), and seed coat may, directly or indirectly, to varying degrees, change conductivity values. Isolated embryonic axes (Powell and Matthews 1978) and cotyledons have a much higher leakage level of solutes, thus they yield high conductivity values.

5 Vigour and Viability Determination of Standard Seed Lots

Germination, the viability parameter, of seed samples is usually determined with a standard test under optimal growth conditions according to ISTA (1985) or AOSA (1986) rules in the laboratory or under glasshouse or field conditions. However, the task of measuring vigour is often compounded because laboratory tests (AOSA 1983) do not always accurately predict performance (Burgass and Powell 1984; Hepburn et al. 1984, 1986; Matthews 1985). The laboratory germination tests may correlate with field emergence under favourable and unfavourable field conditions as in sugar beet, flax and onion. In some cases, e.g. maize, the good emergence predicted from laboratory tests may be obtained only under favourable conditions in the field, whereas there may be drastic differences in laboratory germination and field emergence. Thus, laboratory germination is a basic ultimate test for seed viability in general and vigour in some cases. The field emergence and performance parameters such as emergence, emergence rate, seedling and plant growth rate and yield and related measurements quantify the manifestation of vigour. Practically, the time required for 50% emergence (t_{50}) under field conditions is probably one of the most sought indices of seed vigour. Laboratory germination and growth evaluation may give information on germination and performance of a seed lot. However, in practice, the field performance vigour test is essential for the assessment of seed quality for crop performance and for the evaluation of physiological seed treatments to enhance performance, thus leading to methods to maintain and improve seed quality. Consideration of the seed moisture level with respect to its possible effects on vigour and viability may be essential, as very dry seeds, when brought into contact with water with a high water potential, may suffer imbibition injury.

6 Evaluation of Seed Quality

6.1 Vigour and Viability

6.1.1 Current Conduction Through Hydrated Testas

Individual seed selection for potential viability by measuring the electric current passing through the seed has been reported for several varieties of French bean (*Phaseolus vulgaris*), for groundnut (*Arachis hypogea*) and cotton (*Gossypium hirsutum*; Levengood et al. 1975). The findings of both laboratory germination and field experiments showed an association of small current values with more extensive growth, higher yields, and fewer defective plants. The authors suggested that the seeds of low viability marked by high current values may be removed to upgrade the germination and vigour. Critical levels of current can be determined and used to separate seeds according to vigour and viability. The seeds graded by current levels possessed markedly different vigour, viability and respiration. Thus, by comparing the current levels of a standard seed, whose vigour and viability status was tested under standard conditions, the vigour and viability status of a test seed can be predicted. The method has the advantage that after selection the seeds may be returned to the quiescent state without affecting viability. However, this method has not been investigated further and remains at an experimental stage.

6.1.2 Current Conduction Through the Seed Tissue

This method has been described for the tree seed of long-leaf pine (*Pinus palustris*; Cimino and Belcher 1986). The electrical current passed through the seed of known viability was correlated with germination. Germination percent versus weighted electrical readings (in mA) were obtained by multiplying the

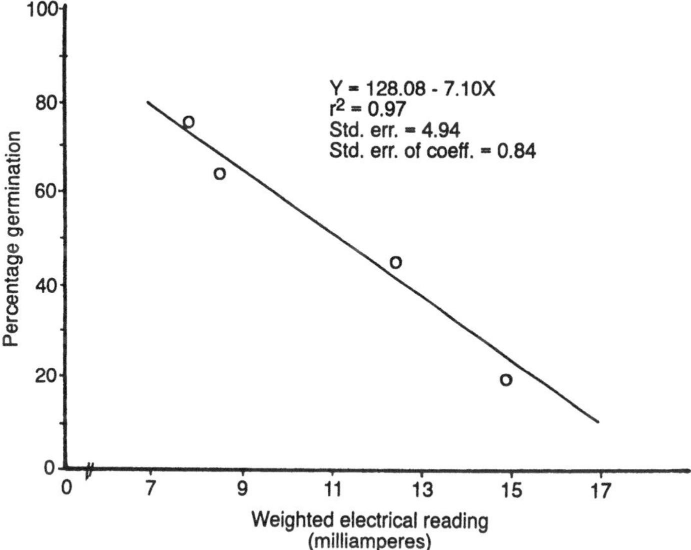

Fig. 2. Linear regression of weighted electrical readings (milliamperes vs germination of long-leaf pine seed. (After Cimino and Belcher 1986)

number of seeds at each current level by the mean of the appropriate electrical category and showed linear regression (Fig. 2). A coefficient of determination of 0.97 was obtained and, based on the data, the formula $Y = 128.08 - 7.10\,X$ was developed to estimate the viability of the long-leaf pine with an average deviation of $\pm 3\%$, where X was calculated as the sum of the weighted average of 400 seeds. The method has yet to be explored further and has rarely been applied to seed quality assessment. Moreover, this method is destructive as the seeds used for the test cannot be preserved unharmed.

6.1.3 Conductivity Measurements on Seed Steep Water

This is widely used as an indirect method to detect vigour and viability. Parrish and Leopold (1978) found an increase in steep water conductivity prior to a decline in seed vigour. Provided mechanical or imbibition injury or membrane rupture is not present, the lower the membrane integrity, the greater the electrolyte leakage into steep water, thus greater conductivity and vice versa (Powell et al. 1984). The conductivity was related inversely to germination (Matthews and Bradnock 1967) and emergence (Matthews and Bradnock 1968). Seeds with damaged or cracked coats and ruptured membranes may show extensive leakage of solutes, which often caused turbidity in the steep water (Duke and Kakefuda 1981), thus yielding very high conductivity (Matthews and Rogerson 1976; Powell and Matthews 1978, 1979). Though the steep water conductivity test is rapid, precise and easy to carry out and is physiologically informative, it is prerogative that the conductivity results should sufficiently correlate with laboratory germination and field emergence and performance. For precise and reliable results individual cultivars or, if possible and feasible, seed lots with a particular ageing regime with respect particularly to moisture and temperature during storage, may need calibration.

6.1.3.1 Single Seed Steep Water Current Measurement

This method has been investigated in several species (Table 1) and has been suggested for seed vigour and viability assessment. This requires a relatively expensive instrument. The current flow through seed steep water is determined in a desired sample size, e.g. 400 seeds in batches of 100 seeds at a time, by a multi-cell electrode assembly similar to the automatic seed analyzer described earlier. Viability and vigour of the seed lot are determined by standard methods. By comparing the current values with viability, a single partition point of current is determined for each seed species to separate viable and poorly viable seeds. The seeds with a steep water current less than the partition value indicate good quality and values greater than the partition point indicate poor quality seeds. Provided the total number of seeds, whose current values are smaller than the partition point, is correlated to the germination test, the total number of good quality seeds compared to 100 seeds tested would be the percent or germination rate.

To detect vigour, a population of good versus poor quality seed lots and the location of this population with respect to the partition point, are compared. The germinable seeds with current values close to the partition point are less vigorous than those germinable seeds with current values much smaller than

Table 1. List of some of the species in which single seed steeping, using an automatic seed analyzer, has been investigated for conductivity testing of vigour and viability

Species	Reference
Triticum aestivum − wheat	Heslehurst (1988)
Zea mays − corn	Waters and Blanchette (1983); Smith and Grabe (1985); Tracy and Juvik (1988)
Glycine max − soybean	McDonald and Wilson (1979, 1980); Steere et al. (1981); Oliveira et al. (1984); Moore et al. (1988)
Phaseolus vulgaris − French bean	Steere et al. (1981)
Brassica oleracea − cabbage	Hill et al. (1988)
− Brussels sprouts	Thornton et al. (1990)
Lycopersicum lycopersicon − tomato	Argerich and Bradford (1989)
Lactuca sativa − lettuce	Moore et al. (1988)
Helianthus annus − sunflower	Anfinrud and Schneiter (1984)
Gossypium spp. − cotton	Steere et al. (1981); Perl and Feder (1983)

the partition point. Similarly, vigour can be detected by determining a vigour partition point using established laboratory vigour tests, which correlate to field performance, or can be detected directly by using the values of field performance parameters of measured current values for calibration.

Regression methods using probit analysis of conductivity and germination data from aged seeds are described by Heslehurst (1988). A somewhat better way to use single seed steep water conductivity measurements to estimate seed quality has been developed by Moore et al. (1988). This method involved finding the derivative (d) of the cumulative frequency (CF) distributions with respect to a class interval of 1 µA electrical current with the natural logarithmic form of Richards function and to make an effective estimation of the slopes [(d CF/d µA) Max] of hypothetical lines tangent to inflection points of the respective sigmoidal cumulative frequency distribution curves. This maximum (Max) internal slope is used as a seed viability index. The greater the internal slope, the less variation there is among individual seed conductivities and the higher the seed quality.

6.1.3.2 Single Seed Steep Water Conductance

This test is based on a high negative correlation between individual seed conductivity and the percent normal seedlings of different seed lots of French bean under field conditions (Mullett 1978; Siddique and Goodwin 1985). The test gives an indication of the membrane integrity of single seeds. A few bad seeds do not prejudice the evaluation of bulk samples. The test involves the determination of a level of conductivity to judge the quality of a seed lot through population studies. Siddique and Goodwin (1985) have investigated and standardized the optimum conditions for predicting germinability of French bean seeds. Using individual seed steep water conductivity and field emergence, a critical conductivity level is determined for each cultivar. The critical conductivity is determined as the value of electrical conductivity between good and poor quality seed lots (Fig. 3). The seeds showing a conductivity value less than the critical level were classed as members of the low conductivity group. The percent seeds in the low conductivity group and the percent normal seedlings obtained in the seedling evaluation test were highly correlated (Siddique and Goodwin 1985). The use of a critical conductivity level in classifying the seeds into low and high conductivity groups appeared to be highly effective in predicting the quality of seeds in terms of the seed to produce normal seedlings. However, the critical level of electrical conductivity needs to be predetermined for each cultivar. This test must be explored further for other large seeded crops. The optimum steeping period of 16 h can probably be minimized to about 0.5 h by steeping seeds in hot water, e.g. at 58 °C (Pandey 1988 b).

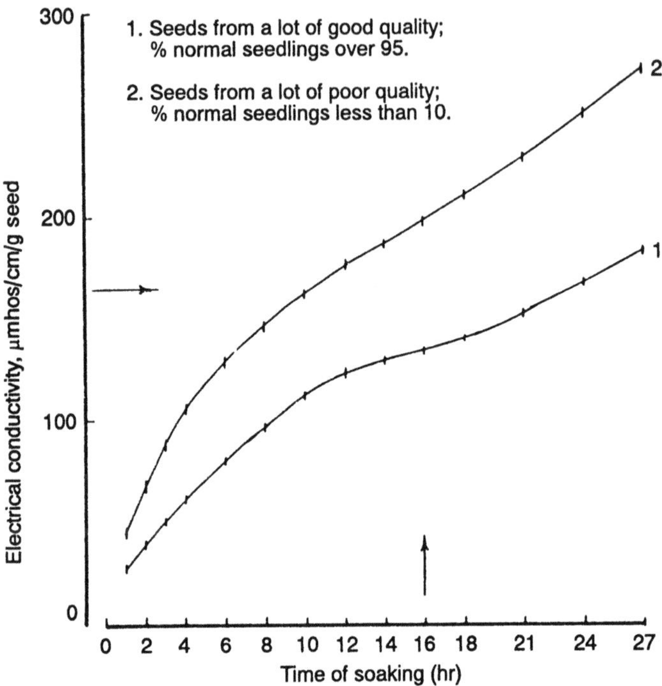

Fig. 3. Increase in seed conductivity in Apollo bean with time of soaking in 80 ml of distilled water at 25 °C. Each point is a mean of three replications, each with 20 observations. *Vertical arrow* denotes the critical conductivity level used in classifying the seeds in the low or high conductivity groups. *Vertical bars*, SE. (After Siddique and Goodwin 1985)

6.1.3.3 *Bulk Seed Steep Water Conductance*

Though suitably developed for peas and French bean (Matthews and Bradnock 1967, 1968), the test has since been evaluated and suggested for use in a wide range of species (Table 2). Contrary to single seed steeping, bulk steeping may reduce the variation in conductivity and thus for the test. Moreover, it simultaneously sustains the risk of prejudice by a few damaged or bad seeds that cannot be manually eliminated. This test meets most of the standards envisaged for a suitable test; it is simple, relatively inexpensive, rapid to perform and highly predictive of field performance. The test requires a conductivity meter which is relatively much less expensive than automatic seed analyzers, and the method is more versatile.

With this method the extent of deterioration in membrane integrity due to ageing is discernible in the magnitude of electrolyte efflux into hot water and it is almost as good at predicting the extent of membrane integrity, seed vigour and viability as cold water steeping (Pandey 1988b). The hot water steeping test is quicker than the cold water method and may be of practical importance due to its rapidity in seed quality assessment, where bulk steeping in cold water

Table 2. List of some of the species in which bulk seed steeping, using a conductivity meter, has been investigated for conductivity testing of vigour and viability

Species	Reference
Triticum aestivum – wheat	Don et al. (1981); Sheppard et al. (1989)
Hordeum vulgare – barley	Sheppard et al. (1989)
Zea mays – maize	Waters and Blanchette (1983)
Oryza sativa – rice	Kuo (1986)
Lolium multiflorum – Italian grass	Marshall and Naylor (1985)
Glycine max – soybean	Yaklick et al. (1979); AOSA (1983); Loeffler et al. (1988); Adam et al. (1989)
Pisum sativum – pea	Matthews and Bradnock (1967, 1968); Bustamante et al. (1984)
Phaseolus vulgaris – French bean	Matthews and Bradnock (1968); Brouwer and Mulder (1982)
Allium cepa – onion	Dearman et al. (1986); Pandey (1988b)
Lycopersicum lycopersicon – tomato	Anfinrud and Schneiter (1984); Pandey (1988b)
Ricinus communis – castor	Thomas (1960)
Gossypium spp. – cotton	Anderson et al. (1964); Bishnoi and Delouche (1980)
Pinus spp. – loblolly and slash pines	Bonner (1986)

is correlated with field emergence, and may prove as good as bulk or single seed steeping in cold water. However, the conductance results may vary with experiments due to small changes in temperature during incubation or in the duration of incubation. This can be overcome either by running controls of a high vigour seed lot with each test or by closely monitoring the temperature of the incubation medium and the duration of incubation. The method could probably be further refined.

6.1.3.4 Limitations of Steep Water Conductivity Testing of Vigour and Viability

This test depends on the degree of negative relationship between conductivity and seed vigour and viability under field conditions. It cannot be applied to seeds in which conductivity values have poor or no relationship with field emergence. Basically, all major factors that are known to influence vigour and viability of seed lots, including genotype, nutrition and growth conditions of the mother plant, physiological maturity of seed at harvest, physical handling of the seed during processing, and seed moisture content and temperature during storage (Perry 1981) along with any applied treatment, may affect the ability to repair and reorganize cell membrane during imbibition, which in turn may influence conductivity values and vigour and viability. Furthermore, vigour and viability are subject to influence by variation in the planting environment. Differences among seed lots due to these attributes, which cause significant variation in the results, may affect the evaluation of seed quality. Thus, details of the effects of variables on results for individual cases may have to be worked out to determine the reliability and versatility of the test. Standard vigour tests described by Fiala (1987a) often, but not always, correlate better with field emergence than the standard germination test (Perry 1984; Fiala 1987b), with the exception of the conductivity test for peas, which fulfills the requirements of a reliable vigour test, accepted internationally. The conductivity test is routinely used to rank pea seed lots since, in general, it gives a good correlation with field emergence of high ranking seed lots (Hampton and Coolbear 1990). Although the conductivity test has the tremendous advantages of simplicity and rapidity and meets most of the requirements for a good vigour test, there have been limitations in extending the test to other species. Provided initial seed moisture is similar, the mineral ion content and damage to the seed or seed coat, if any, does not prejudice – or significantly affect – the conductivity results measured under standard conditions. The restriction of comparisons within cultivars and, if possible, within seeds with an identical condition of variables, may hold promise for a range of other species with greater reliability. For instance, in soybean seeds, comparison within cultivars may be more reliable (Hampton and Coolbear 1990). Tests involving the assessment of individual seeds in a population using an ordinary (Siddique and Goodwin 1985) or multi-cell (McDonald and Wilson 1980) conductivity meter have been preferred.

The conductivity test cannot discriminate among the seeds, e.g.: (1) viable but damaged seeds from non-viable deteriorated or damaged ones; (2) heat or chemically killed seeds from viable ones; (3) damaged or deteriorated seeds from insect- or pathogen-infested ones; (4) dead or viable seeds from dormant; (5) normal seeds from developmentally, anatomically or physiologically defective ones; and so on, which therefore may sometimes yield comparable conductivity values.

6.2 Membrane Integrity Detection

More leakage of solutes, thus higher conductivity values, from less viable or vigorous seeds (Simon and Raja Harun 1972; Leopold 1980; Powell and Matthews 1981) may represent aged, damaged, or non-functional cellular membranes (Simon and Raja Harun 1972; McKersie and Stinson 1980; Murphy and Noland 1982) and cellular rupture caused by imbibition damage (Powell and Matthews 1979; Duke and Kakefuda 1981). While the loss of membrane integrity may elicit physiological and biochemical deteriorative changes in macromolecules, particularly the membrane lipids, the manifestation of membrane integrity deterioration together with the ensuing loss of essential cell constituents, which may serve to attract microorganisms, including pathogens, may form the basis of vigour and viability loss. As reviewed by Wilson and McDonald (1986) and Benson (1990), the conductivity measurement is used as an indirect approach to determine predisposed membrane integrity deterioration through ageing, especially involving oxidative and free radical stresses, both in recalcitrant and orthodox seeds. The detection of membrane integrity through conductivity measurements is useful in studying the physiological and biochemical aspects of seed ageing and seed treatments involving cellular membranes. However, the conductivity results need careful interpretation, since a possible contribution of cells suffering mechanical/imbibition injury and/or dead cells or tissues may lead to erroneous and spurious conclusions which cannot be distinguished from the manifestation of membrane integrity per se.

Membrane integrity is often tested by steeping seeds as such (Tilden and West 1985) or by steeping excised embryonic axes (Ferguson et al. 1990) individually or in bulk, measuring conductivity with a single or multi-cell electrode assembly. Solute leakage into steep water has been accepted as a test for membrane integrity (Parrish and Leopold 1978) and is widely used to study the effects of natural and accelerated ageing (McKersie and Tomes 1980; Givelberg et al. 1984; Wilson and McDonald 1985; Senaratna et al. 1988; Benson 1990), high and low temperature stresses (Leopold 1980; McKersie and Tomes 1980; Gray and Steckel 1983; Givelberg et al. 1984; Vertucci 1989), desiccation (Biddington et al. 1982; Argerich and Bradford 1989; Herter and Burris 1989), and physiological treatments, e.g. antioxidants for retarding ageing (Gorecki and Harman 1987) and as a physiological basis for seed vigour and viability (Powell et al. 1986).

6.3 Priming, Membrane Reconstitution and Repair Study

Conductivity measurement has been used to study membrane reconstitution and repair during priming and to elucidate the physiological basis of such processes in aged seeds (Tilden and West 1985; Pandey 1988a, 1989a).

6.4 Solute Leakage

Conductivity measurement of seed steep water has been adopted as a laboratory procedure to assess solute leakage from the seeds (AOSA 1983; Powell 1986) and is used to study the various aspects of leakage and loss of solutes.

6.5 Imbibition Injury/Damage

Damage to seeds during imbibition following the rapid inrush of water (imbibition injury), which was discussed earlier, is monitored by measuring the conductivity of the steep water with steeping time (Powell et al. 1986; Wann 1986). Similarly, a conductivity test has been employed for studying dehydration effects on imbibitional leakage from desiccation-sensitive seeds (Becwar et al. 1982).

6.6 Permeability Characteristics

The phenomenon of hardseededness and its extent may be determined by conductivity measurements. Hard seeds may imbibe very slowly or not at all in a given period of time. For instance, hard seeds of French bean may not imbibe in 24 h, as evidenced by the lack of increase in the conductivity of the water in which the seeds are placed. Growing significance of permeability characteristics is seen (an observation common to a large number of grain legume crops) in the association of rapid imbibition and poor emergence in legume crops (Powell 1989) and in the association of the delayed permeability of seed coat with good quality (Kuo 1989). The electrical conductivity test has proved to be an effective screening method to detect the delayed permeability of coats in soybean, which is suggested to be a promising factor in producing soybean seeds of good quality (Kuo 1989). Further, the change in the electrolyte leaching rate during imbibition was found to be similar to the water absorbing rate. Thus, an indirect test of water absorption by measuring conductivity is possible.

6.7 Mechanical Damage/Injury

The conductivity test may provide immediate information on the quality of seed before and after processing with respect to detecting mechanical damage or injury. Single seed steep water conductivity may prove to be a useful method for this purpose. High conductivity values produced by injured or damaged seeds due to excessive solute leakage may conspicuously distinguish such seeds from normal ones.

7 Conclusion

Within the known limitations, the conductivity test has been widely investigated and used as a destructive method to asses seed quality and to provide physiological information related to membrane integrity in seeds. The possibility of comparing performance of an individual seed or seed lot in both the conductivity test and in the seedling evaluation test needs to be investigated. In many species seeds with good vigour do not lose viability when soaked in cold water for the duration required for the test. At least, in some cases, the precious planting material used for the conductivity test could be utilized subsequently if the vigour of the seeds has not declined appreciably. This would make the test semi- or non-destructive. However, the soaked seeds are often used for the enzyme assay. Conductivity testing of seeds appears to be one of the most promising areas for continued study for better assessment of the planting value and/or storage potential of seed lots and for the evaluation of physiological seed treatments to maintain or enhance seed quality involving membrane integrity and seed coat characteristics. Generation of information on the extent of membrane integrity deterioration under different sets of conditions in a cultivar of a species in relation to seed quality status under desired and/or varying environmental conditions and the use of such information to predict seed quality may make the test more precise and reliable. The relationship of membrane integrity deterioration (discernible by conductivity under different conditions) to physiological and biochemical changes and seed quality, and the interdependence of the deteriorative processes, provide the basis for the test. This helps to eliminate fortuitous correlations possibly arising from the coincidental interaction of different variables.

References

Abdul Baki AA, Anderson JD (1972) Physiological and biochemical deterioration of seeds. In: Kozlowski TT (ed) Seed biology, vol 2. Acad Press, London, pp 283–315

Adam NK (1962) Physical chemistry. Oxford University Press, Amen House, London, pp 355–440

Adam NM, McDonald MB Jr, Henderlong PR (1989) The influence of seed position, planting and harvesting dates on soybean seed quality. Seed Sci Technol 17:143–152

Anderson AM, Hart JR, French RC (1964) Comparison of germination technique and conductivity tests of cotton seeds. Proc Int Seed Test Assoc 29:81–96

Anfinrud MN, Schneiter AA (1984) Relationship of sunflower germination and vigour tests to field performance. Crop Sci 24:341–344

Argerich CA, Bradford KJ (1989) The effect of priming and ageing on seed vigour in tomato. J Exp Bot 40:599–607

Association of Official Seed Analysts (AOSA) (1983) Seed vigour testing handbook. Publ No 32, AOSA, USA

Association of Official Seed Analysts (AOSA) (1986) Rules for testing seeds. J Seed Technol 6:1–125

Barber RF, Thompson JE (1980) Senescence-dependent increase in the permeability of liposomes prepared from bean cotyledon membranes. J Exp Bot 31:1305–1313

Becwar MR, Stanwood PC, Roos EE (1982) Dehydration effects on imbibitional leakage from desiccation-sensitive seeds. Plant Physiol 69:1132–1135

Benson EE (1990) Free radical damage in stored plant germplasm. International Board for Plant Genetic Resources, Rome

Berjak P, Dini M, Pammenter NW (1984) Possible mechanisms underlying the differing dehydration responses in recalcitrant and orthodox seeds: desiccation-associated subcellular changes in propagules of *Avicennia marina*. Seed Sci Technol 12:365–384

Berjak P, Farrant JM, Mycock DJ, Pammenter NW (1990) Recalcitrant (homoiohydrous) seeds: the enigma of their desiccation-sensitivity. Seed Sci Technol 18:297–310

Bevilacqua LR, Forrati F, Dondero G (1987) 'Callos' in the impermeable seed coat of *Sesbania punicea*. Ann Bot 59:335–341

Bewley JD, Black M (1978) Physiology and biochemistry of seeds in relation to germination, vol 1. Springer, Berlin Heidelberg New York

Bewley JD, Black M (1982) Physiology and biochemistry of seeds in relation to germination, vol 2. Springer, Berlin Heidelberg New York

Biddington NL, Dearman AS, Thomas TH (1982) Effects of temperature and drying rate during dehydration of celery seeds on germination, leakage and response to gibberellin and cytokinin. Physiol Plant 54:75–78

Bishnoi UR, Delouche JC (1980) Relationship of vigour tests and seed lots to cotton seedling establishment. Seed Sci Technol 8:341–346

Bonner FT (1986) Measurements of seed vigour for loblolly and slash pine. For Sci 32:170–178

Borochov A, Halevy AH, Shinitzky M (1982) Senescence and the fluidity of rose petal membranes. Plant Physiol 69:296–299

Bradnock WT, Metthews S (1970) Assessing field emergence potential of wrinkle seeded pea. Hortic Res 10:50–58

Braunstein J, Robbins GD (1971) Electrolytic conductance measurements and capacitive balance. J Chem Educ 48:52–59

Brouwer HM, Mulder JC (1982) Reduced steeping time for the conductivity vigour test of *Phaseolus vulgaris* L. J Seed Technol 7:84–96

Burgass RW, Powell AA (1984) Evidence for repair processes in the invigoration of seeds by hydration. Ann Bot 53:753–757

Bustamante L, Seddon RD, Rennie WJ (1984) Pea seed quality and seedling emergence in the field. Seed Sci Technol 12:551–558

Chapman D, Wallach WFH (1968) Recent physical studies of phospholipids and natural membranes. In: Dennis Chapman (ed) Biological membranes – physical fact and function. Acad Press, New York, pp 139–144

Cimino DG, Belcher EW (1986) Estimating tree seed viability by electromotive measurements. Seed Sci Technol 14:169–175

David R (1989) Dynamics of imbibition in *Phaseolus vulgaris* L. in relation to initial seed moisture content. Plant Physiol 89:805–810

Dearman J, Brocklehurst PA, Drew RLK (1986) Effect of osmotic priming and ageing on onion seed germination. Ann Appl Biol 108:639–648

Delmer DP (1979) Dimethyl sulfoxide as a potential tool for analysis of compartmentation in living plant cells. Plant Physiol 64:623–629

Don R, Rennie WJ, Tomlin MM (1981) A comparison of laboratory vigour test procedures for winter wheat seed samples. Seed Sci Technol 9:641–653

Duke SH, Kakefuda G (1981) Role of the testa in preventing cellular rupture during imbibition of legume seeds. Plant Physiol 67:449–456

Ellis RH, Roberts EH (1980) Improved equations for the prediction of seed longevity. Ann Bot 45:13–30

Farrant JM, Pammenter NW, Berjak P (1986) The increasing desiccation sensitivity of recalcitrant *Avicennia marina* seeds with storage time. Physiol Plant 67:291–298

Farrant JM, Pammenter NW, Berjak P (1988) Recalcitrance – a current assessment. Seed Sci Technol 16:155–166

Farrant JM, Pammenter NW, Berjak P (1989) Germination-associated events and the desiccation sensitivity of recalcitrant seeds – a study on three related species. Planta 178:189–198

Fensom DS (1985) Electrical and magnetic stimuli – hormonal regulation of development. In: Pharis RP, Reid DM (eds) Encyclopedia of plant physiology, vol 11. Springer, Berlin Heidelberg New York, pp 625–652

Ferguson JM, Tekrony DM, Elgi DB (1990) Changes during early soybean seed and axes deterioration. I. Seed quality and mitochondrial respiration. Crop Sci 30:175–179

Fiala F (1987a) Handbook of vigour test methods, 2nd edn. Int Seed Testing Assoc, Zürich, Switzerland

Fiala F (1987b) Report of the vigour test committee 1983–1986. Seed Sci Technol 15:507–522

Givelberg A, Horowitz M, Poljakoff-Mayber A (1984) Solute leakage from *Solanum nigrum* L. seeds exposed to high temperatures during imbibition. J Exp Bot 35:1754–1763

Gorecki RJ, Harman GE (1987) Effects of antioxidants on viability and vigour of ageing pea seeds. Seed Sci Technol 15:109–117

Gray D, Steckel JRA (1983) Freezing injury during seed germination: its influence on seedling emergence in the onion (*Allium cepa*). Seed Sci Technol 11:317–322

Halmer P, Bewley JD (1984) Physiological perspective on seed vigour testing. Seed Sci Technol 12:561–575

Hampton JG, Coolbear P (1990) Potential versus actual seed performance – can vigour testing provide an answer? Seed Sci Technol 18:215–228

Harman GE, Granett AL (1972) Deterioration of stored pea seeds; changes in germination, membrane permeability and ultrastructure resulting from infection by *Aspergillus ruber* and from ageing. Physiol Plant Pathol 2:271–278

Hepburn HA, Powell AA, Matthews S (1984) Problems associated with the routine application of electrical conductivity measurements of individual seeds in the germination testing of peas and soybeans. Seed Sci Technol 12:403–413

Hepburn HA, Goodman BA, Mc Phail DB, Matthews S, Powell AA (1986) An evaluation of EPR measurements of individual seeds in the non-destructive testing of seed viability. J Exp Bot 37:1675–1684

Herter U, Burris JS (1989) Evaluating drying injury on corn seed with a conductivity test. Seed Sci Technol 17:625–638

Heslehurst MR (1988) Quantifying initial quality and vigour of wheat seeds using regression analysis of conductivity and germination data from aged seeds. Seed Sci Technol 16:75–85

Hill HG, Taylor AG, Huang XL (1988) Seed viability determinations in cabbage using sinapine leakage and electrical conductivity measurements. J Exp Bot 39:1439–1447

Holler FJ, Enke CJ (1984) Conductivity and conductometry. In: Kissinger PT, Heineman WR (eds) Laboratory techniques in electroanalytical chemistry. Marcel Dekker, New York, pp 235–266

Hoy DJ, Gamble EE (1985) The effect of seed size and seed density on germination and vigour of soybeans. Can J Plant Sci 65:1–8

International Seed Testing Association (ISTA) (1985) International rules for seed testing. Seed Sci Technol 13:356–513

Kermode AR, Bewley JD (1986) Alteration of genetically regulated syntheses in seeds by desiccation. In: Leopold AC (ed) Membranes metabolism and dry organisms. Cornell University Press, Ithaca, pp 59–84

Kermode AR, Bewley JD (1988) The role of maturation drying in the transition from seed development to germination. V. Responses of the immature castor bean embryo to isolation from the whole seed: a comparison with premature desiccation. J Exp Bot 39:487–497

King MW, Roberts EH (1979) The storage of recalcitrant seeds. Report for the International Board for Plant Genetic Resources Secretariat, Rome, p 96

Kuo WHT (1986) An improved electrical conductivity test for predicting viability of rice seeds. J Agric Assoc China 136:1–5

Kuo WHT (1989) Delayed permeability of soybean seeds: characteristics and screening methodology. Seed Sci Technol 17:131–142

Lange NA (1967) Handbook of chemistry. McGraw Hill, New York, pp 1222

Leopold AC (1980) Temperature effect on soybean imbibition and leakage. Plant Physiol 65:1096–1098

Lerner HR, Ben-Bassat D, Reinhold L, Poljakoff-Mayber A (1978) Induction of 'pore' formation in plant cell membranes by toluene. Plant Physiol 61:213–217

Levengood WC, Bondie J, Chen C (1975) Seed selection for potential viability. J Exp Bot 26:911–918

Levitt J (1980) Response of plants to environmental stresses, vol 2. Acad Press, London

Linder P, Neumann E, Rosenheck K (1977) Kinetics of permeability changes induced by electric impulses in chromaffin granules. J Membr Biol 32:231–254

Lindsay K, Jones MGK (1987) The permeability of electroporated cells and protoplasts of sugarbeet. Planta 172:346–353

Lingane JJ (1958) Electroanalytical chemistry, 2nd edn. Wiley-Interscience, New York, p 669

Lockwood GB (1984) Alkaloids of cell suspensions derived from four *Papavar* spp. and the effect of temperature stress. Z Pflanzenphysiol 114:361–363

Loeffler TM, TeKrony DM, Egli DB (1988) The bulk conductivity test as an indicator of soybean seed quality. J Seed Technol 12:37–53

Lynch DV, Steponkus PL (1987) Plasma membrane lipid alterations associated with cold acclimation of winter rye seedlings (*Secale cereale* L. cv. Puma). Plant Physiol 83:761–767

Marshall AH, Naylor REL (1985) Seed vigour and field establishment in Italian ryegrass *Lolium multiflorum* Lam. Seed Sci Technol 13:781–794

Matthews S (1985) Physiology of seed ageing. Outlook Agric 14:89–94

Matthews S, Bradnock WT (1967) The detection of seed samples of wrinkle-seeded peas (*Pisum sativum* L.) of potentially low planting value. Proc Int Seed Test Assoc 32:553–563

Matthews S, Bradnock WT (1968) Relationship between seed exudation and field emergence in peas and French beans. Hortic Res 8:89–93

Matthews S, Rogerson NE (1976) The influence of embryo condition on the leaching of solutes from pea seeds. J Exp Bot 27:961–968

McDaniels C (1983) Transport of ions and organic molecules. In: Evans DA, Sharp WR, Ammirato PV, Yamada Y (eds) Handbook of plant cell culture, vol 1. Techniques for propagation and breeding. MacMillan Co, New York, pp 696–714

McDonald MB Jr, Wilson DO (1979) An assessment of the standardization and ability of the ASA-610 to rapidly predict potential of soybean germination. J Seed Technol 4:1–11

McDonald MB Jr, Wilson DO (1980) ASA-610 ability to detect changes in soybean seed quality. J Seed Technol 5:56–66

McKersie BD, Stinson RH (1980) Effect of dehydration on leakage and membrane structure in *Lotus corniculatus* (L.) seeds. Plant Physiol 66:316–320

McKersie BD, Tomes DT (1980) Effects of dehydration treatments on germination, seedling vigour, and cytoplasmic leakage in wild oats and birsfoot trefoil. Can J Bot 58:471–476

McKersie BD, Lepock JR, Kruuv J, Thompson JE (1978) The effects of cotyledon senescence on the composition and physical properties of membrane lipid. Biochim Biophys Acta 508:197–212

Moore FD, Jolliffe PA, Stanwood PC, Roos EE (1988) Use of the Richards function to interpret single seed conductivity data. HortScience 23:396–398

Mugnisjah WQ, Nakamura S (1986) Methanol and ethanol stress for seed vigour evaluation in soybean. Seed Sci Technol 14:95–103

Mullett JH (1978) Conductivity testing of bean seed. In: Ballentyne BJ (ed) Bean improvement Workshop, Department of Agriculture, NSW, Australia, pp 105–110

Mullett JH, Wilkinson RI (1979) The relationship between amounts of electrolyte lost on leaching seeds of *Pisum sativum* and some parameters of plant growth. Seed Sci Technol 7:393–398

Murphy JB, Noland TL (1982) Temperature effects on seed imbibition and leakage mediated by viscosity and membranes. Plant Physiol 69:428–431

Nelson OE, Burr HS (1946) Growth correlates of electromotive forces in maize seeds. Proc Natl Acad Sci USA 32:73–84

Oliveira MA, Matthews S, Powell AA (1984) The role of split seed coats in determining vigour in commercial seed lots of soybean, as measured by the electrical conductivity test. Seed Sci Technol 12:659–668

Osborne DJ (1980) Senescence in seeds. In: Thimann KV (ed) Senescence in plants. CRC Press, Boca Raton, pp 13–38

Pandey DK (1988a) Electrolyte efflux into hot water as a test for predicting the germination and emergence of seeds. J Hortic Sci 63:601–604

Pandey DK (1988b) Priming induced repair in French bean seeds. Seed Sci Technol 16:527–532

Pandey DK (1989a) Ageing of French bean seeds at ambient temperature in relation to vigour and viability. Seed Sci Technol 17:41–47

Pandey DK (1989b) Priming induced alleviation of the effects of natural ageing derived selective leakage of constituends in French bean. Seed Sci Technol 17:391–397

Parrish DJ, Leopold AC (1978) On the mechanism of ageing in soybean seeds. Plant Physiol 61:365–368

Pauls KP, Thompson JE (1980) In vitro stimulation of senescence-related membrane damage by ozone-induced lipid peroxidation. Nature 283:504–506

Pauls KP, Thompson JE (1981) Effects of in vitro treatment with ozone on the physical and chemical properties of membranes. Physiol Plant 53:255–262

Pauls KP, Thompson JE (1984) Evidence for the accumulation of peroxidized lipids in membranes of senescing cotyledons. Plant Physiol 75:1152–1157

Penny P, Penny D (1978) Rapid responses of phytohormones. In: Letham DS, Goodwin PB, Higgins TJV (eds) Phytohormones and related compounds: a comprehensive treatise, vol 2. Elsevier/North Holland, Amsterdam

Perl M, Feder Z (1983) Cotton seed quality predection with the automatic seed analyzer. Seed Sci Technol 11:273–280

Perry DA (1981) Handbook of vigour test methods. International Seed Testing Association, Zürich, Switzerland

Perry DA (1984) Commentary on International Seed Testing Association Vigour Test Committee Collaborative Trials. Seed Sci Technol 12:301–308

Phodes D, Stewart GR (1974) A procedure for the in vivo determination of enzyme activity in higher plant tissue. Planta 118:133–144

Powell AA (1986) Cell membranes and seed leachate conductivity in relation to the quality of seed for sowing. J Seed Technol 10:81–100

Powell AA (1989) The importance of genetically determined seed coat characteristics to seed quality in grain legumes. Ann Bot 63:169–175

Powell AA, Matthews S (1978) The damaging effect of water on dry pea embryos during imbibition. J Exp Bot 29:1215–1229

Powell AA, Matthews S (1979) The influence of testa condition on the imbibition and vigour of pea seeds. J Exp Bot 30:193–197

Powell AA, Matthews S (1981) A physical explanation for solute leakage from dry pea embryos during imbibition. J Exp Bot 32:1045–1050

Powell AA, Matthews S, Oliveira MA (1984) Seed quality in grain legumes. Adv Appl Biol 10:217–285

Powell AA, Oliveira MA, Matthews S (1986) The role of imbibition damage in determining the vigour of white and coloured seed lots of dwarf French beans (*Phaseolus vulgaris*). J Exp Bot 37:716–722

Priestley DA (1986) Seed ageing – implications for seed storage and persistence in the soil. Cornell University Press, Ithaca

Roberts DR, Dubroff EB, Thompson JE (1986) Exogenous polyamines alter membrane fluidity in bean leaves – a basis for potential misinterpretation of their true physiological role. Planta 167:395–401

Roberts EH (1973) Predicting the storage life of seeds. Seed Sci Technol 1:499–514

Roberts EH, Black M (1989) Seed quality. Seed Sci Technol 17:175–185

Roberts EH, Ellis RH (1984) The implications of the deterioration of orthodox seeds during storage for genetic resources conservation. In: Holden JHW, Williams JT (eds) Crop genetic resources conservation and evaluation. George, Allen and Unwin, London, pp 18–36

Roberts EH, Ellis RH (1989) Water and seed survival. Ann Bot 63:39–52

Roberts EH, King MW (1980) Storage of recalcitrant seeds. In: Withers LA, Williams JT (eds) Crop genetic resources, the conservation of difficult material. IBPGR/IUBS Series B 42, pp 39–48

Rolston MP (1978) Water impermeable seed dormancy. Bot Rev 44:365–389

Samad IMA, Pearce RS (1978) Leaching of ions, organic molecules, and enzymes from seeds of peanut (*Arachis hypogea* L.) imbibing without testas or with intact testas. J Exp Bot 29:1471–1478

Schmidt DH, Tracy WF (1988) Endosperm type, inbred background, and leakage of seed electrolytes during imbibition in sweet corn. J Am Soc Hortic Sci 113:269–272

Schmidt DH, Tracy WF (1989) Duration of imbibition affects seed leachate conductivity in sweet corn. HortScience 24:346–347

Senaratna T, Gusse JF, McKersie BD (1988) Age-induced changes in cellular membranes of imbibed soybean seed axes. Physiol Plant 73:85–91

Seyedin N, Burris JS, Flynn TE (1984) Physiological studies on the effects of drying temperatures on corn seed quality. Can J Plant Sci 64:497–504

Sheppard SC, Alder V, Evender NG, Rossnagel BG (1989) Relationship between seed vigour and sensitivity of ionizing radiation. Seed Sci Technol 17:205–222

Siddique MA, Goodwin PB (1985) Conductivity measurements on single seeds to predict the germinability of French beans. Seed Sci Technol 13:643–652

Simon EW (1984) Early events in germination. In: Murray DR (ed) Seed physiology, vol 2. Germination and reserve mobilization. Acad Press, Australia, pp 77–115

Simon EW, Raja Harun RM (1972) Leakage during seed imbibition. J Exp Bot 23:1076–1085

Singer S, Nicholson G (1972) The fluid mosaic model of the structure of cell membranes. Science 175:720–731

Smith AJ, Grabe DF (1985) Radiographic density measurements for determination of viability and vigour in corn (*Zea mays*) seeds. Seed Sci Technol 13:759–768

Steere WC, Levengood WC, Bondi JM (1981) An electronic analyzer for evaluating seed germination and vigour. Seed Sci Technol 9:567–576

Steponkus PL (1984) Role of plasma membrane in freezing injury and cold acclimation. Annu Rev Plant Physiol 35:543–584

Styer RC, Cantliffe DJO (1983) Changes in seed structure and composition during development and their effects on leakage in two endosperm mutants of sweet corn. J Am Soc Hortic Sci 108:721–728

Tanaka H, Hirao C, Semba H, Tozawa Y, Ohmomo S (1985) Release of intracellularly stored 5-phosphodiesterase with preserved cell viability. Biotechnol Bioeng 27:890–892

Tao KLJ (1978) Factors causing variations in conductivity test for soybean seeds. J Seed Technol 3:10–18

Thomas CA (1960) Permeability measurements of castor-bean seeds indicative of cold-test performance. Science 131:1045–1046

Thornton JM, Powell AA, Matthews S (1990) Investigation of the relationship between seed leachate conductivity and the germination of *Brassica* seed. Ann Appl Biol 117:129–135

Tilden RL, West SH (1985) Reversal of the effects of ageing in soybean seeds. Plant Physiol 77:584–586

Tracy WF, Juvik JA (1988) Electrolyte leakage and seed quality in shrunken-2 maize selected for improved field emergence. HortScience 23:391–392

Tran VN, Cavanagh AK (1984) Structural aspects of dormancy. In: Murray DR (ed) Seed Physiology, vol 2. Germination and reserve mobilization. Acad Press, Australia, pp 1–44

US Salinity Laboratory Staff (1954) Diagnosis and improvement of saline and alkali soils. US Dep Agric Handb 60, 160 pp

Vertucci CW (1989) Relationship between thermal transitions and freezing injury in pea and soybean seeds. Plant Physiol 90:1121–1128

Wann EV (1986) Leaching of metabolites during imbibition of sweet corn. Crop Sci 26:731–733

Waters L, Blanchette B (1983) Prediction of sweet corn (*Zea mays*) field emergence by conductivity and cold tests. J Am Hortic Sci Soc 108:778–781

Weaver JC, Powell KT, Mintzer RA, Sloan SR, Ling H (1984) The diffusive permeability of bilayer membranes: the contribution of transient aqueous pores. Bioelectrochem Bioenerg 12:405–412

Willard HH, Merritt LL Jr, Dean JA (1974) Conductance methods. Instrumental methods of analysis, 5th edn. Litton Educational Publ, New York, pp 740–770

Wilson DO Jr, McDonald MB Jr (1986) The lipid peroxidation model of seed ageing. Seed Sci Technol 14:269–300

Yaklick RW, Kulik MM, Anderson JD (1979) Evaluation of vigour tests in soybean seeds: relationship of ATP, conductivity and radioactive tracer multiple criteria laboratory tests to field performance. Crop Sci 19:806–810

Yoshida S, Niki T (1979) Cell membrane permeability and respiratory activity in chilling-stressed callus. Plant Cell Physiol 20:1237–1242

Zimmerman U, Pilwat G, Riemann F (1974) Reversible dielectric breakdown of cell membranes by electrostatic fields. Z Naturforsch 29b:304–305

Determination of the Surface Areas of Seed

H. A. VAN DE VENTER and S. DE MEILLON

1 Introduction

A knowledge of the surface area of plant organs such as leaves and fruits is often required in studies on, for example, growth, photosynthetic and transpiration rates, spray coverage, chemical residues, heat transfer and design of peeling systems. Considerable attention has consequently been paid to the development of methods to determine surface area of these organs (Mohsenin 1970).

Determination of the surface area of seeds has found limited application in seed science. Various studies on water uptake and loss (during drying operations) by seeds have, however, required that their surface area be known, and this has prompted the development of a number of methods to determine this physical characteristic. It is assumed that the main interest of plant scientists is in the determination of total seed surface area and these methods are described in some detail in this chapter. Only brief reference to the determination of specific surface area and projected area is made.

2 Total Surface Area

2.1 Coating Methods

Measurement of the surface area of fruits, such as apple, can be performed by peeling in narrow strips and determining the planimeter sum of the areas of tracings of the strips (Baten and Marshall 1943). In the case of objects such as seeds, where peeling is impossible or impractical, an external coat of flexible collodion or silicon rubber solution can be applied to the surface. After the material has set, it can be peeled off. The total area of the material corresponds to the surface area of the seed.

Another approach is to apply a coating in such a manner that it can be expressed as a uniform gain in mass per unit surface area.

2.1.1 Determination of Coating Area

Kamffer et al. (1989) used flexible collodion as a coating to determine surface area of maize kernels. The kernels were suspended on a string, using bent paper clips as clamps, and dipped in flexible collodion. Upon removal, the collodion polymerized in air (within 15 min) to form a flexible, non-elastic, transparent film. The film was coloured by dipping each kernel, with its coating, in a 1% Sudan Black solution for ca 1 min. After a further 10-min drying period, the collodion film was carefully bisected by cutting it with a scalpel. The two film halves were removed from each kernel and the two small openings caused by the clamp tips covered with tiny pieces of masking tape. By making appropriate small cuts, the film was flattened for area determination by means of a calibrated photocell planimeter. For each determination the coatings of ten kernels were used and readings checked three times. Undoubtedly, coating area can also be determined by means of computer image analysis.

To test the accuracy of the above method, steel spheres with a diameter of 7 mm were coated with flexible collodion. To simulate the effect which indentations (e.g. the concave area of the pericarp above the embryo in maize kernels) may have on the results, depressions of specified dimensions were drilled into the steel spheres. Values for the surface area of the indented spheres, as determined experimentally by collodion coating, were compared with values derived by calculation. The highest difference between experimental and calculated values, determined in three experiments, with three replicates each, was 1.32%. It was concluded that the flexible collodion coating method provides highly reliable estimates of the seed surface area of maize kernels.

The values obtained by the flexible collodion coating method for kernels of six maize cultivars were also compared to surface area calculations according to Nemenyi and Szodfridt (1985; see Sect. 2.2). The latter method gave consistently greater estimates (12.4% − 55.8%, depending on the specific cultivar) than the flexible collodion method.

The authors have found (unpubl. results) that the flexible collodion obtained from certain suppliers does not polymerize satisfactorily and is therefore unsuitable for use in the procedure described. It is therefore advisable to test available supplies of flexible collodion for suitability before attempting application of this method.

Although Jindal et al. (1974) suggested the use of a silicon rubber solution as coating, no reference to the use of this material for the determination of surface area of seeds could be found in the available literature.

2.1.2 Determination of Coating Mass

In a method first introduced by Hedlin and Collins (1961), the seed and control objects are coated with an adhesive and a single, randomly packed layer of small metal particles. The attendant change in mass, compared to a change in the mass of control objects with known surface area, gives a measure of the surface area of the seed. Although Hedlin and Collins (1961) developed the method for research on cereal grains, their experiments were conducted on a variety of regular objects differing in shape and density. The procedure has subsequently been used (with modifications) to determine seed surface areas of navy beans, maize, barley, oats and wheat (Bakker-Arkema et al. 1971) and alfalfa, buckwheat, clover, oats, rye and sorghum (Jindal et al. 1974). The following description provides alternatives suggested by modifications in the latter papers. The mesh sizes indicated are US Standard Sieve. The corresponding width of opening (mm) is indicated for each mesh size. The data were obtained from Table 1 of Judson (1927).

Before applying adhesive, the seeds are weighed on an analytical balance.

Varnish appears to be a satisfactory adhesive. Hedlin and Collins (1961) obtained best results using a slow-drying varnish (Berry Bros. Lionoil floor varnish). Spar varnish was used by Bakker-Arkema et al. (1971), while Jindal et al. (1974) used an unspecified varnish.

The seeds are dipped in varnish, after which the excess is removed. This can be achieved by rolling the seeds over a paper towel (Bakker-Arkema et al. 1971), after which they are air dried on a wire screen for about 30 min (with turning to ensure a uniform coating). In the more elaborate method of Hedlin and Collins (1961), the seeds are dipped and then shaken vigorously in about 500 g sand (-4 to $+8$ mesh; $4.76-2.38$ mm) until no more sand adheres to them ($10-15$ s should suffice). Since dry sand removes too much of the varnish, a varnish-coated sand should be used. This is prepared in advance by thoroughly mixing 10 g varnish with 500 g sand and spreading the coated sand to dry in air for about 20 min at room temperature. After the adhesive-coated seeds are removed from the sand, they are allowed to dry in air on a fine screen for $2-3$ min. A uniform layer of adhesive about 0.005 mm thick is obtained in this way.

Jindal et al. (1974) used a technique to apply adhesive coating which eliminated the need for air drying of the seeds. The adhesive-applying material was prepared by sieving and water washing 2.04 kg (4.5 lb) of -8 to $+16$ mesh ($2.38-1.19$ mm) sand and 1.25 kg (2.75 lb) of -14 to $+28$ mesh (1.41-ca. 0.64 mm) sand. About 20 g of varnish was added to the dried sand in each group and thoroughly mixed. The two sand groups were then allowed to air dry for 24 h. Selection of sand size was made according to the size of the seeds. The coarser sand was used for seeds of maize and soybean, while the relatively fine sand was used for smaller seeds such as those of wheat and sorghum.

The sand was placed in a wide-mouth 3.8 l (1 gallon) glass jar. Three glass rods, 13 mm in diameter and 100 mm long, were cemented to the inner wall of each jar to promote agitation. Twenty g of varnish was then added to the varnish-coated sand in the jar. After securing the lid, the jar was rotated for 5 min at 84 rpm on a ball mill tumbling unit. The seed sample was subsequently placed in the jar and rotated with the wet varnished sand for 5 min. As a result of this, a thin, uniform coating of varnish was applied over the entire surface of the seed.

The seed was separated from the sand by using an appropriate sieve to let the sand pass through. Sieving required vigorous shaking because of the sticky state of the contents. Care was taken so that no sand particle remained attached to the seed. No more than 2 min was required for sieving.

Jindal et al. (1974) maintain that a relatively large quantity of the coating medium and sample minimize experimental errors. A sample of about 454 g (1 lb) seed was coated using the above procedure. Their adhesive coating procedure was found to be impractical in the case of small seeds, such as those of clover and alfalfa, and direct varnish application was performed on seeds of these species.

Following application of adhesive, the seeds are coated with metal powder. Hedlin and Collins (1961) listed the selection criteria for a suitable powder. The particle size should be small compared to the object to be measured, in fact, should be small compared to the smallest radii of curvature that commonly occur on the object; the particles should have a high density to provide an easily measured change in mass, and, they should not adhere to each other. In all three of the cited papers a pure nickel powder was used. The powder is described by Bakker-Arkema et al. (1971) as "Sherritt Gordon Mines Limited Nickel Powder No. 843, Grade SF 270×500 screened to 270×325 mesh" ($0.053 - 0.044$ mm). Application of the powder was performed by Jindal et al. (1974) in a jar similar to the one described for adhesive application, by rotating on a ball mill tumbler for 2 min. Hedlin and Collins (1961) used a variety of shaking methods and reported a shaking time of 4 or 5 min to be satisfactory. Bakker-Arkema et al. (1971) swirled their seeds in nickel powder for about 5 s. The seeds can be separated from the nickel powder by rolling them on a wire screen. The coatings are securely attached and the seeds can be handled with forceps without damage. Bakker-Arkema et al. (1971) reported a nickel coating of approximately 0.023 g cm^{-2} seed surface.

After nickel coating, the seeds are weighed. To calculate seed surface area from the mass increase, standard objects such as spheres, having approximately the same size, shape and density as the seeds, are subjected to the same coating procedures. Their surface areas are calculated on the basis of shape and dimensions. The mass increase of the control objects allows calculation of seed surface area. The method permits measurement of the surface area of single objects down to about 0.64 mm^2 (Hedlin and Collins 1961).

It is important to note that for determinations of seed surface area, using the coating mass method, the seeds should be in moisture equilibration with the air in the laboratory at the time of determination. If the seeds are at a moisture content other than equilibrium, mass change due to moisture migration during the test period cannot be distinguished from that due to the added coating. According to Bakker-Arkema et al. (1971), the resulting error can be as high as 50%.

Jindal et al. (1974) conducted tests on the coating mass method using nylon spheres of different dimensions. In each size class, a group of spheres was selected as control while another was treated as sample. The largest difference between calculated and experimental surface areas was of the order of 6.5% with the mean difference being 2.8%.

2.2 Calculation Methods

Various attempts have been made to calculate seed surface areas on the basis of relationships between particular seed dimensions (or other properties such as mass) and surface area, as well as on the basis of the resemblance of their shape to geometric bodies.

2.2.1 Calculation from Seed Dimensions

In a study with kernels of eight maize cultivars, Nemenyi and Szodfridt (1985) found the following relationship between surface area and linear dimensions of the kernels:

$$A = -177.8 + 17.1614\, l + 8.6017\, b + 26.8866\, t \; ,$$

where A = surface area (mm^2), l = length (mm), b = breadth and t = thickness of kernels (mm).

Calculated surface areas were compared to surface areas obtained by removal of pericarps with subsequent area determination on graph paper. Differences between the two values ranged between -3.5 and $+5.4\%$ for kernels of the eight cultivars.

Application of the above formula by Kamffer et al. (1989) to determine surface areas of kernels of six maize cultivars gave estimates consistently greater than those obtained from the flexible collodion method (see Sect. 2.1.1). These authors adjusted the formula of Nemenyi and Szodfridt (1985) by altering the constants for each of the cultivars to arrive at similar values to those obtained by using flexible collodion. They ascribed the fact that the values of the constants differed for different cultivars to variation in shape of the kernels. It can therefore be expected that the values of the constants will also differ for different seed lots of the same cultivar showing variation in kernel shape.

A geometric method was developed for determining the surface area of rough rice by Morita and Singh (1979). The presence of many small ridges on the hull surface (28.802 ridges cm^{-2}) necessitated an elaborate procedure. Precise dimensions and shape of the ridges were determined from electron microscopic pictures of the side, end and top views of the surface. The surface area of a ridge was calculated by applying a mathematical function of the shape of the side and end views. The average density of the ridges per unit basic area of the hull was determined by visually counting the number of ridges in the unit basic area selected on several parts of the electron microscopic pictures of the top view of the hull. The total basic area of the hull was determined by a method similar to that used by Hosokawa and Motohashi (1975). The hull was removed from the kernel and flattened between slide glasses and its area measured by means of an image analyzing computer. The total surface area of the rice hull was calculated from the following relationship:

Total surface area = (basic area)×(density of ridges)
×(surface area of one unit ridge).

Morita and Singh (1979) obtained an estimate which was 80% to 100% larger than values for the surface of rough rice reported by several other researchers and claimed that their estimate should be closer to the actual value than previously published data.

2.2.2 Calculation on the Basis of Resemblance to Geometric Bodies

The resemblance of the shape of several fruits, eggs, vegetables and seeds to geometric bodies such as spheres, prolate or oblate spheroids and ellipsoids can be employed to calculate the surface area using the appropriate formulas applicable to standard shape objects (Jindal et al. 1974). Due to irregularities of seed profiles, it can be assumed that only approximate values for seed surface area will be obtained.

An example of this approach is the calculation of the surface area of gram (*Cicer arietinum* L.) seed performed by Dutta et al. (1988). Taking the shape of the seed to be closest to a prolate spheroid, the surface area was calculated by using the mathematical expression:

$$S = 2\pi b^2 + \frac{2\pi ab}{e} \sin^{-1} e \ ,$$

where a = semi-major axis (mm), b = semi-minor axis (mm) and

$$e = \left[1 - \left(\frac{b}{a}\right)^2 \right]^{1/2} .$$

Wratten et al. (1969) determined seed surface area of rough rice grains on the assumption that they are elliptical in shape. Individual grains were imbedded in wax after which they were sliced into 15 sections by means of a microtome. The major and minor axes of each section were measured through a pre-calibrated microscope. The perimeter of each section was calculated by using the approximate formula:

$$P = \pi \sqrt{\frac{a^2 + b^2}{2}} \ ,$$

where P = perimeter, a = major axis, and b = minor axis.

The average perimeter of two consecutive sections was multiplied by the thickness between the two sections to obtain surface area. Summation of these areas produced the area of the grain, less the area of the two end portions. The end portions were then considered as right circular cones and the surface area of each portion was calculated using the equation:

$$\pi r \sqrt{r^2 + h^2} \ ,$$

where r = one-half the average of the major and minor axis of an end section and h = length.

End portion areas were added to body area to give the total area of one entire grain of rice.

Chuma et al. (1982) developed a method for simultaneous measurement of surface area, volume and size of seeds. The grains and seeds used were those of wheat, rough rice, milled rice and soybeans. The instruments which were used included a shadowgraph, camera, projector, digitizer and a micro-computer. Photographic pictures of the grains and soybean seeds screened on a shadowgraph were taken after being magnified ten times. Cartesian coordinates of the image profiles were read by a digitizer and logged into a micro-computer. The size, surface area and volume were calculated from the coordinates using Langrange's interpolation formula and Simpson's rule. The accuracy of the method was not tested on the grains and soybeans but on specifically shaped objects, viz. a cylinder, sphere, ellipse and a sector. The error in area estimation was found to be 0.5% for the sphere and was largest in the sector at 1.6%.

2.2.3 Calculation from Seed Mass or Volume

Nemenyi and Szodfridt (1985) and Kamffer et al. (1989) have reported relationships between maize kernel mass and surface area.

The relationship between mass and surface area is described by the following exponential formula (Kamffer et al. 1989):

$$A = k_1 k_2^{k_3 \cdot m},$$

where A = surface area, k_1, k_2, k_3 = constants, and m = mass.

The values of k_1 and k_2 differed for each of the cultivars examined (probably because of variation in kernel shape), whereas the value of k_3 was 1 throughout.

Chuma et al. (1982) found high correlation coefficients (r = 0.993 to 1) between volume and surface area of various grains and soybean seeds, indicating that volume may be used to predict surface area.

It should be borne in mind that both seed mass and volume vary with moisture content.

3 Specific Surface Area

An air permeability method, which should serve as a useful tool to calculate the specific surface areas of granular materials such as seeds under packed conditions, was described by Jindal et al. (1974). They defined specific surface as the surface of the pores in a porous medium exposed to fluid flow per unit mass or volume. Their technique, which employs the principles of air flow through porous media, takes into account only the non-contacting surface area available in a packed bed and specific surface values obtained by this method were consequently consistently lower than those obtained by the coating method (described in Sect. 2.1.2).

4 Projected Surface Area

During research on the electrical measurement of moisture content of individual kernels of maize (Kandala et al. 1987), a need arose for the rapid, accurate determination of the kernel projected area. Lawrence et al. (1988) adapted a video-imaging system to measure the projected area of single maize kernels in a multiple-instrument control program. The control program required interpretive BASIC or MS-Pascal. Because these languages were too slow for the video-data processing, an assembly-language subroutine was written to perform the task in satisfactory time. Precision of area measurements was determined to be 0.8% at the 95% confidence level.

A computer vision system was used by Paulsen et al. (1989) to determine the lengths, widths and projected areas of maize kernels. Projected areas agreed within 1.8% of calculated areas.

5 Concluding Remarks

A limited number of methods for the determination of surface area of seeds is available. While the collodion coating method appears to offer the greatest degree of precision, this procedure is not suitable for seeds of small size. Choice of method will be determined by factors such as seed size, seed coat morphology, degree of precision required, intended application and available resources.

References

Bakker-Arkema FW, Rosenau JR, Clifford WH (1971) The effect of grain surface area on the heat and mass transfer rates in fixed and moving beds of biological products. Trans ASAE 14:864–867

Baten WD, Marshall RE (1943) Some methods for approximate prediction of surface areas of fruits. J Agric Res 66:357–373

Chuma Y, Uchida S, Shemsanga KHH (1982) Simultaneous measurement of size, surface area, and volume of grains and soybeans. Trans ASAE 25:1752–1756

Dutta SK, Nema VK, Bhardwaj RK (1988) Physical properties of gram. J Agric Eng Res 39:259–268

Hedlin CP, Collins SH (1961) A method of measuring the surface area of granular material. Can J Chem Eng 2:49–51

Hosokawa A, Motohashi K (1975) Constant drying rate of a single paddy grain. Nogyo-kikai Gakkaishi 37:326–330

Jindal VK, Mohsenin N, Husted JV (1974) Surface area of selected agricultural seeds and grains. Trans ASAE 17:720–725, 728

Judson LV (1927) Sieves and screens. In: Washburn EW (ed) International critical tables of numerical data, physics, chemistry and technology. McGraw-Hill, New York, pp 329–333

Kamffer FH, De Meillon S, Van de Venter HA, Gaigher HL (1989) A simple method for the determination of the surface area of seeds. Plant Varieties Seeds 2:105–108

Kandala CVK, Leffler RG, Nelson SO, Lawrence KC (1987) Capacitive sensors for measuring single-kernel moisture content in corn. Trans ASAE 30:793–797

Lawrence KC, Leffler RG, Nelson SO (1988) Software control of video imaging system for area determination. Int Summer Meet ASAE, St Joseph, MI, Pap 88–3037

Mohsenin NN (1970) Physical properties of plants and animal materials. Gordon and Breach Science, New York

Morita T, Singh RP (1979) Physical and thermal properties of short-grain rough rice. Trans ASAE 22:630–636

Nemenyi M, Szodfridt G (1985) Die die Oberfläche der Körner des Hybridmaises (*Zea mays* L.) beeinflussenden Faktoren. Z Acker- Pflanzenb 154:217–221

Paulsen MR, Wigger WD, Litchfield JB, Sinclair JB (1989) Computer image analysis for detection of maize and soybean kernel quality factors. J Agric Eng Res 43:93–101

Wratten FT, Poole WD, Chesness JL, Bal S, Ramarao V (1969) Physical and thermal properties of rough rice. Trans ASAE 12:801–803

The Discrimination of Seeds by Image Processing

D. G. MYERS

1 Introduction

Digital image processing is a computer-based discipline with a rich theoretical foundation and many subdivisions. One of the most important of these is machine vision; the problem of creating an autonomous visual response for a machine such as a robot. While machine vision covers numerous topics, much of it is concerned with pattern recognition. That is, in techniques for discriminating between various objects in the field of view.

Discriminating seeds by visual observation is a traditional but largely qualitative practice dependent on the skills of the observer concerned and the seeds being assessed. Digital image processing offers a means to making it a fast, quantitative procedure. Until recently, a machine vision approach was theoretically possible but quite impractical. Image processing demands both a vast number of calculations and considerable memory, thus a large mainframe computer was required. Now, however, low-cost personal computers more than meet all the requirements.

2 An Overview

An image is a two-dimensional representation of a three-dimensional scene. Within that scene is some foreground object of interest and a background. If radiation such as sunlight is reflected off the scene and sensed by an appropriate instrument coupled to a computer, then a digital image may be formed. A digital image is one or more arrays of numbers. Each of the numbers − termed a picture element or pixel − represents the energy of the reflected radiation at some point or points in time over some given spectral band width and over some small area.

Processing algorithms used for digital images may be oriented to extracting information on the spatial properties of the image, its spectral, its temporal or any combination of these properties. With regard to seeds, a spatial orientation usually implies a study of the shape of the seed. The assumption is that regardless of distortions to the seed shape caused by environmental, growth and other factors, there remains some intrinsic set of properties which identifies that particular seed. Thus, it is a taxonomic approach to discrimination. A spectral orientation, in contrast, will gain chemical information. However, as spectral measurements are usually confined to broad bands within the visible and infrared spectra, these measurements are only of the bulk chemistry of the visible surface of the seed. Nevertheless, just the measurement of colour alone is sufficient to easily discriminate between many categories of seeds.

The goal of most image processing work with respect to seed discrimination has been to create a "black box" classifier; an instrument which accepts as an input an image of one or more seeds and provides as an output a decision to some desired level of confidence. That decision can be either an identification of the seed or a true/false output if the object is simply to identify whether the seed is of a particular type or not. Being computer-based, such an instrument is very fast – personal computers can generate a result within milleseconds – and it is non-destructive. While an operator can control the process, and that requires an individual with only the most basic computer skills, it is just as easy to program the system for fully automated operations. Because it is programmed, it is very flexible and very easily upgraded or expanded. It is also low-cost. Personal computers are quite common, thus creation of such an instrument often just means some additions to the hardware of an existing unit plus the image processing software.

There are three prime disadvantages. The procedure described is essentially oriented at analyzing just one specimen or a small number of seeds at a time even though it can do this at very high speed. Although spectral properties do relate to the bulk surface chemistry of the seed, nothing is gained on the detailed chemistry, especially biochemistry, of the seed. Finally, it is not always possible to achieve the confidence level desired in the decision.

In very general terms, a machine vision system is not only superior to a human observer for seed discrimination, but around three orders of magnitude faster, more sensitive in many important respects and it gives a quantitative result. The cost also strongly favours the machine approach. However, human observation is more flexible. Machine vision is not sufficiently advanced to be able to deal easily with occluded, that is, overlapping, or closely packed seeds. Consequently, a digital image processing approach requires some means whereby the seeds are spaced and viewed either singularly or in small numbers.

Even with this caveat, image processing has considerable potential in seed discrimination. In particular, as a sampling technique for identifying the constituents of mixtures. In this role, it may be used to identify, for example, the proportion of grass or weed seeds in a sample of cereals, the number of broken nuts following shelling or a potentially new cultivar in a sample from a breeding program. Thus, it has considerable promise in the practical sphere in a variety of quality control applications within both the producing domain and the industrial.

3 Image Processing Systems for Seed Discrimination

A typical image processing system for seed discrimination is shown in Fig. 1. It has at least four subsystems; a sensor to detect the radiation from the scene, a processor to actually perform the processing, a digitizer to convert the sensor output to a form suitable for the processor and a display system to show results. Also needed, of course, is software to control the system.

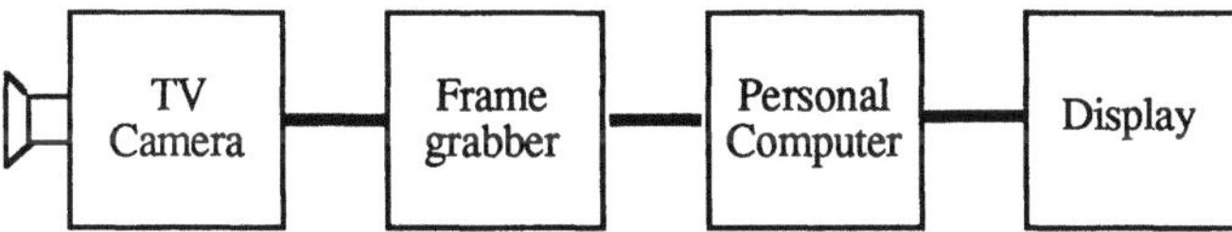

Fig. 1. An image processing system for seed discrimination

Sensor systems are invariably a television camera. If colour is important, then a colour television camera. If a detailed spectral analysis is required, then it must be a scanning spectrometer, but as these tend to be far more expensive than a colour television camera, there is a strong incentive to restrict attention to colour alone. Virtually all modern television cameras will satisfy the requirements of image processing for seed discrimination. However, there is some advantage in having a geometrically stable sensor to ensure that measurements will be nearly identical from one processing session to the next. CCD (charge-coupled device) cameras such as those used in most consumer television camera systems are of this type. The older vidicon cameras are a scanning electron beam device and are not of this type. In addition, it is prudent to avoid interlaced television cameras wherever possible. Broadcast television operates in an interlaced mode. Throughout the world, television transmits 25 or 30 pictures per second in order to give the appearance of smooth motion, but each picture is transmitted as two half-pictures. One of these consists of all the even lines of the picture and the other all the odd. The effective picture rate is therefore 50 or 60 which is necessary to overcome flicker problems. Although the time betweeen scans is small, minor variations will occur within the system and the two half-pictures may not mesh perfectly at the micro-scale. Thus, edges can be slightly shifted and levels slightly different. This creates difficulties for image processing algorithms. Non-interlaced cameras are available from most professional television products companies.

Digitizers to convert television signals to a computer-compatible form are often termed frame grabbers. A frame is simply the technical term for a television image. Thus when activated, the unit waits until it detects the start of the next frame in the signal and then converts it to a stream of numbers. These are usually stored within a memory internal to the unit so that a computer can read the information out with ease. Frame grabbers are, electronically, very simple units and becoming more so with each year. Thus, the choice here is based on what features are desired, what software is supplied with the unit and its compatibility with the available computers. Of these, it is the software which is of most interest. Prospective purchasers need to look for a good library of subroutines which are either accessible to programs written in standard high level languages such as C or which can be readily combined to perform more complex processing tasks.

The processor is usually a personal computer. This needs to be a moderately powerful machine; at least an AT-compatible and preferably better. The reasons for this are based on the nature of image processing itself. Typically, each output image point is derived by processing a small neighbourhood of points about the corresponding point in the input image. An image is a quite large data structure; usually about 256 kbytes, or three times this for colour images. Thus, each image processing operation requires of the order of N/4 million operations and M/4 million memory retrievals, where N is the number of computations performed within each neighbourhood to derive the single output point and M is the number of points in the neighbourhood. Given M is usually 9 and N can be of the order of 20, and that several such operations may be needed in order to derive a result, image processing is a computationally intensive task.

If memory is not sufficiently large to hold both the source image and result, then processing time will be dominated by the time taken to retrieve information off disc. This is likely to be very substantial indeed. Thus memory needs to be of the order of 2 to 4 megabytes for most machines. If the computer has a cache − not essential but certainly useful − then there will be a notable improvement in speed. A co-processor is not necessary as most computations are in integer arithmetic, but it should give some improvement in performance. Given the size of images and that in seed discrimination a large number of images would normally be processed, a hard disc is virtually mandatory. For most applications, this should be at least 40 megabytes and preferably much more.

The one problem area in setting up a system is software. That supplied with the digitizer or available with common packages tends to be low level functions only. However, seed discrimination requires a good repertoire of high level image processing functions. Unfortunately, high level packages tend to be both expensive and targeted at specific applications such as medical image processing. Therefore, some custom software development is usually necessary.

The hardware configuration described seems perfectly adequate for most applications envisaged. The future should see a reduction in costs, some improvement in speed and better integration of subsystems. The issues in the development of these instruments, then, are the algorithms used to reach the decisions and to a lesser extent, the programming of these algorithms for the computer. However, what is also needed is some automated transport mechanism to move the seeds in front of the sensor at speed.

4 Data Acquisition

There are two choices for imaging seeds. If only the shape is required, then the most covenient approach for many seeds is simply to place them on a light table or some similar structure with backlighting. The light can then be adjusted to gain maximum contrast. If more information is desired or an alternative view, then the seeds must be foreground objects in a scene. The scene background, illumination and the viewing geometry are all parameters under the control of the user and can be manipulated to advantage to reduce problems in the image processing computations which follow. The background in this case should be some non-reflective material with a brightness significantly different from the seed. Black is the best choice for most circumstances. The scene illumination needs to eliminate all shadows and that suggests a ring lamp around the camera. If colour is important, then the light source should have a colour temperature to match that of colour television. That is, a white equivalent to 6500 K for PAL systems.

The viewing geometry depends very much on what view of the seed is to be acquired. Because of limitations in the focal lengths of common lenses and so problems of focussing, the usual views considered are either a side view of the seed or a top view with the seed resting under gravity. The latter is much easier to acquire as there is no necessity to align the seed for viewing and backlighting may be used. In addition, of course, the nature of some seeds makes this the only appropriate view. There is also the option of stereoscopy in viewing; the acquisition of multiple views of the scene. This would allow three-dimensional measurements to be made.

Problems with sensors are, on the whole, relatively minor. Most television cameras will accept a variety of lenses, thus focussing properly on the seed is not an issue. The cameras can provide an image of about 512×512 points and that is generally more than adequate. The dynamic range and signal to noise ratio are equally well above the minimum requirement. The usual imaging problems are reflections and shading effects due to non-uniform illumination. Incorrect colour temperature can also occur if an improper light source is used. However, there can be errors due to contrast distortion, spectral distortion and due to interlaced scanning. These first two are easily corrected.

Converting a television signal to a computer-compatible form introduces two subtle distortions. A television signal is a continuous electrical signal. A computer, however, can only process a finite set of numbers. The process of conversion can be conceptually regarded as a three-stage process of sampling, quantizing and coding. In sampling, the continuous signal is converted to a discrete signal. That is, a signal which exists only at particular points in time or space, but whose magnitudes exactly match the original signal at those points. Quantizing now restricts those signal magnitudes to a finite range of values. Finally, coding expresses those magnitudes as binary numbers. Quantizing clearly introduces a distortion. Although it is not random, it is convenient to treat it as such and so it is termed quantization noise. Most modern

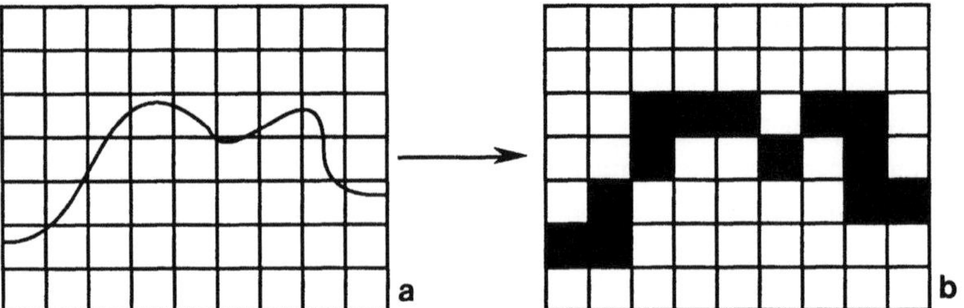

Fig. 2a, b. Aliasing effects in images. **a** True edge of an image; **b** Quantized edge of an image

video digitizers usually quantize to 8-bits or 256 levels. In this case, quantization noise can be safely ignored.

Sampling also introduces errors. A famous result, Nyquist's theorem, indicates that distortions will be produced in sampling if the sampling rate does not exceed a rate $1/2\omega$, where ω is the maximum frequency of the signal. The distortion is termed aliasing. For one-dimensional signals, aliasing is easily described and easily prevented. However, images are two-dimensional and the situation is much more complex. An illustration of the problem is shown in Fig. 2. Aliasing results in curved objects being represented with a certain degree of "chunkiness". Thus the effect, in general terms, is to change the shape of viewed objects around those portions of the object where there is fine detail or curves with a high gradient and to affect the apparent texture of those objects. Unfortunately, visual inspection is still the best approach for assessing the impact of aliasing in images.

Unless a very high resolution camera is used, it can be safely assumed any image of a seed will have some aliasing present. Measurements based on broad properties of the seed will only be slightly unaffected. Those dependent in any way on texture, fine detail or curvature can be expected to exhibit some error. This is a potentially serious problem only in cases where the seeds to be discriminated are quite similar in shape and size.

5 The Theory of Image Processing

The scope and character of image processing can be judged from good introductory texts such as those by Ballard and Brown (1982), Gonzalez and Wintz (1987) and Schalkoff (1989). More advanced treatments of many of the topics are found in Agrawala (1977) and Chellapa and Sawchuk (1985).

That part of image processing most applicable to seed discrimination is pattern recognition. This may be described as a three-stage process of conditioning, feature extraction and feature evaluation.

Conditioning is concerned with eliminating or reducing errors resulting from the acquisition phase. Thus, it is concerned with reducing or eliminating noise, correcting contrast distortion, correcting geometric distortion, if any, correcting any shading distortion and correcting spectral distortions.

A feature is anything of interest within an image which can be quantified. In general, a number of features are determined to form an attribute set. Spatial features generally relate to the shape of an object or the texture of its surface. Spectral features describe the domains within the image with similar hue, saturation or spectral characteristics.

The features of greatest interest in seed discrimination relate to shape. These can be categorized into two broad groups. Edge-directed features encapsulate edge information in some way. Many directly relate to curvature. Morphological features are more concerned with deriving a value which relates to the overall form of the object. Many morphological measures therefore directly or indirectly include area. One important subdivision, topological measures, essentially capture some geometrical feature of the object. This is a very popular class of features and very widely used in most applications areas.

Feature evaluation is primarily based on three broad approaches. By far the most popular for seed discrimination is technically described as supervised statistical classification. The measured attribute set is taken to be the coordinates of a vector within an n-dimensional decision space. A training procedure is executed prior to any classification on known seeds. This determines the decision boundaries for that space and divides it into a set of classes. Then the classification of the seed involves taking measurements and observing into which of these classes the measured vector falls. Statistical pattern recognition of this type is very successful and easily applied.

The training process is moderately complex. Because the known seeds are subject to natural variation, the decision space cannot be categorized by a set of points. Rather, it must be categorized by a set of multivariate probability distributions. In general terms, these would be Gaussian with, ideally, small variances and zero covariances, but there is little guarantee of this. Thus, the possibility exists for the distribution of a given class to overlap those of others. The decision boundaries must therefore be defined in such a way as to minimize error according to some statistical criterion. It is rarely obvious whether overlap exists or which parameters contribute most towards it. Thus, it is invariably the case that the decision boundaries are defined as a result of a number of experiments on the known classes of seeds. Further, via discriminant analysis, it is common to try and ascertain what constitutes a best parameter set. This is very easily done with one of the many statistical packages available for personal or mainframe computers.

A point to stress it is extremely ill-advised to select a feature set based on visual observation. A machine vision system and the eye are radically different and the cues the eye uses to identify are not available to the former. To illustrate, it is perfectly obvious to an observer whether a grain is wheat or barley, but to a machine vision system, they both have similar major shape parameters. Thus, a feature like aspect ratio could very easily classify them as identical. Continuing with this argument, it is always prudent to observe the performance of any classifier with all the seeds likely to be encountered in its work.

In very general terms, it is also good practice to use discriminant analysis no matter how obvious the classification problem may seem and to test as many representative seeds as possible. That is, to create the discriminator using seeds taken from several growing seasons and several sites and to test with their common contaminants.

The decision boundaries are defined according to some criterion such as minimum distance or maximum likelihood. These give rise to hyperplanes within the decision space which are attractive because of their simplicity and ease of implementation. If these should prove inadequate and more complex structures are needed, then a possibility to consider is neural networks (Vemuri 1988; I.E.E.E. Computer 1988; Wasserman 1989). Whereas a classifier based on minimum distance involves a set of weighted sums of the feature values, neural networks employ a non-linear and possibly layered structure. They belong to the class of unsupervised systems which can automatically learn to classify new patterns. Adaptive or learning classifiers, in general, are quite complex and need some skill to design. Neural networks, however, offer some promise of allowing quite simple adaptive systems to be created.

The second approach to feature evaluation is syntactical and draws on formal language theory (Gonzalez and Thomason 1978; Schalkoff 1989). A formal language has an alphabet of symbols and a grammar which, amongst other things, defines how symbols may be linked to form strings or sentences. This part of the definition is described by a set of rules or primitive string operations. Given an arbitrary string of symbols drawn from the alphabet, it is possible to test whether a particular grammar could produce that stirng or not via these rules. In more common terms, it is possible to parse a given sentence to see if it belongs to the language. For an image, a finite set of primitive features such as short lines and arcs of different curvature can be created and treated as an alphabet. A shape then, can be described as a sentence. If a grammar exists for different objects to be recognized, this string may be tested to determine to which it conforms. Creating the grammar clearly requires testing a large number of known objects of the type to be recognized. Syntactic methods are complex but they allow some variation in the shape of the object, as a well-defined grammar should only focus on the intrinsic properties of that shape. Consequently, syntactic methods offer a rather sophisticated approach to image understanding and one with some possibility of recognizing damaged seeds.

The third approach to feature evaluation is to use rule-based systems. These are often termed expert systems due to a common application area. Expert systems are the focus of considerable attention at present. Numerous books describe the basic technology and related concepts (see, for example, Patterson 1990 and Schalkoff 1989 for a brief description on their application to image processing) and a number of journals such as I.E.E.E. Expert are devoted to their advancement. Rule-based systems focus on the relationhips which exist between data and use logical principles to systematically search for the information required. That search is guided by a set of rules expressed in IF< >THEN format. Rule-based systems are very well suited to circumstances where dependencies exist between data that are best, or can only be, described by heuristic relationships. There are many ways to use rule-based systems in image processing. One is to create lists of assertions using the measured features. These are of the form that the measure of some feature is, for example, less than some numeric value. Thus, the outcome of the assertion is true or false. The attribute set is therefore transformed to a set of logical values. Now the rules are applied to identify the object. In general terms, these would be oriented to first discovering the broad class of objects to which the particular pattern belongs, then the specific class member.

6 Attribute Sets in Seed Discrimination

6.1 Introduction

The attribute sets commonly chosen for seed discrimination are usually derived from shape features. That is, from a two-dimensional view of the seed. Of course, several views may be taken and features determined from them all, but the research conducted so far has not shown any great advantage in doing this. Further, contrary to what might be imagined, the number of such features considered is usually quite small. So far, no research has been reported on using three-dimensional parameters in any machine vision system for seed discrimination.

6.2 Colour Features

Discriminating seeds via their spectra has some theoretical appeal but not much practical. While the cost of scanning spectrometers is undoubtedly a major reason for this, in very general terms it seems a highly complex approach to a problem where far simpler methods can give the same result. Thus, most interest has centred on discriminating via colour alone. This begins by deriving the saturation or hue, or both, of every pixel in the image. These are easily calculated. Given the red, green and blue outputs of a camera, it can be shown that the saturation S and hue H are given by:

$$S = \sqrt{(R-Y)^2 + (B-Y)^2}$$

$$H = \tan^{-1}\left(\frac{R-Y}{B-Y}\right),$$

where $Y = 0.299R + 0.587G + 0.114B$.

Note, too, that the colour signal in broadcast television is in terms of hue and saturation, thus these can be derived by directly digitizing the composite colour signal.

6.3 Edge Extraction

Many features in image processing either directly or indirectly depend on edges. Therefore, the first stage in feature extraction is to derive those edges. This appears to be a simple task, but can be surprisingly difficult.

There are several approaches to edge extraction. One is the following. If the background has been chosen carefully, then the distribution of pixel values within the image will be similar to that shown in Fig. 3. If a histogram like this, showing pixel magnitudes within an image, is observed – and it should be relatively easy to ensure it will occur – then any pixel below some magnitude is background and any above is seed. It becomes a simple matter to detect an edge point and then trace around the contour.

If there is no clear separation between background and foreground, then a more complex edge extraction algorithm will be needed. Under this circumstance, it would be best to use one of several algorithms based on the difference between two low-pass spatial filters (see, for example, Chellapa and Sawchuk 1985; Canny 1986; Nalwa and Binford 1986; Torre and Poggio 1986).

One or two problems can arise in edge extraction. Edges can be "noisy" due to natural factors, to slight variations in the image due to interlaced scanning and due to aliasing. Edge tracing algorithms can become unstable if this noise is too great and they may limit cycle. That is, they may continuously cycle about a small cluster of points. A spatial smoothing algorithm such as curve fitting can overcome this. A simple yet quite effective alternative is to restrict the tracing algorithm such that is searches for the next point within some limited angular range of the tangent of the current point or the average tangent of some small neighbourhood about the current point. A more serious problem concerns seeds such as wheat. The brush end of the grain is a region of uncertainty and it is very difficult, if not impossible, to determine the exact contour of the seed. It is also difficult to assess the degree of error when a contour is extracted. As a consequence, it must be assumed any measure based on area will have errors and possibly quite substantial errors.

One reason for the importance of edge extraction is that it offers a very efficient data compression technique for storing images. After edge extraction, the image is represented by only 1000 or so coordinate pairs compared to the

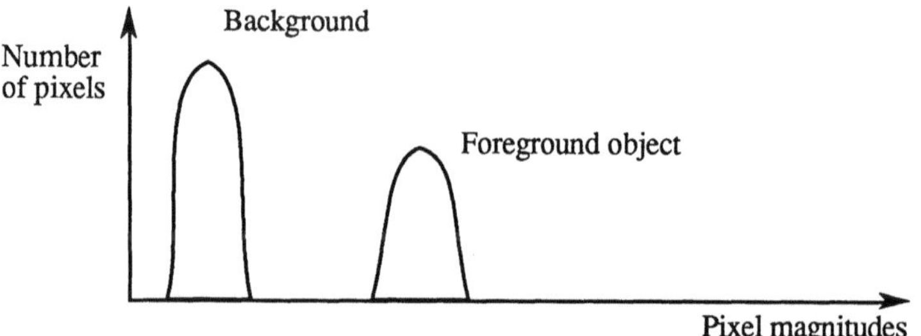

Fig. 3. A typical image histogram

256 kbytes of data needed for the full image. Thus, literally thousands of images can be stored on a disc. Reconstruction of the shape at a later time is very easy, as is the extraction of parameters for analysis.

6.4 Topological Features

For shape analysis, the features derived are overwhelmingly topological. The obvious choices here are length, width, area and perimeter. Since it is difficult to calibrate camera systems, it is most convenient to use normalized features. Many of these are oriented to defining the deviation of the object's shape from a circle. The three most commonly used features which meet this requirement are shown in Fig. 4. They are:

1) The Aspect Ratio
The major axis of a seed is the longest axis through the centroid. The minor axis is the perpendicular to the major axis through the centroid. Then the aspect ratio is the ratio of the major to minor axis. It is a measure of the roundness of the object.

2) The Compactness
The compactness of an object is defined as the area divided by the square of the perimeter. If the object is circular, the compactness is 0.25.

3) Minimum Enclosing Rectangle
The minimum enclosing rectangle is the smallest rectangle within which the object will fit. It usually, but not always, has sides equal to the major and minor axes. To normalize, it is usual to take the ratio of the area of the object to the area of the minimal enclosing rectangle.

The alternatives to these features are often far more complex to derive. Therefore, some slight variations are often chosen if additional complexity is needed. For example, dividing the major axis into a number of sections and taking the aspect ratio for each of these. In wheat, due to the uncertainty associated with

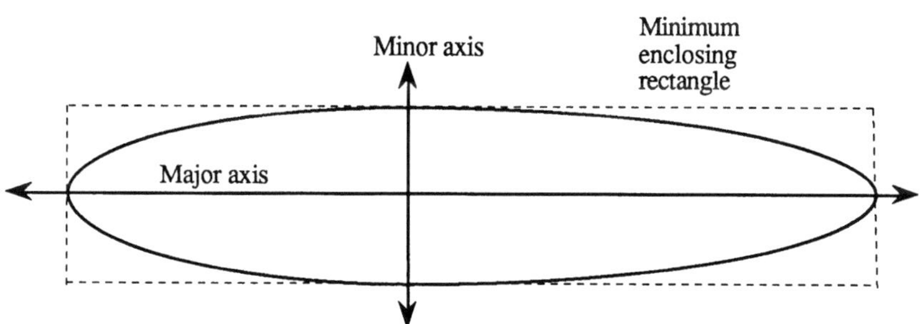

Fig. 4. Common topological features

the brush end, the section width is some integral division of the germ end-centroid spacing. An alternative is to take radii from the centroid to the edge at fixed angular spacings and normalize to the major axis. In this case, any radii in the brush region can be ignored.

6.5 Edge-Directed Features

The simplest edge-directed feature is a chain code. To develop the code, a finite number of direction angles are defined, and then the curve traced within some bound according to these direction angles. Thus, the edge is translated to a string of symbols. Unless the edge is well defined or if some acceptable comparison criterion is evolved, chain codes can be difficult to use in applications such as seed discrimination. Hence, there is no research reported in the public domain on this.

A more complex set of features are Fourier descriptors. Consider some curve in a digital image. It will consist of a finite number of points. There are a number of ways to define this curve, but the most common are either by its two-dimensional coordinates or by radii from some given point to each point of the curve. Then the well-known discrete Fourier transform (DFT; Brigham 1988) may be applied to this data. This results in a set of numbers − the Fourier descriptors − giving the magnitudes of a set of sinusoids plus a constant which define that curve. Since sinusoids are orthogonal, this representation is unique. Another possibility for applying Fourier descriptors is discussed in Zahn and Roskies (1972).

The first Fourier coefficient describes the average value of the curve. However, the next and all subsequent terms give the magnitude of a sinusoid needed to give the best approximation to that curve. That is, they encapsulate the curvature of the shape. Note that the upper $N/2$ coefficients are similar to the lower. Further, that in an application like this, the values around $N/2$ will predominantly describe aliasing rather than any useful image property.

Fourier coefficients are very easily derived and are widely used, but they do have some disadvantages. A curve reconstructed by only taking some of the Fourier components and their corresponding sinusoids is notably different in the sense that there is significant over- and undershoot about any sharp transition in the curve. This is termed Gibbs phenomena. It suggests that while a set of Fourier coefficients may uniquely describe a curve, the individual components may not quite capture the curvature in the manner desired. For that reason, some thought has been given to alternative orthogonal expansions in which the coefficients relate more directly to curvature. An obvious choice here is Chebyshev polynomials, but so far no results with these have been reported in the literature.

6.6 Other Features

The literature records numerous other features which are used in a wide variety
of applications. Most have never been applied to seed discrimination, although
comments have been made on the possibility of using them from time to time.
Two warrant some attention.

One set of features widely used in many areas of image processing are moments. There are a number of these, but the usual ones are the geometric moments given by:

$$M_{PQ} = \sum_{i=0}^{N-1} \sum_{j=0}^{N-1} x^P y^Q f(x,y) \ .$$

Here, $f(x, y)$ is 1 for image pixels and zero otherwise. The central moments are
obtained by substituting $(x-\mu)^P$ for x^P and $(y-\mu)^Q$ for y^Q respectively.

Moments derive from the concept of the moment of inertia in mechanics
(Weijak 1983; Teh and Chin 1988). They are easily derived and quite widely
used, but are not without their problems. These moments are not invariant to
geometric transformation. A modified form is, but there are only seven such
moments. It is not known whether moments are a complete representation, but
it is known that around ten are needed in order to generate a reasonable reconstruction of a shape. Moments are not orthogonal, thus there is considerable
redundancy. The higher moments are also very prone to noise and should be
used with considerable caution.

Orthogonal polynomials can be used to decompose a one-dimensional
structure − a curve − into the sum of a set of known curves. While this can
be extended to two dimensions, it is better suited to encapsulating texture information and the like rather than area. However, the idea of decomposing an
object into known forms has appeal and part of morphological analysis is concerned with the logical union of simple shapes into complex forms (Schalkoff
1989, Pitas and Venetsanopoulos 1990).

7 Seed Discrimination by Image Processing

7.1 Introduction

Image processing has been applied to various seed discrimination problems
(see, for example, Travis and Draper 1985; Keefe and Draper 1986; Berlage et
al. 1988). In addition, a number of systems have been developed (Travis and
Draper 1985; Saperstein et al. 1987; Keffe and Draper 1988; Westerlind 1988).
Most interest centres on the very general issue of quality control and to a lesser
extent as an aid in breeding programs. In the quality control role, image processing has some definite advantages over alternative methods (see, for example, Draper and Keefe 1987).

As a conceptual issue, the application of image processing may be divided into just two problems:

1) discriminating between dissimilar classes of seeds;
2) discriminating between similar classes of seeds.

The former is almost trivial; the latter can present some major difficulties.

7.2 Discrimination Between Dissimilar Classes

Many seed discrimination problems are characterized by significantly different measures in the features of interest between the classes. For example, there are problems such as identifying the proportion of weed seeds in cereal samples or the proportion of broken nuts. While the feature distributions may be broad within the decision space, the classes are so widely spaced that this is of no practical consequence. Under these circumstances, creating an image processing classifier is simplicity itself. One or more parameters are chosen based on the size, shape or colour parameters, which differ strongly, an initial trial is conducted to derive the decision boundaries for a statistical classifier and then that classifier is implemented. Many problems of this kind simply involve discriminating seeds into two classes and this can often be done by just aspect ratio alone, or aspect ratio and one colour parameter, or even one or more absolute parameters such as length plus some relative parameters.

While caution needs to be exercised, a careful visual observation is often sufficient to determine an acceptable feature set. Naturally, which features to choose is strongly dependent on the problem itself. An early paper illustrates what may be expected. Travis and Draper (1985) investigated the general problem of seed discrimination, examining a range of seeds including cereals, fodder plant seeds, weed seeds, vegetable seeds and rape seeds. Using just the size parameter and length, they were easily able to achieve a 95% confidence level in classification.

7.3 Discrimination Between Similar Classes

The other extreme in terms of a discrimination problem is when most of the features are similar from class to class. While decision boundaries can be set, the discrimination error is almost certainly high and the classifier of diminished practical value. This circumstance can arise for a number of reasons, but two important cases need to be highlighted. One is where the number of classes becomes large but the seeds are not substantially different. The other is when the problem is to discriminate between varieties of the one seed when those varieties have little genetic diversity.

An obvious approach to overcoming these problems is to increase the size of the attribute set. However, there are only a finite number of attributes which can be measured. The choice therefore becomes critical and the more obvious may not be the best. As a result, much of the published literature on the discrimination of seeds by image processing and much of the current research in the field tend to concern investigating more complex featues than those discussed earlier. Much of this work is directed at the discrimination of varieties, particularly wheat varieties. In addition, the interest in more complex techniques such as syntactic and rule-based methods is largely focussed on this same problem.

While the complexity of these problems should not be understated, it is also important to keep them in perspective. The essential machine vision problem being referred to there is one of accurately identifying a single seed. The essence of this problem is achieving high accuracy while minimizing the complexity of the classifier. Apart from a range of problems in breeding, however, most practical situations are not restricted to testing single seeds. Rather, any number of samples can be taken and tested. Clearly, if enough tests are conducted, then a statistically significant result can be obtained. Furthermore, while a machine vision system may experience difficulties, it will in most circumstances still outperform a human observer. In most tasks, too, it will perform significantly better.

7.4 The Discrimination of Wheat Varieties

Although it is conceptually only a small part of the seed discrimination problem, the identification of wheat varieties is of considerable practical importance. Because it is typical of the most difficult seed discrimination problems, it is worthwhile examining it in more detail and reviewing some representative work. As might be expected, the literature on the discrimination of wheat varieties tends to fall into three subclasses; North American, European and Australian. Where there is significant genetic diversity between the varieties, then excellent results have been reported. Where there is very little, results are on the whole good, but not of a standard acceptable in many practical applications. Almost all of the research in this area has been concerned with the attribute set for statistical classifier and most has been directed to parameters derived from the crease down view.

Keefe and Draper (1986) examined the problem of discriminating between five common British varieties. In contrast to almost all other work, they examined the side view rather than the top view of the grain. Ten measurements were made of each kernel in the experiment and a number of measures derived from these, including many commonly used in shape analysis. Although the results on the whole were reasonably acceptable, they commented that compactness and aspect ratio seemed poor parameters. A more recent paper (Draper and Keefe 1987) summarizes the authors' philosophy.

Zayas et al. (1986) have considered the problems of discriminating American wheats. Although 15 varieties were considered in their tests, these were grouped into 11 pairs and the analysis confined to a procedure for identifying the pairs. For each kernel, the major and minor axis was determined from the crease down view and a number of parameters derived from these. Using discrimintant analysis, they were able to create a classifier with an accuracy in excess of 80%. However, results were not uniform and in some poor classifications were achieved. In more recent work (Zayas et al. 1989), the authors have divided the major axis into a number of equal segments and measured the grain axis at each of these points. This allows a number of terms like aspect ratios to be derived. This feature set has given excellent results.

A Canadian team (Saperstein et al. 1987; Neuman et al. 1987) has been investigating the problem of creating a classifier which will identify wheats to Canadian standards. From the crease down view, the grain profile was computed and processed. A number of measures were derived, but a discriminant analysis suggested the best were, in the following order, length, perimeter, the first Fourier component, a normalized moment N_{12} and the compactness. A classifier constructed on these alone gave mixed results. For some varieties accuracy was very high, but for others it was quite disappointing. Nevertheless, the authors felt that construction of a classifier for the Canadian wheat industry was feasible, although more work was needed. They commented that some of the problems encountered could be attributed to genetic similarity between the wheats. A later study (Neuman et al. 1989a, b) investigated colour as a means of discriminating wheats. Some local factors encouraged this, and the results obtained were encouraging.

Australian wheats have been investigated by Myers and Edsall (1989). Difficulties were encountered because of the limited genetic diversity. In addition, as the shape is more spherical than other wheats, shape and length parameters in general do not discriminate as strongly as elsewhere. Their study had several objectives which included and assessment of some aspects of experimental procedures. It was found that side views were of little importance. The seven most significant parameters found were, in the following order, the compactness, the second Fourier coefficient, the fourth Fourier coefficient, the third Fourier coefficient, packing and the aspect ratio. The Fourier coefficients were derived from an analysis of the germ end of the grain in an arc bounded by the minor axis. The packing is the ratio of area to the square of the minor axis. Like other results, the classifier derived worked well in some circumstances but not all. An interesting feature of the results was that where a kernel was incorrectly classified, then that error classification was very uniform. As a result, if it was known that a sample of wheat was uniform or a relatively simple admixture, it would be possible to classify in with high accuracy by noting the overall results. This led the authors to conclude that a rule-based classifier should prove quite successful.

8 Conclusions

In large measure, it can be stated the problem of discriminating seeds by image processing has been solved. The issues remaining are practical, particularly how to present seeds for imaging (see, however, Chen et al. 1989). In addition, as mentioned earlier a machine system will generally outperform a human observer within some quite broad limits. The most significant problem remaining is that machine-based systems do not perform as well as desired when the seeds are very similar. In practical terms, this usually means a problem of discriminating between varieties. While the performance achieved is not as good as researchers would desire, it is nonetheless quite acceptable. Further, most researchers would be confident that more complex procedures will soon achieve the results desired.

References

Agrawala AR (ed) (1977) Machine recognition of patterns. I.E.E.E. Computer Society Press, Los Alamitos, CA

Ballard DH, Brown CM (1982) Computer vision. Prentice Hall, Englewood Cliffs, NJ

Berlage PG, Cooper TM, Aristazabel JF (1988) Machine vision identification of diptoid and tetraphoid ryegrass seed. Trans Am Soc Agric Eng 31:24−27

Brigham EO (1988) The fast Fourier transform and its application. Prentice Hall, Englewood Cliffs, NJ

Canny J (1986) A computational approach to edge detection. I.E.E.E. Trans Pattern Anal Machine Intell 8:679−698

Chellapa R, Sawchuk AA (eds) (1985) Digital image processing, vols I and II. I.E.E.E. Computer Society Press, Los Alamitos, CA, USA

Chen C, Chiang YP, Pomeranz Y (1989) Image analysis and characteristics of cereal grains with a laser range finder and camera contour extractor. Cereal Chem 66:466−470

Draper SR, Keefe PD (1987) Electrophoresis of seed storage proteins and whole-seed morphology using machine vision: alternative or complementary methods for cultivar identification. Int Symp Int Seed Test Assoc, Leningrad, USSR, pp 27−35

Gonzalez RC, Thomason MG (1978) Syntactic pattern recognition: an introduction. Wiley, New York

Gonzalez RW, Wintz P (1987) Digital image processing. Addison Wesley, New York

I.E.E.E. Computer (1988) Special issue on artifical neural systems. Washington, DC

Keefe PD, Draper SR (1986) The measurement of new characteristics for cultivar identification in wheat using machine vision. Seed Sci Technol 14:715−724

Keefe PD, Draper SR (1988) An automated machine vision system for the morphometry of new cultivars and plant gene bank accessions. Plant Varieties Seeds 1:1−11

Myers DG, Edsall KJ (1989) The application of image processing techniques to the identification of Australian Wheat Varieties. Plant Varieties Seeds 2:109−116

Nalwa VS, Binford TD (1986) On detecting edges. I.E.E.E. Trans Pattern Anal Machine Intell 8:699−714

Neuman M, Saperstein HD, Shwedyk E, Bushuk W (1987) Discrimination of wheat class and variety by digital image analysis of whole grain samples. J Cereal Sci 6:125−132

Neuman MR, Saperstein HD, Shwedyk E, Bushuk W (1989a) Wheat grain colour analysis by digital image processing. I. Methodology. J Cereal Sci 10:175−182

Neuman MR, Saperstein HD, Shwedyk E, Bushuk W (1989b) Wheat grain colour analysis by digital image processing. II. Wheat class discrimination. J Cereal Sci 10:183–188

Patterson DW (1990) Introduction to artificial intelligence and expert systems. Prentice Hall, Englewood Cliffs, NJ

Pitas I, Venetsanopoulos AN (1990) Morphological shape analysis. I.E.E.E. Trans Pattern Anal Machine Intell 12:38–45

Saperstein HD, Neuman M, Wright EH, Shwedyk E, Bushuk W (1987) An instrumental system for cereal grain classification using digital image analysis. J Cereal Sci 6:3–14

Schalkoff RJ (1989) Digital image processing and computer vision: an introduction to theory and implementation. Wiley, New York

Teh C, Chin RT (1988) On image analysis by the method of moments. I.E.E.E. Trans Pattern Anal Machine Intell 10:496–513

Torre V, Poggio T (1986) On edge detection. I.E.E.E. Trans Pattern Anal Machine Intell 8:147–163

Travis AJ, Draper SR (1985) A computer-based system for the recognition of seed shape. Seed Sci Technol 13:813–820

Vemuri V (1988) Artificial neural networks; theoretical concepts. I.E.E.E. Computer Society Press, Los Alamitos, CA

Wasserman PD (1989) Neural computing: theory and practice. Van Nostrand Reinhold, New York

Weijak JS (1983) Moment invariants in theory and practice. Image Vision Computing 1:79–83

Westerlind E (1988) Seed scanner; a computer-based device for the determination of other seeds by number in areal seed. Food Sci Technol 16:287–297

Zahn CT, Roskies RZ (1972) Fourier descriptors for plane closed curves. I.E.E.E. Trans Computers C-21:269–281

Zayas I, Lai FS, Pomeranz Y (1986) Discrimination between wheat classes and varieties by image analysis. Cereal Chem 63:52–56

Zayas I, Pomeranz Y, Lai FS (1989) Discrimination of wheat and non-wheat components in grain samples by image analysis. Cereal Chem 66:233–237

Physicochemical Analysis of Wheat Starch

T. R. Noel, S. G. Ring, and M. A. Whittam

1 Introduction

Starch occurs as water-insoluble granules which are partially crystalline and bi-refringent when viewed between crossed polars. Wheat grains contain two main populations of granules, the larger A-type granules are formed early in seed development and have a major dimension ranging from $10-35\,\mu m$, the smaller B-type granules are formed later in development and range in size from $1-10\,\mu m$. The smaller B-granules can contribute up to 30% by weight of the starch (Evers et al. 1974; Dengate and Meredith 1984). The granules contain two macromolecules, amylose and amylopectin.

Amylose is an essentially linear polymer of $1-4$ linked α-D-glucose which can be characterized by its iodine-binding behaviour of $19.5\% \pm 0.5\%$ w/w iodine under standard conditions (Banks and Greenwood 1975). While some amylose fractions are linear as judged by their quantitative hydrolysis to maltose by the exo-acting enzyme β-amylase, many amylose preparations contain branched species (Hizukuri et al. 1981). In a recent study it was found that wheat amylose was lightly branched with, on average, around five chains per molecule (Takeda et al. 1987). The branch points in amylose can be removed by treating with enzymes specific for the $\alpha\,1-6$ linkage (Banks and Greenwood 1975). Most wheat starches which have been examined contain between 20% and 30% w/w amylose, the amylose content depending on granule size, maturity and botanical origin (Matheson 1971; Galliard and Bowler 1987; Morrison and Gadan 1987).

Amylopectin is characterized by its low iodine binding behaviour of $<1\%$ w/w under standard conditions and is a highly branched structure. A wheat amylopectin had an average chain length of 19 (Hizukuri and Maehara 1990). Enzymic debranching of the amylopectin followed by gel permeation chromatography revealed a polymodal weight distribution of constituent chains with peak apices at degree of polymerization (d.p.) 11.5, 17.7, 40 and 1600 and peak shoulders at d.p. 13.5 and 80. The amylopectin was classified as having an intermediate iodine-binding behaviour (for an amylopectin) of $0.76\,g/100\,g$ polysaccharide.

Granular wheat starch as isolated, also contains non-polysaccharide components such as lipid, particularly lysophospholipid, which can represent 1% of the dry weight (Morrison and Gadan 1987; Morrison 1988). This lipid is thought to occur within the internal structure of the granule and is not an accidental contaminant. The granular starch may also contain associated protein (Lowy et al. 1981). Both of these non-starch components can modify starch behaviour (Galliard and Bowler 1987).

Chemical and enzymic methods have been used to determine the chemical composition of starch granules and the fine structures of the constituent polysaccharides. Physicochemical methods have found use in the characterization of starch polysaccharides through their iodine-binding behaviour and their behaviour in aqueous solution including the determination of molecular size, shape, weight and polydispersity. The functional behaviour of starches has also been examined including gelatinization and retrogradation behaviour. In this way physicochemical methods are helping to establish relationships between structure, composition and functional behaviour. The range of physicochemical techniques available for characterizing starch is potentially wide. In this chapter we consider methods which can be applied to analysis of seeds, particularly the characteristics of wheat starch and its variation with botanical origin.

2 Starch Isolation

Prior to any physicochemical analysis of the starch it must first be isolated in such a way that avoids inadvertant physical, chemical or enzymic modification of the starch granule. Methods are available which substantially achieve these objectives either for extraction of starch from bulk samples (Adkins and Greenwood 1966) or from a single grain (South and Morrison 1990). It is usual to adapt methods, depending on the source of the starch and the proposed subsequent research. For these reasons a preferred method is not given, rather general comments have been made on aspects of starch purification.

The dry milling of cereal grain exposes the starch granule to localized heat and shear which damages the ordered granular structure of some of the starch, with a result that the starch polysaccharides become soluble in water at room temperature. Wet homogenization of the grain in a blender at room temperature achieves cellular disruption with minimum granule damage. The grain is first softened by steeping overnight in water at 2 °C to minimize enzyme activity; it is preferable to further inhibit enzyme action by the addition of an agent such as mercuric chloride (0.01 M) (Adkins and Greenwood 1966). The softened grain is homogenized and cell wall debris is removed on 150- and 60-µm screens. This residue can be re-extracted by homogenization to increase the yield of starch. With care, an 80% – 90% recovery of the starch is achievable. The starch fraction passing through the screen is collected by sedimentation and at this stage is invariably contaminated with cell wall debris, cytoplasmic proteins and lipids.

Cell wall debris can be separated from granular starch by exploiting the more rapid sedimentation of the latter from water. Alternatively, centrifugation of the crude starch through 80% w/v aqueous caesium chloride at 14000 g results in a sediment free of contaminating cell debris and is particularly appropriate for small samples (South and Morrison 1990). Contamination of starch preparations with cell wall debris is readily observed in the light microscope. Cytoplasmic proteins can be removed by first washing the starch in 0.1 M NaCl followed by repeated shaking of aqueous starch dispersions with toluene. The shaking denatures proteins which then, after mild centrifugation (2000 g, 30 min), separate at the toluene/water interface. Care is needed to avoid loss of small starch granules in this proteinaceous layer. Gentle mechanical agitation, for example by bubbling nitrogen through this layer, aids the recovery of the small granules. Other possible methods for the removal of protein contaminants include extraction of the starch with aqueous solutions of sodium dodecanoate (South and Morrison 1990) or sodium dodecyl sulphate (Gough et al. 1985). While protein extraction is achieved, both of these compounds can bind to and complex with starch, particularly amylose, and therefore modify properties. Caution in the use of these compounds is therefore advised. Proteolytic enzymes have also been used to hydrolyze and therefore extract protein contaminants, although it is preferable to avoid their use as they may also interact with and bind to the starch granule.

Surface lipid can be extracted by washing the starch in methanol; for a more detailed review of starch lipids and their extractability, see Morrison (1988). Carefully purified wheat starches typically contain 25 – 30 mg of lipid nitrogen and, by difference 10 – 20 mg of protein nitrogen per 100 g of starch. The lipid of wheat starch is primarily lysophospholipid at a level of 800 – 900 mg/100 g of starch. Complete extraction of lipid and protein components requires disruption of the native granular structure and it is proposed that some lipid and protein are an integral part of the native granule of wheat starch.

In many ways it is best to store the starch as hydrated granules in water, covered with toluene to inhibit microbial growth, at 2 °C. The drying of the starch granule can encourage associations between chains which may prove difficult to reverse. The isolated starch may be dried by first washing with acetone on a sintered funnel followed by vacuum drying at 20 °C. Drying the starch at higher temperatures, particularly above 40 °C, is to be avoided as the drying process may modify the crystallinity and gelatinization behaviour of the granular starch.

In the isolation of starch from single grains (South and Morrison 1990) dissection out of the germ prior to soaking and homogenization is a convenient method to avoid much of the inadvertant contamination of starch with non-starch components. The yield of starch from a single grain is typically 25 – 35 mg.

3 Properties of Isolated Polysaccharides

3.1 Purification

Isolated wheat starch granules contain lipid which has the potential to complex about ten times its weight of amylose (Karkalas and Raphaelides 1986). As the lipid content of wheat starch is of the order of 1% w/w, a major fraction of the amylose present can be complexed. This complexation will interfere with subsequent chemical and physico-chemical studies on the isolated polysaccharide and it is essential, before further study, to remove this contaminating lipid. Dimethyl sulphoxide (Banks and Greenwood 1975), particularly 90% aq. DMSO, is a good solvent for both starch and lipid. Quantitative dissolution is usually achieved by stirring a 2% w/w dispersion overnight at room temperature, more rapid dissolution may be achieved by heating to $60-100\,°C$. For many studies, e.g. determination of molecular weight, dissolution under milder conditions is to be preferred (Everett and Foster 1959). Other solvent systems based on DMSO, e.g. urea/DMSO (Morrison and Laignelnet 1983), have also been used to dissolve successfully granular starch.

Once dissolved the starch polysaccharides may be precipitated by the addition of 9 vol of alcohol leaving lipid in the supernatant. A 1-butanol precipitate is soluble in boiling oxygen-free water and is a starting material for the purification of amylose from starch by successive precipitation from aqueous solution using thymol and 1-butanol, leaving amylopectin in the supernatant (Banks and Greenwood 1975). For the determination of the amylose content of the starch, through its iodine binding behaviour, ethanol is used as a precipitant instead of 1-butanol. The ethanol precipitate can be redissolved in DMSO for amylose determination using either colorimetric or physico-chemical methods.

3.2 Amylose Determination

Amylose forms a blue-coloured complex with iodine/potassium iodide in aqueous solution through its interaction with a polyiodide ion. Amylose contents can then be determined either from the absorbance of the blue complex (Gilbert and Spragg 1964), or from a more direct measure of the amount of iodine bound by the polysaccharide as a function of concentration of free iodine. The colorimetric method has the advantage of convenience and is perhaps best calibrated with amyloses of known purity.

Physico-chemical methods allow the determination of amylose content and are independent of the availability of amylose standards. The iodine-binding behaviour of starch, amylose and amylopectin fractions can be determined using amperometric (BeMiller 1964) and potentiometric methods (Schoch 1964a). A semi-micro differential potentiometric method has been developed (Banks et al. 1971) which has proved useful in determining the iodine-binding

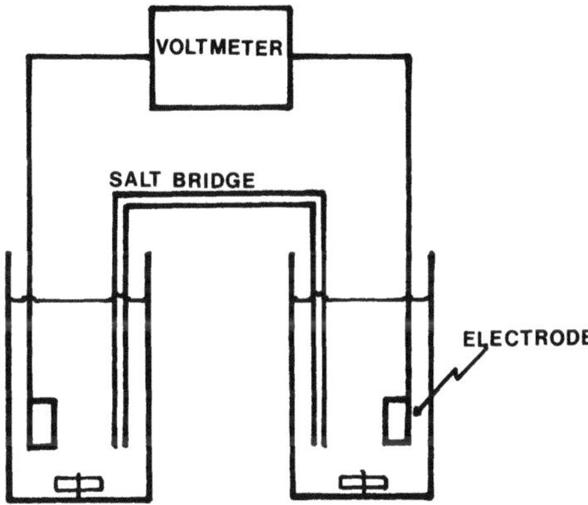

Fig. 1. Schematic diagram of apparatus for semi-micro differential potentiometric titration

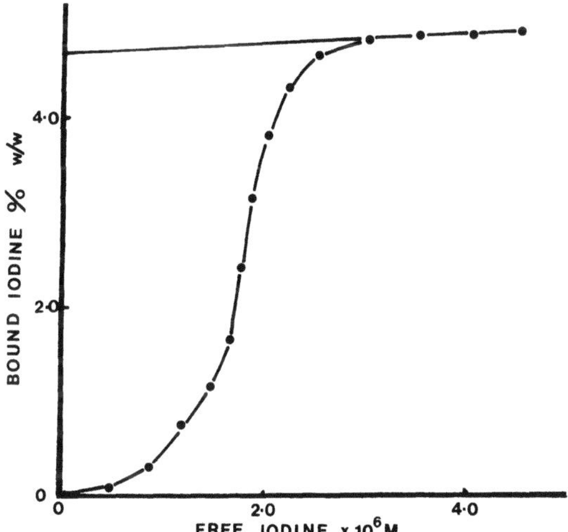

Fig. 2. Plot of iodine-binding behaviour of a starch as a function of the free iodine concentration

behaviour of starch preparations containing 3–6 mg of amylose. The apparatus is shown schematically in Fig. 1. Two half-cells containing 830 ml of 0.01 M KI are connected by a salt bridge. Each half-cell contains a platinum electrode which is connected to a digital voltmeter through a length of 1.5-mm diameter silver wire. One ml of a solution of starch in dimethyl sulphoxide containing between 3 and 6 mg of amylose is added to one half-cell and an equivalent amount of DMSO added to the other. A known amount of 0.005 M I in 0.01 M KI is added to the sample cell (usually 0.1 ml). Some of the iodine will complex with the amylose leaving the remainder "free" in solution. The free iodine will give rise to a potential difference between the half-cells which is measured in the millivoltmeter. Iodine/potassium iodide is added to the other half-cell using a micrometer pipette until a null potential is obtained. From the differences in volume of iodine added to sample and blank half-cells it is possible to calculate the amount of iodine bound as a function of free iodine concentration as shown in Fig. 2 for pea starch. Following convention, the extrapolated iodine-binding capacity of the starch at a zero concentration of free iodine is 4.6% w/w, indicating an amylose content of 24% ± 1% w/w, assuming that under the same conditions the non-amylosic polysaccharides bind no iodine at an extrapolated zero concentration. In addition to determining the amylose content of starches and the purity of amylose fractions, this method has found use in characterizing different amylopectins, and fractions of starch which have an iodine-binding behaviour intermediate between that of amylose and amylopectin.

3.3 Solution Properties of Amylose and Amylopectin

Physical methods have been used to characterize the behaviour of purified amylose and amylopectin fractions in aqueous solution (Blanks and Greenwood 1975). The principal methods adopted in recent years include viscometry, total intensity light scattering and photon correlation spectroscopy. For success these methods depend on true solution being achieved. A number of solvents have been used including 1 M KOH, DMSO and water. One M KOH and DMSO are both good solvents for both amylose and amylopectin. With alkali care has to be taken to avoid inadvertant depolymerization or chemical modification. This is usually achieved by performing the dissolution at low temperature (2 °C) in an inert nitrogen atmosphere, followed by dilution to 0.1 M. DMSO is also a good solvent; variability in the water content of this solvent can lead to experimental variability. Water is not a good solvent for the starch polysaccharides, as concentrated solutions of amylose rapidly retrograde at room temperature and even dilute solutions (1 mg ml^{-1}) are only stable for a limited but sufficient time (4−5 h) for measurements to be made. Once amylose has precipitated from aqueous solution, it is not possible to redissolve it quantitatively in water at temperatures < 140 °C. Neutral aqueous solutions of amylose may be prepared by direct aqueous leaching of granular starch, from the 1-butanol complex or by neutralization of an alkaline solution. Using capillary viscometers, it is possible to determine the intrinsic viscosity of amylose and amylopectin solutions. Both amylose and amylopectin adopt conformations in solution which can be approximated to an equivalent sphere (Banks and Greenwood 1975). Differences in intrinsic viscosities represent differences in size. For a series of different amyloses or amylopectins this represents a useful indicator of likely differences in degree of polymerization and molecular weight. Total intensity light scattering has been used to determine the average molecular weight and size of amylose and amylopectin in aqueous solution through measurements of the intensity of scattered light as a function of scattering angle (Banks and Greenwood 1975). Its advantage as a technique is that it is an absolute instrument, but it is not well suited to the screening of different samples. More recently, another light scattering technique, photon correlation spectroscopy, based on scattering of a coherent light source, has been used to determine the translational diffusion coefficient D_t of starch polysaccharides (Ring et al. 1985). From Stokes' law it is possible to calculate a size parameter for the polysaccharide

$$D_t = k\,T/6\,\pi\,\eta\,r,$$

where k is Boltzmann's constant, T temperature, η solution viscosity and r the radius of the equivalent sphere which represents the diffusional behaviour of the polysaccharide in aqueous solution.

More recently, gel permeation chromatography with the column eluant being monitored by refractive index and low angle light scattering detectors has been used to determine the molecular weight distribution and structural heterogeneity of amyloses from different botanical sources (Hizukuri and Takagi 1984; Takeda et al. 1984). The amylose was first purified by repeated crystallization from water as a 1-butanol complex. Gel permeation chromatography was performed on a cross-linked hydrophilic vinyl polymer (Toyopearl) at 45 °C. At this temperature precipitation of the amylose did not occur within the time course of the experiment.

A refractive index detector continuously monitored the column eluant, its output being proportional to amylose concentration. The eluant was also monitored by a low angle light scattering photometer. In a normal light scattering experiment the scattered intensity is measured as a function of scattering angle in the range $30 - 135\,°$. From appropriate extrapolation of the data to zero scattering angle it is possible to calculate an average molecular weight. In the low angle light scattering experiment the scattered light is measured at an angle of $5\,°$ and it is assumed that the difference in scattered light intensity at $5\,°$ and an extrapolated $0\,°$ is small. By appropriate calibration with samples of known molecular weight it is possible to obtain reliable values for the molecular weight of the unknown species. An advantage of the light scattering experiment performed after gel permeation chromatography in a closed system is that any dust or aggregated material is either filtered out or is well separated from the material of interest.

From the combined outputs of the refractive index detector and low angle light scattering detector, it is possible to calculate a molecular weight, and from the gel permeation chromatography a molecular weight distribution. As the gel permeation separates polysaccharide on the basis of molecular size in solution, the relationship between size and molecular weight can be probed. This can provide information on structural heterogeneity, e.g. branching, as branched species occupy a smaller hydrodynamic volume in solution than the corresponding linear chain.

Most studies have been limited to the examination of amylose; this may, in part, be due to the lack of a suitable gel permeation packing for examining a large molecule such as amylopectin. The significance of differences in molecular weight or size to the functional behaviour of starch or to its biosynthesis and assembly into the starch granule is difficult to assess at this stage. The limited information available at present indicates that fine structure is more important.

4 Granule Properties

4.1 Gelatinization

The native starch granule is both partially crystalline and birefringent. It swells reversibly to a limited extent in water at room temperature; this change is reversed by drying. On heating the starch granule in water, at a characteristic temperature known as the gelatinization temperature, the granule loses its ordered structure and irreversibly swells to many times its original size. At the same time amylose is preferentially solubilized. The gelatinization temperature for wheat starch ranges from $60-70\,^\circ$C. On continued heating further granular swelling occurs with further solubilization of starch. Even at $100\,^\circ$C complete dissolution is not achieved; the dispersion consists of swollen gelatinized granules in a predominantly amylose solution. Characteristics of the gelatinization process include the temperature of gelatinization, the heat required to disrupt the native structure and the extent of swelling and solubilization after gelatinization.

The most direct way of determining gelatinization temperature is to use a light microscope equipped with a heating stage (Watson 1964). An aqueous dispersion of starch 0.2% w/w is mounted in a cavity slide; care is taken to avoid fractionation of the granule population. A thermocouple can also be mounted within the cavity. The slide is heated at a constant rate $2-3\,^\circ$C min^{-1}. The temperature at which sudden irreversible swelling of the granule coupled with loss of birefringence defines a gelatinization temperature. For an individual granule the disappearance of birefringence occurs over a small temperature range $<1\,^\circ$C. For the granule population birefringence loss occurs over a range of a few $^\circ$C. With care the gelatinization temperature, expressed as the mid-point of the range, can be defined to better than $0.5\,^\circ$C. The starch granule is a non-equilibrium structure, characteristics such as the gelatinization temperature will therefore depend on the time scale of the measurement and heating rate. For most practically used heating rates, e.g. $1-10\,^\circ$C min^{-1}, this dependence is small. Very slow heating rates, e.g. $<1-2\,^\circ$C h^{-1} should be avoided as further crystallization of the starch polysaccharides can occur during heating, leading to an increase in gelatinization temperature.

Differential scanning calorimetry (DSC) has also been used to probe starch gelatinization (Donovan 1979; Biliaderis et al. 1986; Slade and Levine 1988). This technique has been widely used to study the melting of crystalline synthetic polymers and protein denaturation (Privalov and Gill 1988). In the more usual form of this technique, the heat flow required to raise the temperature of two cells, a reference cell and a sample, at a constant rate, e.g. $10\,^\circ$C min^{-1},

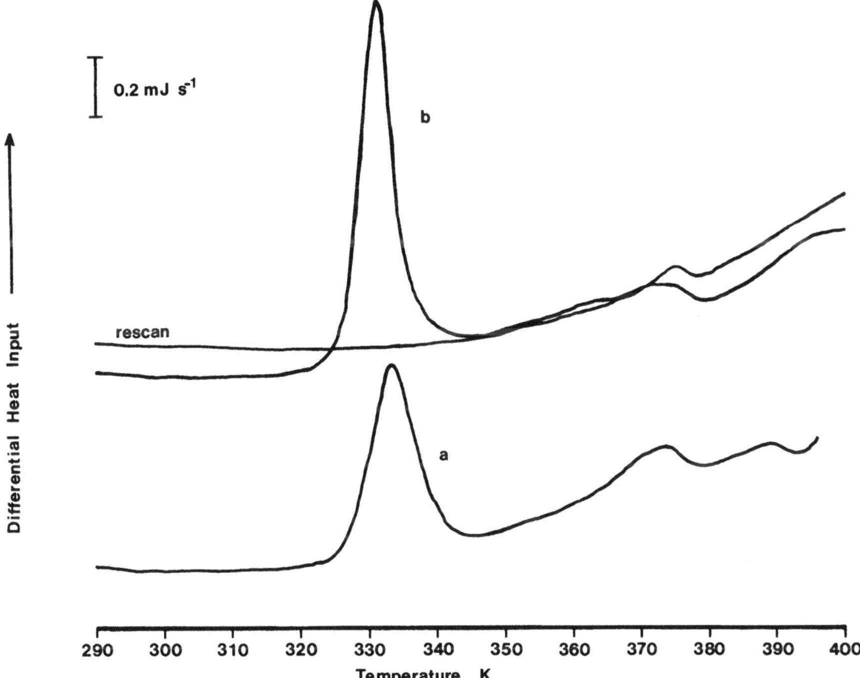

Fig. 3. Plot of differential heat flow versus temperature showing the gelatinization of small (*a*) and large (*b*) wheat starch granules

is compared. From the differential heat flow it is possible to calculate the heat capacity of the sample and the enthalpy change for the melting process. The gelatinization of starch is an endothermic process occurring over a temperature range with a mid-point temperature, Tm. Figure 3 shows a plot of differential heat input in mJ s^{-1} versus temperature in K and compares the gelatinization behaviour of A- and B-type granules from wheat. Gelatinization occurs over 15−20 K and the Tm for A- and B-granules is 331 and 334 K respectively. The size of the endotherm is a measure of the gelatinization enthalpy which can be calculated after calibration of the instrument from the enthalpy of melting of a reference compound such as indium. For the A- and B-granules the gelatinization enthalpies were 10.9 ± 0.2 mJ mg^{-1} and 10.4 ± 0.2 mJ mg^{-1} polysaccharide respectively and are comparable with other reported data on wheat starch (Russell 1987; South and Morrison 1990). The heat capacity of the solubilized and partially solubilized starch polysaccharides is greater than that of granular starch. This is shown as the difference in differential heat input prior to gelatinization of the original scan and an immediate rescan (Fig. 3).

Another characteristic of thermal analysis of wheat starch is the presence of further endothermic transitions in the vicinity of 370 K. These are attributed to the melting of complexes formed between amylose and the integral starch lipid (Kugimiya and Donovan 1981; Donovan et al. 1983). This complex may be present in the native granule or may form during the gelatinization process. In this case it must be argued that the expected exotherm for complex formation is masked by the gelatinization endotherm. The temperature range of the gelatinization endotherm for native granules corresponds very well to the range obtained for loss of granular birefringence (Donovan et al. 1983). DSC therefore provides a convenient means of monitoring gelatinization, a typical experiment lasting less than 1 h and being largely automated. Its greatest value lies in probing and assessing changes in the granule which arise as a result of processing treatments. For example, heat moisture treatment (Colonna et al. 1987) of the native granule can change both the shape of the endotherm and Tm and can decouple parts of the endothermic process from loss of granular birefringence (Donovan et al. 1983). DSC has also found widespread use in monitoring the reassociation of starch chains during starch retrogradation (Mciver et al. 1968; Miles et al. 1985; Russell and Oliver 1989). Other applications include the examination of the sensitivity of the gelatinization process to variations in composition of water (Donovan 1979) and the presence of other solutes (Spies and Hoseney 1982; Eliasson 1986). In these studies DSC has been shown to have proven value. DSC also has potential for studies on the variability of starch with botanical origin. In this case, while differences in Tm and gelatinization enthalpy may well be observed, their significance may be more difficult to assess, as the relationships between starch structure, granule organization and gelatinization behaviour are not well understood.

4.1.1 Experimental Conditions

There are a number of commercial differential scanning calorimeters which could be used for these experiments. One consideration is the design and use of the sample cell. Starch readily sediments from aqueous suspension, if a starch sediment is heated; when some of the granules within the population start to gelatinize, they "remove" water from other granules. The water distribution within the sediment becomes more non-uniform. As gelatinization is influenced by water content, particularly at water contents < 50%, the observed gelatinization behaviour may not represent that expected for starch in excess water. This problem is less severe when the depth of sediment is small and can be overcome by using cells which have a relatively large basal area compared to their height or by performing experiments on small quantities of material, i.e. a few mg.

A further consideration of cell design is the presence of vapour space above the sample. Ideally, there should be no vapour space above the liquid in a DSC cell as heat effects can be observed owing to liquid vaporization or to changes in pressure. In practice, the vapour space is kept as small as possible and these effects are assumed to be insignificant.

4.2 Swelling Behaviour

After gelatinization the starch granule swells to many times its original size. The swelling of wheat starch takes place in two stages: a significant swelling occurs within the range 60–70°C followed by a second step at 80–90°C (Doublier 1987). Solubilization of starch polysaccharide, primarily amylose, chiefly occurs during the second stage. As a result of granule swelling, the volume fraction occupied by the starch granules increases, resulting in an increase in dispersion viscosity. Particle volume fraction is the main determinant of the low shear viscosity of starch pastes, although other factors including particle shape, deformability and size distribution are also important (Doublier 1987). The amount of amylose solubilized during gelatinization has a greater effect on the mechanical behaviour of starch gels prepared by cooling concentrated dispersions of gelatinized starch to room temperature. The amylose forms a matrix gel within which swollen granules are embedded (Miles et al. 1985). The stiffness of the amylose matrix gel shows a very strong power law dependence on concentration (Ellis and Ring 1985).

Chemical procedures for total carbohydrate (Dubois et al. 1956) and amylose (Gilbert and Spragg 1964) can be used to measure the solubilization of starch during gelatinization. The swelling of the starch granule can be assessed in several ways. For example, it is possible to measure the length of the major axis of the granule using image analysis techniques (Ellis et al. 1988) or manually from enlarged photomicrographs. While this information is useful, a measure of the volume occupied by the granules is more relevant to a discussion of the functional behaviour of starch as a thickening agent. The most widely used method to assess granule swelling is to gelatinize the starch at the required temperature and then to centrifuge the gelatinized suspension at 2000–3000 g for 15 min (Schoch 1964b). A measure of the extent of swelling is then obtained from the weight of the sediment. A criticism of this method is that it does not measure the volume fraction of gelatinized granules directly. It assumes that gelatinized granules from different sources, irrespective of size distribution and shape, pack in the sediment in a similar way with comparable volume fractions. The method is attractive for its convenience and simplicity, but it is not well suited to measurements on small quantities of starch.

Granule swelling may also be determined by measuring the exclusion of a high molecular weight dextran from the gelatinized granule (Evans and Haisman 1979). A commercially available blue dextran (Pharmacia) is a high molecular weight water-soluble dextran (MW 2×10^6) which is excluded from gelatinized starch granules. If a solution of this dextran 0.1% w/w is mixed with an equal volume of a concentrated, washed dispersion of gelatinized starch, the volume of water not occupied by the gelatinized granules will lead to a dilution of the blue dextran. From measurements of the absorbance of the blue dextran at 630 nm it is possible to calculate this dilution, the volume of water not occupied by the granules and hence by difference the volume occupied by the gelatinized granules. Usually it is possible to sediment the gelatinized granules with a mild centrifugation 2000–3000 g for 15 min leaving the blue dextran in a clear supernatant. Occasionally, for some impure starch preparations the supernatant may be faintly turbid; this turbidity interferes with the measurement. In this case the supernatant may be clarified by a more vigorous centrifugation at 10000 g for 30 min.

To determine gelatinization and swelling behaviour a 1% – 2% w/w aqueous suspension of starch is usually appropriate. Gelatinization may be performed by heating a gently stirred suspension to the required temperature or by pouring a more concentrated suspension into stirred water at the required temperature. Either of these methods can avoid clumping. Gelatinization behaviour is also dependent on the rate of heating. For example, it was shown (Doublier 1987) that when wheat starch was gelatinized by heating at $6\,^{\circ}\mathrm{C}$ min^{-1}, granule swelling was greater than when it was heated at $1\,^{\circ}\mathrm{C}\ \mathrm{min}^{-1}$. Similar conclusions were made in another comparative study of the gelatinization behaviour of wheat and maize starches (Ellis et al. 1989).

5 Concluding Remarks

Physicochemical methods can provide valuable information on the properties of starch and are sufficiently sensitive so that analysis can be performed on the starch extracted from a single grain. In combination with improved chemical and enzymic methods for examining the fine structure of amylose and amylopectin, they can help provide links between fine structure and functional behaviour. Through investigations of the variability in starch structure with botanical origin, a much better understanding may ultimately be gained of the relationship between the way starch is biosynthesized in different crop plants and functional behaviour.

Acknowledgements. We thank the Ministry of Agriculture, Fisheries and Food for support.

References

Adkins GK, Greenwood CT (1966) Isolation of cereal starches in the laboratory. Staerke 18:213–218
Banks W, Greenwood CT (1975) Starch and its components. Edinburgh University Press, Edinburgh
Banks W, Greenwood CT, Muir DD (1971) The characterization of starch and its components, part 3. The technique of semi micro, differential, potentiometric, iodine titration and factors affecting it. Staerke 23:118–124
BeMiller JN (1964) Iodometric determination of amylose amperometric titration. In: Whistler RL (ed) Methods in carbohydrate chemistry, vol 4. Starch. Acad Press, New York, pp 165–167
Biliaderis CG, Page CM, Maurice TJ, Juliano BO (1986) Thermal characterization of rice starches: a polymeric approach to phase transitions of granular starch. J Agric Food Chem 34:6–14

Colonna P, Buleon A, Mercier C (1987) Physically modified starches. In: Galliard T (ed) Starch: properties and potential. Critical reports on applied chemistry, vol 13. John Wiley, Chichester, pp 79–114

Dengate H, Meredith P (1984) Variation in size distribution of starch granules from wheat grain. J Cereal Sci 2:83–90

Donovan JW (1979) Phase transitions of the starch-water system. Biopolymers 18:263–275

Donovan JW, Lorenz K, Kulp K (1983) Differential scanning calorimetry of heat-moisture treated wheat and potato starches. Cereal Chem 30:381–387

Doublier JL (1987) A rheological comparison of wheat, maize, faba bean and smooth pea starches. J Cereal Sci 5:247–262

Dubois M, Gilles KA, Hamilton JK, Rebers PA, Smith F (1956) Calorimetric method for determination of sugars and related substances. Anal Chem 28:350–356

Eliasson AC (1986) On the effects of surface active agents on the gelatinization of starch – a calorimetric investigation. Carbohydr Polym 6:463–476

Ellis HS, Ring SG (1985) A study of some factors influencing amylose gelation. Carbohydr Polym 5:215–222

Ellis HS, Ring SG, Whittam MA (1988) Time dependent changes in the size and volume of gelatinized starch granules on storage. Food Hydrocolloids 2:321–328

Ellis HS, Ring SG, Whittam MA (1989) A comparison of the viscous behaviour of wheat and maize starch pastes. J Cereal Sci 10:33–44

Evans ID, Haisman DR (1979) Rheology of gelatinised starch suspensions. J Texture Stud 10:347–370

Everett W, Foster JF (1959) The subfractionation of amylose and characterization of subfractions by light scattering. J Am Chem Soc 81:3459–3469

Evers AD, Greenwood CT, Muir DD, Venables C (1974) Studies on the biosynthesis of starch granules, part 8. A comparison of the properties of the small and the large granules in mature cereal starches. Staerke 26:42–46

Galliard T, Bowler P (1987) Morphology and composition of starch. In: Galliard T (ed) Starch: properties and potential. Critical reports on applied chemistry, vol 13. John Wiley, New York, pp 55–78

Gilbert GA, Spragg SP (1964) Iodometric determination of amylose iodine sorption: "blue value". In: Whistler RL (ed) Methods in carbohydrate chemistry, vol 4. Starch. Acad Press, New York, pp 168–169

Gough BM, Greenwell P, Russell PL (1985) On the interaction of sodium dodecyl sulphate with starch granules. In: Hill RD, Munck L (eds) New approaches to research on cereal carbohydrates. Elsevier, Amsterdam, pp 99–108

Hizukuri S, Maehara Y (1990) Fine structure of wheat amylopectin: the mode of A to B chain binding. Carbohydr Res 206:145–159

Hizukuri S, Takagi T (1984) Estimation of the distribution of molecular weight for amylose by the low-angle laser-light scattering technique combined with high performance gel chromatography. Carbohydr Res 134:1–10

Hizukuri S, Takeda Y, Yasuda M, Suzuki A (1981) Multi-branched nature of amylose and the action of debranching enzymes. Carbohydr Res 94:205–213

Karkalas J, Raphaelides S (1986) Quantitative aspects of amylose-lipid interactions. Carbohydr Res 157:215–234

Kugimiya M, Donovan JW (1981) Calorimetric determination of the amylose content of starches based on formation and melting of the amylose-lysolecithin complex. J Food Sci 46:765–770

Lowy GDA, Sargeant JG, Schofield JD (1981) Wheat starch granule protein: the isolation land characterization of a salt-extractable protein from starch granules. J Sci Food Agric 32:371–377

Matheson NK (1971) Amylose changes in the starch of developing wheat grains. Phytochemistry 10:3213–3219

Mciver RG, Axford DWE, Colwell KH, Elton GAH (1968) Kinetic study of the retrogradation of gelatinised starch. J Sci Food Agric 19:560–563

Miles MJ, Morris VJ, Orford PD, Ring SG (1985) The roles of amylose and amylopectin in the gelation and retrogradation of starch. Carbohydr Res 135:271–281

Morrison WR (1988) Lipids in cereal starches: a review. J Cereal Sci 8:1–15

Morrison WR, Gadan H (1987) The amylose and lipid contents of starch granules in developing wheat endosperm. J Cereal Sci 5:263–275

Morrison WR, Laignelet B (1983) An improved colorimetric procedure for determining apparent and total amylose in cereal and other starches. J Cereal Sci 1:9–20

Privalov PL, Gill SJ (1988) Stability of protein structure and hydrophobic interaction. Adv Protein Chem 39:191–234

Ring SG, I'Anson KJ, Morris VJ (1985) Static and dynamic light scattering studies of amylose solutions. Macromolecules 18:182–188

Russell PL (1987) Gelatinization of starches of different amylose/amylopectin content. A study by differential scanning calorimetry. J Cereal Sci 6:133–145

Russell PL, Oliver G (1989) The effect of pH and NaCl content on starch gel ageing. A study by differential scanning calorimetry and rheology. J Cereal Sci 10:123–138

Schoch TJ (1964a) Iodometric determination of amylose potentiometric titration: standard method. In: Whistler RL (ed) Methods in carbohydrate chemistry, vol 4. Starch. Acad Press, New York, pp 157–160

Schoch TJ (1964b) Swelling power and solubility of granular starches. In: Whistler RL (ed) Methods in carbohydrate chemistry, vol 4. Starch. Acad Press, New York, pp 106–108

Slade L, Levine H (1988) Non-equilibrium melting of native granular starch, part 1. Temperature location of the glass transition associated with gelatinization of A-type cereal starches. Carbohydr Polym 8:183–188

South JB, Morrison WR (1990) Isolation and analysis of starch from single kernels of wheat and barley. J Cereal Sci 12:43–51

Spies RD, Hoseney RC (1982) Effect of sugars on starch gelatinization. Cereal Chem 59:128–131

Takeda Y, Shirasaka K, Hizukuri S (1984) Examination of the purity and structure of amylose by gel-permeation chromatography. Carbohydr Res 132:83–92

Takeda Y, Hizukuri S, Takeda C, Suzuki A (1987) Structures of branched molecules of amyloses of various origins, and molar fractions of branched and unbranched molecules. Carbohydr Res 165:139–145

Watson SA (1964) Determination of starch gelatinization temperature. In: Whistler RL (ed) Methods in carbohydrate chemistry, vol 4. Starch. Acad Press, New York, pp 240–241

The Isolation of Wheat Mitochondrial DNA and RNA

D. F. Spencer, M. W. Gray, and M. N. Schnare

1 Introduction

Interest in plant mitochondrial (mt) DNA and RNA has increased dramatically in the last 10 years. The physical complexity of flowering plant mtDNA, first described by Quetier and Vedel (1977), is still a topic of study and is as yet not fully understood. The presence in plant mitochondria of a unique 5 S ribosomal RNA (rRNA) (absent from all other mitochondria) was recognized early and was the topic of several publications from our group (Bonen and Gray 1980; Spencer et al. 1981).

The protocols presented here describe our methods for isolating mtDNA and mtRNA from embryos (the viable germ) of wheat (*Triticum aestivum*). The embryos are obtained by a mechanical fractionation of wheat seed. The pioneering work in the isolation of viable embryos was that of Johnston and Stern (1957). Early versions of the protocols presented here were described briefly in Cunningham and Gray (1977) and Bonen and Gray (1980). Several procedures have been adapted from Kolodner and Tewari (1975).

Studies of wheat mtDNA and mtRNA obtained by such protocols include: the preliminary characterization of rRNA gene organization in mtDNA (Bonen and Gray 1980); the determination of the sequence of the 5S rRNA (Spencer et al. 1981); the determination of the sequence of the gene for the small subunit ("18S") rRNA (Spencer et al. 1984); sequence determination of protein-coding genes (Bonen et al. 1984, 1987; Boer et al. 1985); studies of mtDNA-encoded transfer RNAs (tRNAs) (Gray and Spencer 1983; Joyce et al. 1988a, b; Joyce and Gray 1989a, b); processing of RNA precursors of tRNAs and tRNA-like sequences (Hanic-Joyce and Gray 1990; Hanic-Joyce et al. 1990); and description and characterization of messenger RNA editing in wheat mitochondria (Covello and Gray 1989).

The protocols presented here have been used primarily for preparing wheat mitochondria, but we have used them, with minor modifications, for rye (*Secale cereale*). They should be adaptable to many other flowering plants and, where appropriate, suggested modifications for such purposes are included. A previous article in this series (Jackson 1985) described a simpler but less rigorous method of isolating mtDNA from etiolated shoots of maize (*Zea mays*).

2 Fractionation of Wheat Embryos from Seed

The first steps in the protocol are the grinding of seed and sieving of the resulting fragments to separate the embryos from the bulk of the other seed components. Solvent flotation is used to purify the embryos further. These initial procedures are the most difficult to adapt to seeds other than wheat. Our limited experience with rye seed indicates that embryos can be obtained, with only a few modifications, using the following grinding and sieving methods. The precise grinding times required and the exact sieve combinations can only be determined by testing small lots of the seed of interest. In general, smaller seeds from cereals and other monocotyledonous plants are probably the most easily accommodated, whereas with dicotyledonous plants, where the final product is really embryo axes, the methods are less likely to be usable. Preliminary tests suggest that the protocol described here should be applicable to *Glycine max* (soybean) seeds.

2.1 Grinding and Sieving of the Seed

The seed is generally processed in lots of 20 kg. We purchase untreated seed in 25-kg bags and store it in a cold room (4 °C). Because grinding of the seed generates moderate amounts of dust, it is best done in a fume hood and a dust mask should be used. Ear protectors may be desirable because the grinding step is noisy.

The actual milling is performed on 500-g lots. Weigh 500 g of seed in a 500-ml plastic beaker and note the level of seed (approximately 500 ml), so that subsequent weighings are unnecessary. Add the seed to a 3.8-l stainless steel container of a three-speed Waring commercial blender (Model CB-6), secure the cover and grind at low setting (15 500 rpm) for 20 s. Transfer all of the ground seed to the top sieve of a 20-cm diameter U.S. Series standard sieve set consisting of: upper sieve 1.70 mm (12 mesh); middle sieve 1.18 mm (16 mesh); lower sieve 600 µm (30 mesh); and a solid pan on the bottom. Place the cover on the top sieve and shake the assembled sieves vigorously for 15 to 20 s. Return the contents of the top sieve to the blender and re-grind for 20 s at low speed. Discard the material on the middle (1.18 mm) sieve and transfer the crude embryos, which are retained on the lower sieve (600 µm), to a beaker. Return the contents of the blender to the top sieve and repeat the shaking. Repeat the above steps in 500-g lots until the 20 kg of seed has been processed. The crude embryos can be stored overnight by covering the beaker containing them with aluminum foil or plastic food wrap and placing them at 4 °C.

The next step removes most of the bran from the crude embryos. Pour the crude embryos (in small lots) onto the 600-µm sieve. Using unheated air from a hair dryer (such as an Oster Airjet dryer) or any suitable hand-held air blower, carefully blow the bran out of the sieve and away from the embryos. This is best done by holding the blower at about a 45 ° angle and approximately

30 – 60 cm away from the sieve while shaking the sieve back and forth horizontally. The bran is undesirable and a small loss of embryos is inevitable. If working with dicots there is no bran to remove but small pieces of seed coat are eliminated with this procedure.

2.2 Flotation of the Embryos

The embryos are next separated from endosperm by flotation in a mixture of cyclohexane and carbon tetrachloride. We have found that the optimal ratio of these two solvents varies slightly even for different varieties of wheat, and therefore when working with embryos from other plants, the solvent ratios must be determined empirically. For wheat, the cyclohexane to carbon tetrachloride ratios vary from 10 : 25 to 10 : 27; the latter is the mixture we currently use.

Prepare 3700 ml of the cyclohexane/carbon tetrachloride mixture in a large Erlenmeyer flask. Generally, a technical grade of solvent is adequate but such poorer quality solvents should be filtered through glass wool (packed in a conical funnel) to remove particulate material such as rust. Occasionally, filter paper will have to be used to eliminate fine debris. Be sure the solvents are thoroughly mixed together before use.

Place the embryos in a 2000-ml glass beaker and add solvent mix to a final volume of approximately 1800 ml. Stir the embryos vigorously with a long glass stirring rod, then allow the components to separate. The good embryos should float to the surface while endosperm sinks and dead embryos and embryo/endosperm particles remain suspended. After the mixture has stabilized, suction the embryos off the top using a short section of glass tubing (6 – 8 mm i.d.) connected with Tygon tubing to a filter flask connected to an aspirator filter pump. The embryos are recovered from the flask by filtering the embryo suspension, under low vacuum, through a coarse sintered glass (Buchner) funnel containing a disk of Whatman #1 paper and mounted on a second filter flask. The solvent mix can be recycled by passing it through glass wool followed by filtration through Whatman #1 paper.

Repeat the flotation of the recovered embryos using recycled solvent in a smaller capacity beaker of similar height to the first, and with proportionately less solvent (1000 ml). Recover the embryos as above and repeat the flotations twice with recycled solvent mix, then twice using fresh solvent mix each time. The embryos should be processed as quickly as possible so as to minimize contact with the solvent.

After the final flotation and collection of the embryos, leave the embryos on the sintered glass filter for about 30 min under suction with occasional stirring. Then distribute the embryos in a thin layer on a glass tray and place in a fume hood. Allow the embryos to dry completely, usually overnight or longer, before continuing. Stir the embryos periodically while they are in the tray. All traces of solvent must be eliminated to permit proper embryo imbibition and germination.

2.3 Final Purification of the Solvent-Floated Embryos

The starting material is the dried embryos from the previous section. This is a single sieving step that eliminates broken pieces of embryos as well as embryos with fragments of endosperm attached.

Assemble a sieve set with 850 µm (20 mesh) on top, 710 µm (25 mesh) in the middle and a solid pan on the bottom. Transfer the embryos to the top sieve and, after covering, shake vigorously for approximately 1 min. The material retained on the middle (710 µm) sieve is the final purified embryos. Broken embryos collect on the bottom pan and embryo/endosperm fragments collect on the top sieve.

The embryos must be stored dry and cool (4 °C). Transfer the embryos to a small glass or plastic beaker, cover with Parafilm (American National Can), cut a few holes in the Parafilm, place the small beaker inside a larger beaker containing Drierite (or similar desiccant) and cover the larger beaker tightly with Parafilm. Store at 4 °C. When embryos are needed, allow the beaker(s) to equilibrate to room temperature before opening in order to prevent condensation from accumulating inside.

The yield of wheat embryos is generally 2–2.5 g/kg starting seed. The initial grinding times may have to be adjusted to achieve this yield.

3 Large-Scale Preparation of Wheat Embryo Mitochondria

The protocol as it is presented here assumes 24 g (pre-germination weight) of embryos. It is possible to scale the methods up or down. The protocols that follow can also be used for extractions of shoot tissue provided there is compensation for the dilution effect on buffers caused by the higher water content in such material. All operations are performed at between 0° (ice) and 4 °C. Several of the solutions used contain β-mercaptoethanol and bovine serum albumin (BSA; fraction V, essentially fatty acid-free). These components should be added to buffers on the day mitochondria are prepared.

3.1 Germination/Imbibition of the Embryos

The embryos are distributed in 2.4-g lots onto 150×15 mm Petri dishes containing a circle of tightly fitting Whatman 3 MM paper. The filter paper should be prewetted with 16 ml of 1% glucose before sprinkling the embryos onto it. Cover the dishes and place them in the dark (at room temperature) for 24 h. If the embryos are healthy, there should be little free glucose solution after this time and there should be no detectable signs of fungal or bacterial contamination. We have experienced serious problems with fungal contamination with one variety of rye; these problems are very difficult to eliminate other than by testing different sources and varieties of seed.

3.2 Isolation of Mitochondria

The embryos of many seed plants have tightly associated epiphytic bacteria (Wallace and Lochhead 1951). Therefore it is often desirable, although not essential, to wash the embryos before continuing. Transfer the embryos to a chilled beaker containing 200 ml of buffer A (0.44 M sucrose, 0.05 M Tris-HCl, pH 8, 0.003 M EDTA-Na, 0.001 M β-mercaptoethanol, 0.1% BSA) and stir the embryos with a glass rod. Filter the wash liquid through four layers of cheesecloth. If the filtrate is extremely cloudy, a second wash may be desirable.

Transfer the embryos to a precooled mortar that has been placed in a container of ice. Add 100 ml of buffer A and grind the embryos with a pestle for approximately 5 min. Filter the homogenate through four layers of cheesecloth into a chilled beaker, squeeze excess liquid from the ground residue, then return the ground embryos to the mortar. Add 100 ml of buffer A and repeat the grinding and filtering. Repeat this step a third time. If the tissue used is shoot rather than embryo, the pH probably should be checked to ensure that acids in the cell vacuoles have not exhausted the buffering capacity of buffer A.

Divide the final filtrate between two chilled 250-ml plastic centrifuge bottles. Centrifuge the bottles at 900 g for 6 min in a swinging bucket preparative centrifuge. The pellets from the first low-speed spin are enriched for nuclei, although they contain large amounts of cell debris. If nuclear DNA is required, this fraction should be saved. Pour the supernatant fractions into another two 250-ml bottles and centrifuge at 1900 g for 6 min; then pour the supernatants into two 250-ml bottles. The second low-speed pellet contains nuclei as well as proplastids, or their fragments, and small amounts of any contaminating epiphytes. This fraction is of no value and should be discarded. Although the preparation of nuclear DNA is not a primary focus of this chapter, brief descriptions of further processing of this fraction will be included. For nuclear DNA, the first low-speed spin pellets should be washed twice with buffer A, using 25 – 30 ml each time. Resuspend the pellets in the buffer as described for mitochondrial pellets in the following paragraph, centrifuge at 1900 g for 6 min, discard the supernatants and repeat. Nuclear DNA can be extracted directly at this point by resuspending the nuclei, adding sodium acetate or sodium chloride to 0.3 – 0.5 M final concentration, and performing phenol extractions as outlined later in this protocol. Alternatively, the nuclei can be purified further by centrifuging them through the same sucrose gradients described later for mitochondria. If additional purification of the nuclear DNA is desired, the DNA can be centrifuged in a cesium chloride gradient as outlined later for mtDNA. It should be noted that 24 g of wheat embryos yield approximately 10 mg of nuclear DNA and therefore only a fraction of the total low-speed spin pellet need be retained for nuclear DNA isolation.

To continue the isolation of mitochondria, centrifuge the two bottles containing the second low-speed spin supernatants at 17000 to 24000 g for 20 min. If cytosolic RNA is desired, the top two-thirds (approximately) of the supernatant fraction should be removed and processed as described later for mtRNA; otherwise, the supernatant fraction is poured off and any extremely loose material from the pellet is discarded. Resuspend the crude mitochondrial pellets in 70–75 ml of cold buffer A. This can be accomplished either by stirring the pellet with a glass rod or by flushing the buffer back and forth over the pellet with a 10-ml pipette connected to a motorized pipette filler such as a Pipet-Aid (Drummond Scientific). The hard white material at the bottom of the mitochondrial pellet should be left behind if possible. Transfer the resuspended mitochondria (in two lots) to a chilled 30-ml (or larger) Potter Elvehjem-type tissue grinder. Homogenize (about 10–12 strokes with a Teflon pestle) until the mitochondria are completely resuspended, then transfer the suspension to two chilled 50-ml polypropylene centrifuge tubes and centrifuge at 1900 g for 6 min. Pour the supernatant fractions into two clean tubes and recentrifuge. Transfer the supernatants to two tubes and centrifuge at 17000–24000 g for 20 min. Discard the supernatant fraction and any loose material from the top of the pellet.

At this stage there are two different series of procedures according to whether the final product is to be RNA or DNA. In each case, mitochondria are eventually purified on sucrose gradients.

3.3 Preparation of Mitochondria for DNA Isolation

Resuspend the mitochondrial pellets in a total of 30 ml chilled buffer A containing $MgCl_2$ at a final concentration of 10 mM. The resuspension should be done as described above, using the tissue grinder. Transfer the suspension back to a 50-ml centrifuge tube and add DNase I to a final concentration of 50 µM. (The DNase can be dissolved beforehand at a concentration of 2 mg in 10 ml buffer A and the solution then added to the resuspended mitochondria.) Mix the DNase I thoroughly with the mitochondria but do so carefully because DNase I is particularly sensitive to denaturation. Leave the tube on ice (or at 4 °C) for 1 h.

After the DNase treatment, transfer the mitochondria to a chilled 250-ml flask. Slowly (over about 10 min) add 2 vol cold buffer B (0.44 M sucrose, 0.05 M Tris-HCl, pH 8, 0.02 M EDTA-Na, 0.001 M β-mercaptoethanol and 0.1% BSA). Agitate the mitochondrial suspension slowly during the buffer addition. Divide the final suspension among four 50-ml polypropylene centrifuge tubes. Centrifuge at 17000–24000 g for 20 min. Discard the supernatant fraction, resuspend the pellets in a total of 60 ml buffer B (using a tissue grinder, as above), combine the suspensions into two polypropylene tubes and repeat the high-speed spin. Discard supernatants. At this point the mitochondria are purified on sucrose gradients. Both DNA and RNA preparations are handled identically.

3.4 Purification of Mitochondria on Sucrose Gradients

The mitochondria are next banded in discontinuous sucrose gradients. The specific recipes here are intended for use with a Beckman SW 25.1 swinging bucket rotor but could be adapted to a different ultracentrifuge rotor. Three discontinuous sucrose gradients must be prepared. Each consists of a 7.5-ml lower layer of 1.55 M sucrose in buffer E (0.05 M Tris-HCl, 0.003 M EDTA-Na, pH 8) and a 15-ml upper layer of 1.15 M sucrose in buffer E. The sucrose used should be a nuclease-free grade; on the day the gradients are to be prepared, β-mercaptoethanol should be added to 0.001 M and BSA to 0.1 % final concentration.

Each gradient can accommodate 3–5 ml of sample. Resuspend the mitochondrial pellet from above in a total of 9 ml buffer A (use the tissue grinder as described previously), then load one-third of this on each of the three gradients. If other fractions are to be included at this stage, then individual mitochondrial pellets or nuclei from the first low-speed pellet should be resuspended in 3–5 ml of buffer A and loaded onto individual gradients. Centrifuge the gradients at 22500 rpm for 1 h; set the braking profile to the gentlest recommended by the manufacturer for the rotor in the centrifuge being used.

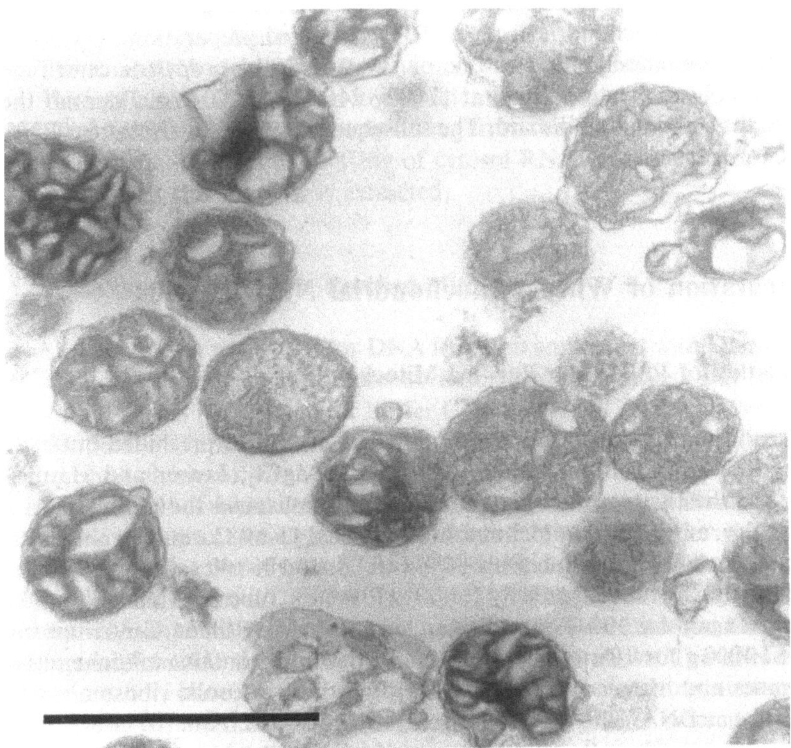

Fig. 1. Electron micrograph of purified wheat mitochondria. The mitochondria were fixed by adding one-tenth volume of pure glutaraldehyde to the diluted mitochondrial suspension following the sucrose gradient step. Subsequent washings, osmium tetroxide postfixation, uranyl acetate prestaining, dehydration, embedding, sectioning and electron microscopy were performed by G. Faulkner (Electron Microscopy Unit, Faculty of Medicine, Dalhousie University). *Bar* = 1 μm

The mitochondria form a fairly compact, concentrated, yellow/brown band at the interface of the 1.15 and 1.55 M layers. There will probably also be a moderately large pellet which contains small quantities of mitochondria, some proplastid fragments, and a large quantity of epiphytic bacteria. Our analyses of the rRNAs from this fraction indicate that in wheat embryos the major epiphyte is a member of the γ-subdivision of the purple bacteria (either *Escherichia coli* or a closely related organism), consistent with early observations of Wallace and Lochhead (1951). This pelleted material should not be permitted to contaminate the mitochondria. Nuclei purified on these gradients will pass through both sucrose layers and collect as a tight pellet on the bottom.

Connect a Pasteur pipette to an aspirator (filter pump) and carefully remove the bulk of the gradients above the mitochondrial bands, without disturbing the lower regions of the gradients. Slowly remove the mitochondrial bands from each tube with a syringe and an 18-gauge needle with a smooth 90° bend. Transfer the mitochondria to a small flask, and slowly (over approximately 10 min) dilute the mitochondria with 2 vol cold buffer C (0.05 M Tris-HCl, 0.02 M EDTA-Na, pH 8). The electron micrograph in Fig. 1 shows purified wheat embryo mitochondria from a DNase I-treated preparation.

Transfer the mitochondria to one or two 50-ml polypropylene centrifuge tubes and collect by centrifuging at 17000–24000 g for 20 min. Pour off the supernatant fraction and discard. The subsequent steps are different for RNA and DNA preparations.

4 Preparation of Wheat Mitochondrial Nucleic Acids

4.1 Isolation of RNA from Purified Mitochondria

Resuspend the combined mitochondrial pellets in 9 ml of prechilled buffer D (0.01 M Tris-HCl, pH 8.5, 0.05 M KCl, 0.01 M $MgCl_2$; Leaver and Harmey 1976). Use the tissue grinder as before to fully resuspend the mitochondria. Transfer the resuspended mitochondria to a chilled 15-ml Corex (Corning) centrifuge tube. Add 1 ml of a chilled 20% (v/v) Triton X-100 solution in buffer D, and vortex the tube vigorously for 30 s. Place the tube back on ice for 30 s and vortex again for 30 s. Repeat this cycle a total of five times. Centrifuge the tube at 10000 g for 10 min. The pellet from this spin contains mitochondrial membranes and material adhering to them, such as cytosolic ribosomes and most of the mtDNA. This material is generally discarded. Pour the clear supernatant into a graduated 40–50-ml centrifuge tube. Add an equal volume 2×detergent mix [2% sarkosyl (sodium N-lauroylsarcosinate), 0.1 M NaCl], and then an equal volume phenol/cresol/8-hydroxyquinoline (500:70:0.5; Parish and Kirby 1966) which has been equilibrated with 10–50 mM Tris-HCl (pH 7.5–8). Cover and seal the tube with either a cap or Parafilm and extract by vigorous shaking in a motorized shaker at 4 °C. Clarify the phases by centrifugation at 1900 g (or higher) at 0 – 4 °C, preferably in a swinging bucket rotor. Carefully remove the upper, aqueous phase, avoiding the debris at the interface, and transfer this phase to a clean, chilled tube. Add solid NaCl to 0.5 M final concentration (0.6 g/20 ml solution), vortex the tube until the salt is dissolved and repeat the phenol extraction with fresh phenol/cresol until no more material is produced at the aqueous-phenol interface. To the final aqueous phase add 2 vol cold 95% – 100% ethanol and after covering the tube mix the contents thoroughly. Place the tube at −15 to −20 °C. The RNA is fairly stable in this form. Generally, the precipitated RNA should be redissolved in a small volume and additional phenol extractions performed. It is necessary to give the final RNA two cycles of ethanol precipitation in order to eliminate all traces of phenol. Cytosol RNA can be isolated in a similar fashion, although the solution volumes will be greater. By delaying the addition of the NaCl until after the first phenol extraction rather than before, most DNA is trapped in the aqueous-phenol interface (complexed with protein) and thus eliminated. Figure 2 shows two profiles of three fractions of RNA from a wheat embryo mitochondrial preparation, resolved either by formaldehyde-agarose gel electrophoresis (A) or by urea-polyacrylamide gel electrophoresis (B).

The final RNA yields from 24 g of embryos are: approximately 1.25 mg total mitochondrial RNA (from the main mitochondrial fraction, the Triton X-100 supernatant); approximately 0.5 mg from the Triton pellet (which is normally discarded); and 150 – 200 mg of cytosol RNA (although normally only about 1/20 of this amount is extracted).

4.2 Isolation of DNA from Purified Mitochondria

The mitochondria to be used for DNA isolation are treated with DNase I prior to fractionation on sucrose gradients. Resuspend the gradient-purified mitochondrial pellets in 3 – 3.5 ml of buffer C using a small tissue grinder to make the suspension uniform, then transfer the mitochondria to a graduated tube. Bring the volume to 4.5 ml with buffer C. Add 0.1 ml buffer C containing 1.25 mg of self-digested Pronase (Calbiochem), or an equivalent amount of Proteinase K, mix carefully, then add 0.4 ml of 25% (w/v) sarkosyl. Mix thoroughly but very carefully to minimize physical degradation (shearing) of the DNA. Place the tube at 25 °C for 30 min. If nuclear DNA is to be purified as extensively as the mtDNA, the nuclei pellets from the sucrose gradients can be resuspended in 4.5 ml of buffer C and processed as above. The above procedures are not, however, completely effective at deproteinizing nuclear DNA and so the preferred method is to perform several phenol extractions on the resuspended nuclei, followed by one or two ethanol precipitations, prior to preparation of the DNA for CsCl gradient purification.

The specifics of banding the DNA(s) in CsCl vary with the ultracentrifuge rotor being used. What follows is designed for the Beckman Type 50 Ti rotor with Quick-Seal polyallomer centrifuge tubes. Bring the solution volumes to

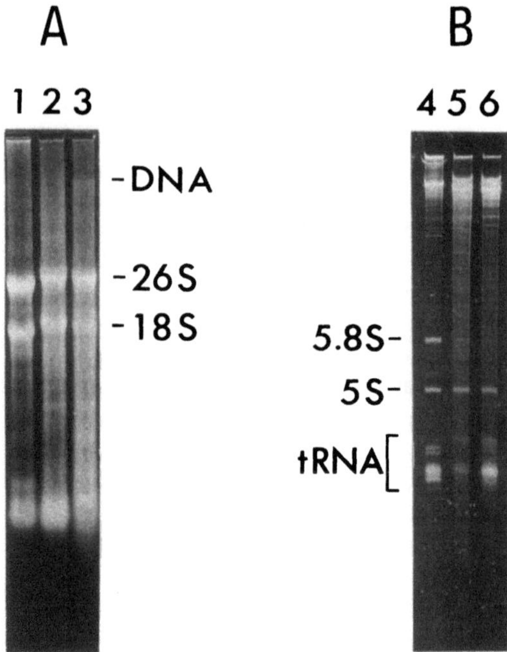

Fig. 2. Electrophoretic analysis of RNA fractions from a wheat mitochondrial preparation. **A** Agarose gel (1%) containing 2.2 M formaldehyde and cast in 20 mM MOPS-Na, 1 mM EDTA-Na (pH 7). Five µg of each RNA preparation were dissolved in 20 µl of loading buffer (50% formamide, 2.2 M formaldehyde, 20 mM MOPS-Na, 1 mM EDTA-Na, pH 7, 0.1% w/v bromophenol blue, 0.1% xylene cyanol) and 10 µg of ethidium bromide were added to each sample. The samples were heated at 65 °C for 10 min, then loaded onto the gel and electrophoresed at 5 V/cm. *Lane 1* Total wheat cytosolic RNA; *lane 2* mtRNA from Triton X-100 supernatant fraction; *lane 3* RNA from Triton X-100 pellet fraction. There is no detectable contamination by cytosolic RNA in the mtRNA from the Triton X-100 supernatant fraction. **B** Polyacrylamide gel (6%; 20×20×0.15 cm) containing 7 M urea and cast in 50 mM Tris-boric acid, 1 mM EDTA (pH 8.3). RNA (2.5 µg of each preparation) was dissolved in 10 µl of NMF/urea loading buffer (60% v/v N-methylformamide, 5 M urea, 10 mM Tris-HCl, pH 7.6, 1 mM EDTA-Na) containing 0.1% (w/v) bromophenol blue, heated to 65 °C for 5 min, and electrophoresed at 10 V/cm. The gel was stained with a solution of ethidium bromide (1 µg/ml). *Lane 4* Total wheat cytosol RNA; *lane 5* RNA from Triton X-100 pellet fraction; *lane 6* RNA from Triton X-100 supernatant fraction. The absence of detectable cytosolic 5.8S RNA in the mitochondrial fraction is an indication of its purity

10.4 ml with buffer C, then add 10.5 g CsCl. Cover the tube(s) and carefully mix the solution by slowly inverting each tube. The high CsCl concentration will precipitate protein (which floats in the dense solution). An optional extra step is to place the tube(s) at 65 °C for 10–15 min to improve deproteinization of DNA. We have found this to be particularly useful when working with nuclear DNA. It is possible to eliminate much of the denatured protein with a centrifugation step (3000 g for 5–10 min); the protein forms a layer on the top and can be left behind when transferring the DNA/CsCl solution to a new tube.

The final step before filling the centrifuge tubes is to add dye to each tube. The traditional dye used in CsCl gradients has been ethidium bromide. Move the DNA/CsCl solution(s) into subdued light, such as in a darkroom with a low intensity incandescent light bulb, and add 0.1 ml of a 20 mg/ml stock solution to each sample. Cover the tube and mix the solutions thoroughly but gently. Ethidium bromide is hazardous and should be handled appropriately. Once the ethidium bromide is added to the DNA, it is important to minimize the exposure of the DNA/dye complex to light, particularly shorter wavelengths. The final DNA/dye/CsCl solution can then be transferred to a centrifuge tube, which is sealed and placed in the rotor. An alternative dye that is at least as good for these purposes is the bisbenzimide dye Hoechst 33258. Add it at the same concentration (0.1 ml of a 20 mg/ml stock to each sample), cover the tube and mix. This quantity of Hoechst dye will not fully dissolve in the CsCl solution, and must be mixed longer (15 – 30 min with periodic gentle agitation) to ensure that the DNA is fully saturated. Because the dye does not intercalate, it is much less likely to damage DNA when the DNA/dye complex is exposed to light, and equally important, it is not considered a health hazard.

Place the sealed tubes in the rotor and centrifuge at 40000 rpm for 44 – 48 h. The centrifuge temperature should be approximately 20 °C. At the end of the run, take the rotor to a room with subdued light, remove the tubes and visualize the DNA band(s) with a hand-held longwave UV (365 nm, black light) source. There are several techniques for removing the DNA band from the tube. We prefer to carefully remove the top region in the gradient (above the band) and then to withdraw the DNA slowly into a 16-gauge cannula connected to a disposable plastic syringe. The dye must then be extracted from the DNA. Although there are several variations of this step as well, we use water-saturated n-butanol as solvent. The easiest method is to do the extractions directly in the syringe as described by Hofmann and Weising (1990). It generally takes at least five and frequently more extractions to eliminate the dye. High-molecular-weight DNA (such as plant nuclear DNA) is quite difficult to extract efficiently. Throughout all of these manipulations the DNA should be handled as gently as possible. Transfer the DNA/CsCl solution to a centrifuge tube, add 2 vol 95% – 100% ethanol (at room temperature) and mix thoroughly by slow inversion. The DNA precipitate can usually be seen immediately but there may also be CsCl precipitation. Centrifuge the tube at approximately 1000 g for 1 – 2 min and examine the pellet. The DNA will generally be at the bottom of the tube (if not, recentrifuge at 2000 g) and the top liquid can be carefully poured off. Any CsCl that precipitates is easily eliminated by repeated washes with 80% ethanol. The pelleted DNA, and the inside surfaces of the centrifuge tube, should be washed several times with 80% ethanol (with a short centrifugation step each time) to remove all CsCl. Any residues of CsCl will make redissolving the DNA difficult or impossible. The final DNA pellet should be dried very briefly in a vacuum desiccator and redissolved. It is often advisable to subject the DNA to a phenol extraction if DNA purity is critical. The trade-off for these additional manipulations is further DNA shearing.

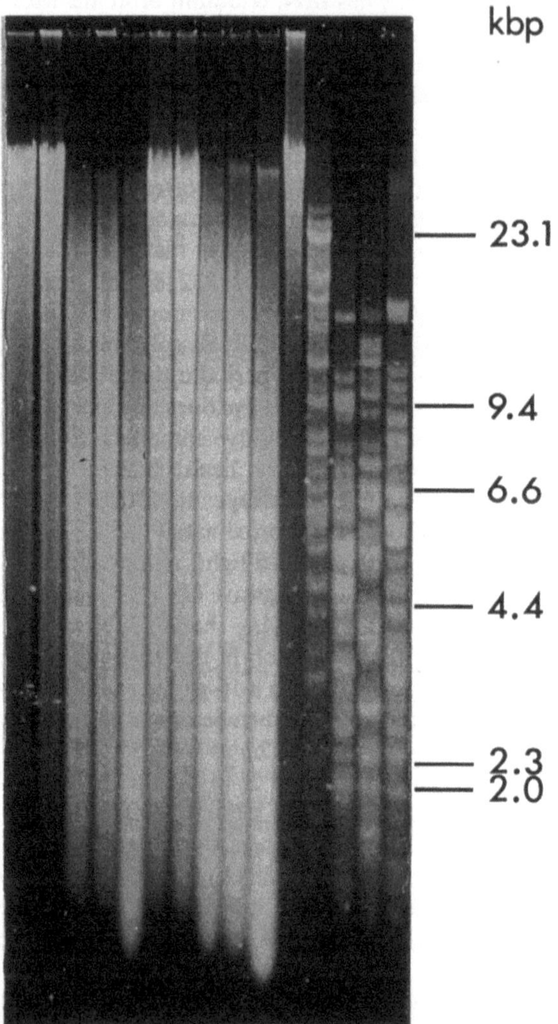

Fig. 3. Agarose gel (0.5%) cast in 100 mM Tris-acetate, 0.1 mM EDTA (pH 8). Approximately 1-µg samples of each DNA were digested in the manufacturer's recommended buffer and incubated at 55 °C (for BclI) or 37 °C (all other enzymes) for 3 h with 5 – 10 units of enzyme. 5× Loading buffer [0.1 M EDTA-Na, pH 7.5, 25% v/v glycerol, 5% w/v Ficoll (Pharmacia), 0.5% w/v bromophenol blue, 0.5% w/v xylene cyanol] was added to each sample, which was heated to 65 °C for 10 min, then loaded onto the gel, with a lane of *Hin* dIII-digested lambda DNA for size markers. The gel was electrophoresed at 0.5 – 1.0 V/cm for 16 h, then stained in a 1 µg/ml solution of ethidium bromide. *Lanes 1 – 5* Purified nuclear DNA from first low-speed pellet; *lanes 6 – 10* purified DNA from second low-speed pellet; *lanes 11 – 15* purified DNA from mitochondrial fraction. *Lanes 1, 6, 11* control; *lanes 2, 7, 12 Sal*I; *lanes 3, 8, 13 Bam*HI; *lanes 4, 9, 14 Eco*RI; *lanes 5, 10, 15 Bcl*I. The *numbers at the right* are the fragment sizes of the lambda reference DNA (in kilobase pairs, *kbp*). The wheat nuclear DNA is nearly completely blocked (by methylation) for cutting by *Sal*I, while *Bcl*I will not cut eubacterial DNAs methylated (in vivo) with the *dam* methylase. These digestions therefore serve as indicators of cross-contamination of various fractions. The mtDNA contains little or no detectable nuclear or epiphytic/eubacterial DNA

The estimated final yield of mtDNA is 150 µg. The yield of nuclear DNA is about 10 mg/24 g embryos; generally only about one-tenth of this amount is actually processed.

Figure 3 is a photograph of an ethidium bromide-stained agarose gel of restriction endonuclease-digested purified DNA from the first low-speed pellet, second low-speed pellet, and mitochondrial fraction. Southern hybridization of this gel (not shown) indicates that the nuclear DNA (first low-speed pellet) has trace contamination with chloroplast DNA, while the principal contaminant of the mtDNA is trace levels of chloroplast DNA, with barely detectable levels of DNA from the bacterial epiphyte.

Acknowledgment. Work in the authors' laboratory on wheat mtDNA and mtRNA is supported by an Operating Grant (MT-4124) from the Medical Research Council of Canada to M.W.G., who is a Fellow in the Evolutionary Biology Program of the Canadian Institute for Advanced Research.

References

Boer PH, McIntosh JE, Gray MW, Bonen L (1985) The wheat mitochondrial gene for apocytochrome b: absence of a prokaryotic ribosome binding site. Nucleic Acids Res 13:2281–2292

Bonen L, Gray MW (1980) Organization and expression of the mitochondrial genome of plants. I. The genes for wheat mitochondrial ribosomal and transfer RNA: evidence for an unusual arrangement. Nucleic Acids Res 8:319–335

Bonen L, Boer PH, Gray MW (1984) The wheat cytochrome oxidase subunit II gene has an intron insert and three radical amino acid changes relative to maize. EMBO J 3:2531–2536

Bonen L, Boer PH, McIntosh JE, Gray MW (1987) Nucleotide sequence of the wheat mitochondrial gene for subunit I of cytochrome oxidase. Nucleic Acids Res 15:6734

Covello PS, Gray MW (1989) RNA editing in plant mitochondria. Nature 341:662–666

Cunningham RS, Gray MW (1977) Isolation and characterization of ^{32}P-labeled mitochondrial and cytosol ribosomal RNA from germinating wheat embryos. Biochim Biophys Acta 475:476–491

Gray MW, Spencer DF (1983) Wheat mitochondrial DNA encodes a eubacteria-like initiator methionine transfer RNA. FEBS Lett 161:323–327

Hanic-Joyce PJ, Gray MW (1990) Processing of transfer RNA precursors in a wheat mitochondrial extract. J Biol Chem 265:13782–13791

Hanic-Joyce PJ, Spencer DF, Gray MW (1990) In vitro processing of transcripts containing novel tRNA-like sequences ('t-elements') encoded by wheat mitochondrial DNA. Plant Mol Biol 15:551–559

Hofmann D, Weising K (1990) Extraction of ethidium bromide from CsCl-purified DNA. Focus 12:29, Life Technologies, Gaithersburg, MD

Jackson JF (1985) Mitochondria. In: Linskens HF, Jackson JF (eds) Modern methods of plant analysis, New Series, vol 1 Cell components. Springer, Berlin, Heidelberg, New York, pp 298–303

Johnston FB, Stern H (1957) Mass isolation of viable wheat embryos. Nature 179:160–161

Joyce PBM, Gray MW (1989a) Chloroplast-like transfer RNA genes expressed in wheat mitochondria. Nucleic Acids Res 17:5461–5476

Joyce PBM, Gray MW (1989b) Aspartate and asparagine tRNA genes in wheat mitochondrial DNA: a cautionary note on the isolation of tRNA genes from plants. Nucleic Acids Res 17:7865–7878

Joyce PBM, Spencer DF, Gray MW (1988a) Multiple sequence rearrangements accompanying the duplication of a tRNA^{Pro} gene in wheat mitochondrial DNA. Plant Mol Biol 11:833–843

Joyce PBM, Spencer DF, Bonen L, Gray MW (1988b) Genes for tRNA^{Asp}, tRNA^{Pro}, tRNA^{Tyr} and two tRNAs^{Ser} in wheat mitochondrial DNA. Plant Mol Biol 10:251–262

Kolodner R, Tewari KK (1975) The molecular size and conformation of the chloroplast DNA from higher plants. Biochim Biophys Acta 402:372–390

Leaver CJ, Harmey MA (1976) Higher-plant mitochondrial ribosomes contain a 5S ribonucleic acid component. Biochem J 157:275–277

Parish JH, Kirby KS (1966) Reagents which reduce interactions between ribosomal RNA and rapidly labelled RNA from rat liver. Biochim Biophys Acta 129:554–562

Quetier F, Vedel F (1977) Heterogeneous population of mitochondrial DNA molecules in higher plants. Nature 268:365–368

Spencer DF, Bonen L, Gray MW (1981) Primary sequence of wheat mitochondrial 5S ribosomal ribonucleic acid: functional and evolutionary implications. Biochemistry 20:4022–4029

Spencer DF, Schnare MN, Gray MW (1984) Pronounced structural similarities between the small subunit ribosomal RNA genes of wheat mitochondria and *Escherichia coli*. Proc Natl Acad Sci USA 81:493–497

Wallace RH, Lochhead AG (1951) Bacteria associated with seeds of various crop plants. Soil Sci 71:159–166

Analysis of Limonoids in Citrus Seeds

Z. HERMAN, C. H. FONG, and S. HASEGAWA

1 Introduction

Limonoids are a group of chemically related triterpene derivatives found in the Rutaceae and Meliaceae families. Among the 38 limonoids reported to occur in citrus and its hybrids, four are known to be bitter in taste. These are limonin, nomilin, ichangin, and nomilinic acid. Limonin is the primary cause of citrus juice bitterness. This problem occurs in a variety of citrus juices, and it is economically important to the citrus industry because bitter juices have a lower market value for producers. Dekker (1988) has reviewed limonoid bitterness and postharvest juice debittering methods. Hasegawa (1989) has reviewed limonoid biochemistry and preharvest approaches to the bitterness problem.

In recent year, limonoids have been shown to occur in citrus as nonbitter glucoside derivatives (Bennett et al. 1989; Hasegawa et al. 1989). At least ten limonoid glucosides have been isolated from citrus seeds, juices, and peel.

Abbreviations Used

CH_2Cl_2	Dichlormethane
CH_3CN	Acetonitrile
C_6H_6	Benzene
H_2O	Water
EtOAc	Ethyl acetate
Me_2CO	Acetone
MeOH	Methanol
NH_4Cl	Ammonium chloride
HAc	Acetic acid
H_2SO_4	Sulfuric acid
HCl	Hydrogen chloride
$KHCO_3$	Postassium bicarbonate
H_3PO_4	Phosphoric acid
$CHCl_3$	Chloroform
R_f	Ratio of the distance traveled by a compound divided by the distance traveled by the solvent front (used in TLC)

The biological function of limonoids in plants is not known. However, there are reports that limonoids inhibit the feeding activity of certain insects (Liu et al. 1990 and references therein). Interest in limonoids has also increased with the finding that feeding certain limonoids to laboratory animals resulted in an inhibition of chemically induced tumors (Lam et al. 1989; Miller et al. 1989). Thus, there is a considerable demand for limonoid compounds, and citrus seeds are an excellent source of both limonoid aglycones and glucosides. This chapter discusses the chemical structures, distribution, extraction, isolation, identification, and quantification of limonoids in citrus seeds.

2 Chemical Structures and Properties

There are basically five structural categories of limonoids (Table 1). The categories are based on structural differences that give rise to solubility characteristics that are important in the processes of extraction and analysis. The section below describes the chemical structures and properties of limonoid aglycones and glucosides. For additional information on the structures and chemistry of limonoids, see Dreyer (1968) and Maier et al. (1980).

2.1 Limonoid Aglycones

Maier and Beverly (1968) first reported that limonoate A-ring lactone (**1**) (Fig. 1), is present in intact citrus fruit tissues. Compound **1** is the monolactone form of limonin (**3**). Limonoid monolactones are water-soluble at neutral or basic pH and are not bitter in taste. However, when the fruit is juiced, a second lactone ring closes due to the acidic conditions as well as the presence of an endogenous enzyme, limonin D-ring lactone hydrolase. This enzyme catalyzes D-ring lactonization under acidic conditions, and the same enzyme opens the

Table 1. Limonoid structural categories and solubilities

Limonoids	Example	Solubility	
		H$_2$O	Organic solvents[a]
Neutral aglycones	Limonin (**3**)		+ + + +
Monocarboxylic aglycones	Limonoate A-ring lactone (**1**)	+ + +	+ + +
Dicarboxylic aglycones	Nomilinoic acid (**4**)	+ + + +	+ +
Monocarboxylic glucosides	Limonin glucoside (**2**)	+ + + +	
Dicarboxylic glucosides	Nomilinic acid glucoside (**5**)	+ + + +	

[a] See text of Section 2 for more details.

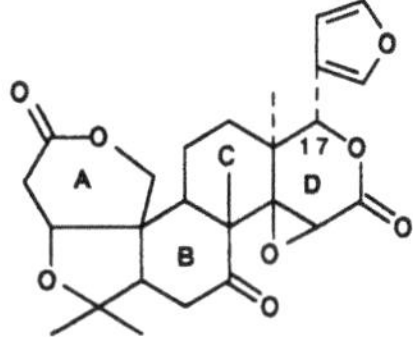

(1) Limonoic acid A-ring lactone: R=H

(2) Limonin 17-ß-D-glucopyranoside:
   R=ß-D-glucose

(3) Limonin

(4) Nomilinoic acid: R=H

(5) Nomilinic acid glucoside:
   R=ß-D-glucose

(6) Nomilinic acid

Fig. 1. Structures of limonoids

D-ring under alkaline conditions (Maier et al. 1969). With D-ring closure during juicing, the resulting dilactones, such as limonin (3) and nomilin are extremely bitter in taste.

Unlike fruit tissue, the limonoid aglycones in citrus seeds are present in several different forms. There are primarily two forms: monolactones, such as limonoic acid A-ring lactone (1) and dilactones, such as limonin (3). The A-ring lactones are usually converted to dilactones prior to extraction or quantification. The dilactones have poor solubility in water, but solubility in citrus juices is higher, due to the presence of pectin and sugars (Chandler 1971). The dilactones are not very soluble in hexane or petroleum ether, but they are soluble in Me_2CO, CH_2Cl_2 or EtOAc. One exception is obacunone, which is somewhat soluble in hexane (Rouseff and Nagy 1982).

Citrus contains a number of acidic aglycones, such as nomilinic acid (6), which have an open A-ring. There are also compounds with both lactone rings open, such as nomilinoic acid (4). These compounds are soluble in a wide range of solvents. A list of all the aglycones found in citrus and its hybrids is given in Table 2.

The limonoid aglycones in seeds are chemically stable. However, limonoids will decompose under strongly acidic conditions, and the lactone rings start opening at pH 9 and above. Nomilin rapidly converts to obacunone at pH 9. Nomilinic acid (5) and obacunone convert to isoobacunoic acid above pH 9.

Table 2. Limonoid aglycones in the seeds of *Citrus* and its hybrids

Neutral	Acidic
1. Limonin	23. Deacetylnomilinic acid
2. Nomilin	24. Nomilinic acid
3. Obacunone	25. Isoobacunoic acid
4. Deacetylnomilin	26. Epiisoobacunoic acid
5. Ichangin	27. Isolimonic acid
6. Deoxylimonin	28. Limonoic acid A-ring lactone
7. Deoxylimonol	29. Deoxylimonic acid
8. Limonol	30. 17-Dehydrolimonoic acid A-ring lactone
9. Limonyl acetate	31. Trans-19-hydroxyobacunoic acid
10. 7α-Obacunol	32. Calaminic acid[a]
11. 7α-Obacunyl acetate	33. Retrocalaminic acid[a]
12. Ichangensin	34. Cyclocalaminic acid[a]
13. Citrusin	35. Isoobacunoic acid diosphenol[a]
14. 1-(10-19)Abeo-obacun-9(11)-en-7α-yl acetate	36. Obacunoic acid[a]
15. Calamin[a]	37. 1-(10-19)Abeo-7α-acetoxy-10β-hydroxy-isoobacunoic acid
16. Retrocalamin[a]	
17. Cyclocalamin[a]	
18. Methyl isoobacunoate diosphenol[a]	
19. Methyl deacetylnomilinate[a]	
20. 6-Keto-7β-deacetylnomilol[a]	
21. Methyl 6-hydroxy isoobacunoate[a]	
22. Isocyclocalamin[a]	

[a] Isolated from calamondin seeds

2.2 Limonoid Glucosides

Citrus seeds contain substantial quantities of limonoid glucosides (Hasegawa et al. 1989). Each of these compounds contains one D-glucose molecule attached via a β-glucosidic linkage to the 17 position of the corresponding limonoid A-ring lactone. The glucoside form of limonin, limonin 17-β-D-glucopyranoside (**2**), is shown in Fig. 1. Compound **2** is also called limonin glucoside.

Table 3. Limonoid glucosides in citrus seeds

Monocarboxylic acids	Dicarboxylic acids
17-β-D-glucopyranosides of:	17-β-D-glucopyranosides of:
1. Limonin	5. Nomilinic acid
2. Nomilin	6. Deacetylnomilinic acid
3. Deacetylnomilin	7. Obacunoic acid
4. Obacunone	8. Trans-obacunoic acid
	9. Isoobacunoic acid
	10. Epiisoobacunoic acid

At least ten limonoid glucosides have been isolated from citrus (Bennett et al. 1989; Hasegawa et al. 1989). More are in the process of being identified. The first ten are listed in Table 3. Four contain one carboxylic acid group, and six contain two carboxylic acid groups, such as nomilinic acid glucoside (5). Due to the glucose moiety, limonoid glucosides are soluble in H_2O and MeOH. At high concentrations, they sometimes crystallize in MeOH. Limonoid glucosides are not very soluble in organic solvents.

Most of the limonoid glucosides appear to be stable at pH 2 to 8, but this has not been well studied. One known exception is nomilin glucoside, which is converted to nomilinic acid glucoside below pH 3.5. Nomilin glucoside is also converted to obacunone glucoside at pH 7.5 and above.

3 Concentrations of Limonoids in Citrus Seeds

3.1 Aglycone Concentrations

Citrus seeds contain the highest concentrations of limonoids of any citrus tissue. Hasegawa et al. (1980) reported that limonoid aglycones comprise approximately 1% of the fresh weight of citrus seeds. Grapefruit seeds contained the highest concentrations and are known to be an excellent source of limonoids. Of the aglycones in citrus seeds, the neutral compounds ranged from 58% – 87% of the total, and the acidic compounds had a range of 13% – 42%. The seeds of kumquat and calamondin contain approximately 0.7% total limonoid aglycones.

Two papers, Hasegawa et al. (1980) and Rouseff and Nagy (1982), reported the analysis of limonoids in the seeds of eight species of citrus. The first paper utilized thin layer chromatography (TLC), the second paper used high pressure liquid chromatography (HPLC). For most species, the predominant limonoid was limonin, followed by nomilin, deacetylnomilin, and obacunone. Rouseff and Nagy (1982) analyzed a number of different cultivars and presented evidence that each species of citrus has a unique limonoid pattern in the seeds. They also reported that aglycones comprise 0.2% – 0.4% of the weight of air-dried seeds.

Ozaki et al. (1991) published a quantitative study of the limonoids in eight varieties of citrus seeds. Table 4 shows the results of the analysis of the neutral aglycones by HPLC. The predominant limonoid in all eight citrus species was limonin. Grapefruit seeds were again found to be the richest source of limonoids. One kg of dried grapefruit seeds contain about 19 g of limonin. The total concentration of limonoid aglycones varied considerably between species.

Herman et al. (1989) reported that the seeds of *C. ichangensis* contain 2900 ppm of the nonbitter limonoid ichangensin. *C. ichangensis* contains relatively little limonin, only 250 ppm. Hashinaga and Hasegawa (1989) reported the concentrations of limonoids in the seeds of *C. sudachi*, and Hashinaga et

Table 4. Concentrations of limonoid aglycones in citrus seeds[a]

Seeds	Limonin	Nomilin	Obacunone	Ichangin	Deacetylnomilin	Total
			(mg/g dry seed)			
Fukuhara	9.77	3.88	0.37	Trace	2.14	15.92
Hyuganatsu	4.68	3.73	0.28	Trace	0.35	9.04
Sanbokan	3.95	1.02	0.30	0.16	1.36	6.79
Shimamikan	7.85	2.01	0.28	0.16	2.01	12.31
Grapefruit	19.06	1.84	1.86	Trace	1.10	23.86
Lemon	8.95	3.03	0.58	Trace	Trace	12.56
Valencia	10.00	2.30	0.08	1.16	1.24	14.78
Tangerine	4.10	1.37	0.35	0.38	3.11	9.31

[a] From Ozaki et al. (1991); concentrations determined by HPLC.

al. (1990) reported the concentrations of limonoids in the seeds of *C. yuzu.*
Limonin was the predominant limonoid in both *C. sudachi* and *C. yuzu.*

In general, acidic limonoids seem to comprise approximately 20% of the total aglycones in citrus seeds. Nomilinic acid, deacetylnomilinic acid, and isolimonic acid are the predominant acidic aglycones (Hasegawa et al. 1980). Herman et al. (1989) reported the concentrations of acidic limonoids in *C. ichangensis.* The papers by Hashinaga and Hasegawa (1989) and Hashinaga et al. (1990) report the acidic aglycone concentrations in *C. sudachi* and *C. yuzu* seeds, respectively.

3.2 Glucoside Concentrations

Limonoid glucosides were first isolated from grapefruit seeds (Bennett et al. 1989; Hasegawa et al. 1989). In these first reports, approximately 15 limonoid

Table 5. Limonoid glucosides in citrus seeds[a]

Seeds	DAG	NG	NAG	OG	LG	DG	Total	TLC[b]
				(mg/g dry seed)				
Fukuhara	0.28	3.22	0.98	1.09	0.51	1.32	7.40	7.92
Hyuganatsu	0.42	1.10	0.76	0.65	Trace	0.37	3.31	2.92
Sanbokan	0.37	1.13	0.55	0.90	0.51	0.89	4.36	3.96
Shimamikan	0.48	1.89	1.29	2.35	0.37	0.69	7.08	6.45
Grapefruit	0.75	2.01	0.89	0.86	1.48	0.68	6.67	7.28
Lemon	0.14	1.53	1.39	1.49	1.44	0.55	6.54	6.20
Valencia	0.13	4.48	0.98	1.06	0.59	1.69	8.94	8.48
Tangerine	1.69	0.42	0.96	0.45	0.90	0.93	5.36	5.15

[a] From Ozaki et al. (1991); concentrations determined by HPLC.
[b] Total limonoid glucosides determined by TLC for comparison.
G = glucoside, DA = deacetylnomilinic acid, N = nomilin, NA = nomilinic acid, O = obacunone, L = Limonin, D = deacetylnomilin.

glucosides were detected using TLC and HPLC. The structures of ten of these compounds were determined. Using TLC, it was estimated that limonoid glucosides comprise nearly 1% of the total fresh weight of grapefruit seeds.

Ozaki et al. (1991) analyzed the limonoid glucosides in the seeds of eight citrus species using HPLC and TLC. The results are given in Table 5. There appear to be six major limonoid glucosides in citrus seeds. Four are monocarboxylic acids; these are the glucosides of limonin, nomilin, deacetylnomilin, and obacunone. Two are dicarboxylic acids; these are the glucosides of nomilinic acid and deacetylnomilinic acid. Whereas limonin is the predominant aglycone in citrus seeds, nomilin glucoside appears to be the predominant glucoside. Limonoid glucosides were found to comprise an average of 0.6% of the dry weight of the seeds that were analyzed.

4 Detection, Identification, and Quantification

4.1 Limonoid Aglycones

A variety of methods have been used to detect, and quantify limonoids. Most methods were developed specifically to detect and quantify limonin in citrus juices. Some of the early methods include chemical derivation followed by spectrophotometry (Chandler and Kefford 1966; Wilson and Crutchfield 1968) or fluorometry (Fisher 1973). Gas-liquid chromatography has also been used (Kruger and Colter 1972).

Carter et al. (1985) compared various methods of detecting and quantifying limonin in citrus juices. These methods are also useful for the analysis of seeds and will be discussed below.

TLC is probably the easiest and most economical method of detecting limonoid aglycones. Basically, the method involves spotting the sample onto a silica gel plate along with the appropriate limonoid standard, developing the plate in a suitable solvent mixture, spraying the plate with Ehrlich's reagent (1% p-dimethylaminobenzaldehyde in EtOH), and developing the plate in a chamber of HCl gas (Dreyer 1965). The HCl gas can be easily generated by mixing about 10 ml H_2SO_4 with 5 g NH_4Cl. The entire analysis takes approximately 1 h. Limonoids produce very distinctive reddish-orange spots on the plate; approximately 0.5 μg can be detected.

Some of the TLC solvent systems used are (1) EtOAc-cyclohexane (3:2); (2) CH_2Cl_2-MeOH (97:3); (3) CH_2Cl_2-EtOAc (3:2); and (4) toluene-EtOH-H_2O-HAc (200:47:15:1). Positive identification of each limonoid aglycone is usually possible using just two of the above solvent systems. R_f values for many aglycones using a solvent system are listed in Table 6.

Acidic aglycones, which are listed in Table 2, do not migrate very much in the above systems. Therefore, diazomethane is often used to methylate the acidic aglycones. The resulting methyl esters can then be analyzed as above.

Table 6. R_fs of limonoid aglycones by TLC[a]

Limonoid	R_f
Ichangin	0.12
Deacetylnomilin	0.12
Deoxylimonin	0.15
Deoxylimonol	0.15
Nomilin	0.19
6-Keto-7β-deacetylnomilol	0.19
Limonin	0.27
Methyl deacetylnomilinate	0.27
7α-Limonol	0.29
Methyl deoxylimonoate	0.33
Calamin	0.35
Limonol	0.38
Methyl nomilinate	0.38
Methyl deoxyisoobacunoate	0.39
Methyl limonoate D-ring lactone	0.40
7α-Limonyl acetate	0.40
Methyl obacunoate	0.42
Methyl isolimonoate	0.42
Methyl epiisoobacunoate	0.49
Retrocalamin	0.53
Obacunone	0.55
Ichangensin	0.55
Methyl isoobacunoate	0.61
Cyclocalamin	0.72
Isocyclocalamin	0.72
Methyl isoobacunoate diosphenol	0.75

[a] Determined by TLC on silica gel plates using EtOAc-cyclohexane (3:2).

Limonoid aglycones can be quantified using TLC. The method developed by Maier and Grant (1970) is precise and can quantify as little as 1 µg of limonin. The method involves extraction, spotting of samples, and then the comparison to a series of standards that are spotted onto the same plate.

HPLC is a popular method used for the detection and quantification of limonoids. Most of the methods have been developed for quantification of limonin in citrus juices and may or may not be useful for seed analysis. The HPLC methods have been reviewed by Carter et al. (1985) and Van Beek and Blaakmeer (1989).

HPLC has been used to analyze limonoids in citrus seeds. One drawback is that limonoids have a maximum UV absorption at 207 nm. This necessitates sample clean up and the use of high-purity solvents when using a UV detector at this low wavelength.

Rouseff and Nagy (1982) quantified limonin, nomilin, obacunone, and deacetylnomilin in the seeds of eight citrus species by HPLC. A Zorbax CN column (4.6×250 mm, Dupont) was used at 40 °C with a flow rate of 1 ml/min. The isocratic solvent system was heptane-isopropanol-MeOH

(1:12:2). Limonoids were detected by UV absorption at 207 nm. Exact retention times were not reported, but the chromatograms showed good peak resolution for the four aglycones.

Herman et al. (1989) reported an HPLC analysis of *C. ichangensis* seeds. A C_{18} reverse-phase column (4.6×250 mm) was eluted isocratically with 49% H_2O, 41% MeOH, and 10% CH_3CN at a flow rate of 1 ml/min. Limonoids were detected by UV absorption at 210 nm. This method quantified limonin, nomilin, deacetylnomilin, obacunone, ichangensin, deacetylnomilinic acid, isolimonic acid, and nomilinic acid. The acidic limonoids were quantified as their methyl ester derivatives.

Hashinaga and Hasegawa (1989) and Hashinaga et al. (1990) reported the HPLC analysis of the limonoids in *C. sudachi* and *C. yuzu* seeds, respectively. For both of these analyses, a 65-min linear gradient was used, beginning with 10% CH_3CN and H_2O and ending with 55% CH_3CN in H_2O. A C_{18} reverse-phase column (4.6×250 mm) was used. The flow rate was 1 ml/min. Detection was by UV absorption at 210 nm. Ozaki et al. (1991) used an isocratic system: H_2O-MeOH-CH_3CN (49:41:10) to quantify limonoid aglycones in eight citrus species.

A radioimmunoassay (RIA) method was developed for the quantification of limonin in citrus tissues (Mansell and Weiler 1980; Weiler and Mansell 1980). The method is fast and extremely sensitive; it can quantify 1 ppb limonin. However, other limonoids can cross-react with the antibody.

An enzyme immunoassay (EIA) method for limonin has been developed (Jourdan et al. 1984; Ram et al. 1988). This method is rapid and sensitive; it can quantify less than 1 ppm limonin. The EIA method is sold commercially as the Bitterdetek Limonin Test Kit (Idetek Inc., San Bruno, CA, USA). This method requires some special equipment. Also, limonoid glucosides cross-react with the antibody (Bert Carter, Ventura Coastal Corp., Ventura, CA, pers. comm. 1989). Both the EIA and RIA methods are designed to be specific for limonin; therefore, they do not quantify other limonoids.

4.2 Limonoid Glucosides

TLC can be used to detect the presence of limonoid glucosides in seed extracts. It can also be used to quantify the total amount of limonoid glucosides (Fong et al. 1989; Ozaki et al. 1991). However, it is difficult to separate individual glucosides by TLC. The TLC solvent system in the above analyses was EtOAc-methyl ethyl ketone-formic acid-H_2O (5:3:1:1). For most citrus seeds, three positive red spots can be observed when the plate is sprayed with p-dimethylaminobenzaldehyde and exposed to HCl gas (Dreyer 1965). The intensity of the spots can be used for quantification by comparison to the spots produced by known amounts of limonin glucoside standards. The method depends on visual analysis of the spots by judges. The result tend to agree closely with HPLC analyses for total limonoid glucosides (Herman et al. 1990; Ozaki et al. 1991).

A HPLC method for the quantification of limonin glucoside was first reported by Fong et al. (1989). Herman et al. (1990) reported an HPLC method for the analysis of five limonoid glucosides. Both methods were developed for the analysis of citrus juices.

Ozaki et al. (1991) reported an HPLC method for the quantification of seven limonoid glucosides in eight different species of citrus seeds. The method employed a C_{18} reverse-phase analytical column (4.6×250 mm) and a linear gradient starting with 15% CH_3CN in 3 mM H_3PO_4 and ending with 26% CH_3CN at 33 min. The flow rate was 1 ml/min, and limonoid glucosides were detected by UV absorption at 210 nm. The retention times for the glucosides of limonin, deacetylnomilinic acid, deacetylnomilin, ichangin, nomilin, nomilinic acid, and obacunone were 13, 16.6, 21, 22.3, 25.8, 27.8, and 30.3 min, respectively.

HPLC can be used during purification procedures as a method to detect the presence and quantity of individual limonoid glucosides. This may be necessary, because TLC does not separate the individual compounds.

4.3 Nuclear Magnetic Resonance Spectroscopy

Nuclear magnetic resonance (NMR) spectroscopy has become the major technique for determining the structures of natural products, and it has been especially useful in the case of limonoids. Both ^1H and ^{13}C NMR spectra provide detailed knowledge about the structural environment of each hydrogen and carbon atom in a molecule. Examples of the information obtainable from such spectra, in addition to the frequencies (chemical shifts) of the various atoms, are the number of hydrogen atoms attached to each carbon atom and the proximity of hydrogen atoms to each other.

Limonoids represent a particularly favorable case for ^1H NMR spectral analysis, since they contain several oxygenated functional groups. Proton signals adjacent to these groups are usually well separated from other signals, in contrast to the situation with many natural products in which most proton signals have similar chemical shifts and thus are not well resolved. Often a proton spectrum alone of a new limonoid is sufficient to reveal its structure, by comparison with spectra of related limonoids. Carbon spectra, which are very sensitive to changes in molecular geometry, provide further structural details. For instance, the position of the C-4 signal immediately differentiates between compounds with a 5-membered A' ether ring, such as limonin (3), those with a 7-membered lactone A-ring, such as nomilin, and those in which the A-ring is open, such as nomilinic acid (6). In addition to simple ^1H and ^{13}C NMR spectra, two-dimensional techniques have become available in recent years. These spectra provide a wealth of further structural information, such as correlations between hydrogen and carbon atoms, and they have proved to be extremely useful for structure determination of limonoids.

NMR spectroscopy can also be used to identify limonoids isolated from citrus, assuming that they are not new compounds, by comparison with spectra of known limonoids. For this purpose, samples can be of a much lower degree of purity than is required for structure determination. In most cases a proton spectrum, which can be run in a few minutes, is sufficient. Citrus limonoids contain at least four methyl groups, and the chemical shifts of these signals in the proton spectra are highly characteristic and usually definitive of a particular compound. Even better evidence of the identity of two compounds can be provided by comparison of their carbon spectra, but this requires several hours to run.

For an example of the use of NMR to identify the structures of limonoid aglycones, see Bennett and Hasegawa (1980). For NMR identification of limonoid glucosides, see Hasegawa et al. (1989) and Bennett et al. (1989).

5 Extraction and Isolation of Limonoids

5.1 Aglycones

Some of the earlier methods for extracting limonoids from citrus seeds are described by Emerson (1948) and Dreyer (1965). Two currently used methods are described below.

In the buffer extraction method, seeds are homogenized in a blender with 10 vol of 0.1 M Tris buffer at pH 7.5. The homogenate is allowed to incubate at approximately 25 °C for 20 h. During the incubation, neutral dilactones, which are insoluble in aqueous solution, become soluble due to hydrolysis of the D-ring lactones by the action of limonin D-ring lactone hydrolase. This enzyme is abundant in citrus seeds (Maier et al. 1969). The mixture is centrifuged and then filtered through Celite filter aid to yield a clear solution. The filtrate is acidified to pH 3 with 1 N HCl to close the D-ring. Limonoids are then extracted twice with CH_2Cl_2. This method is very efficient and quantitative, and the resulting solution contains relatively little nonlimonoid material (Hasegawa et al. 1980; Miyake et al. 1991).

In the solvent extraction method, seeds are dried in a 60 °C oven before grinding with a mill. The meals obtained from dried seeds are washed with petroleum ether or hexane to remove oily materials. Limonoids are then extracted with acetone. The washing and extraction process can be carried out in a Soxhlet or similar type of extractor. Acetone appears to be the best solvent among the solvents used, such as EtOAc, C_6H_6, $CHCl_3$, or CH_2Cl_2 (Miyake et al. 1991). The solvent method is less efficient than the buffer method because the mono- and dicarboxylic acid limonoids present in seeds may not be extracted with organic solvents. The other disadvantage is that the petroleum ether or hexane used for washing may extract some of the very nonpolar limonoids, such as obacunone (Rouseff and Nagy 1982) or possibly methyl isoobacunoate diosphenol and methyl isoobacunoate present in seeds of calamondin (Bennett and Hasegawa 1981). However, the solvent method is convenient and is recommended for large-scale extractions.

Limonin and to a lesser extent nomilin have been of particular interest in the field of food technology since they are involved in limonoid bitterness in citrus juices. These two limonoids can be readily isolated from citrus seeds by fractional crystallization. Grapefruit seeds are ideal for the isolation of these two limonoids.

The extract obtained above is dissolved in CH_2Cl_2 and filtered. Limonin can be easily crystallized from CH_2Cl_2 by simply reducing the volume. The second crop of limonin from the mother liquor can then be crystallized by adding isopropanol until the crystals form. The mother liquor is then brought to dryness and nomilin can be crystallized from acetone.

After removal of most of the limonin and nomilin by crystallization, the other limonoids can be isolated by column chromatography. The residue is first dissolved in CH_2Cl_2, washed twice with 2% $KHCO_3$, and then washed with H_2O to remove acidic limonoids. The remaining limonoids can be fractionated on silica gel columns using solvent systems, such as increasing EtOAc in hexane or increasing MeOH in CH_2Cl_2. Some limonoids are easily fractionated in pure form with one solvent system, while others require two solvent systems. If additional fractionation is needed, increasing diethyl ether (1% – 10%) in CH_2Cl_2 can also be used. For efficient fractionation, a ratio of limonoids to silica gel of 1:100 (w/w) is recommended.

The $KHCO_3$ extract obtained above is acidified to pH 2 and extracted twice with EtOAc to obtain acidic limonoids. The EtOAc fraction is washed several times with H_2O. The acidic limonoids are then methylated with diazomethane. Methylated limonoids can be fractionated on silica gel columns using the procedures described above for the neutral limonoids.

Individual neutral limonoids can be readily crystallized using solvents such as CH_2Cl_2, Me_2CO, MeOH, CH_2Cl_2-isopropanol, and CH_2Cl_2-Me_2CO. For additional information, see Hasegawa et al. (1980), Bennett and Hasegawa (1982), and Bennett et al. (1988). Acidic limonoids have not been crystallized, with the exception of methyl 17-dehydrolimonoate A-ring lactone, which is crystallized from MeOH (Hasegawa et al. 1972).

5.2 Glucosides

Since limonoid glucosides are water-soluble, they can be easily extracted with H_2O. As an example, a quantity of seeds can be homogenized in 15 vol H_2O. Following this, the pH is adjusted to 4 prior to the addition of pectinase (from *Aspergillus niger*, Sigma Chemical Co., St. Louis, MO, USA). The homogenate is then stirred for 20 h at room temperature. This step dissolves pectin in the cell walls and prevents the formation of pectin gels. The mixture is then centrifuged at 13000 g for 15 min, and the supernatant is filtered. The filtrate is then loaded onto an Amberlite XAD column, and the column is washed with several volumes of H_2O. Limonoid glucosides are then eluted with CH_3CN. Following evaporation, the residue is dissolved in H_2O and used for isolation.

For further details, see Bennett et al. (1989). Hasegawa et al. (1989) have reported an alternative extraction method employing 70% MeOH.

It is possible to extract both limonoid aglycones and glucosides from the same batch of seed meal. Following acetone extraction for aglycones (Sect. 5.1), limonoid glucosides can be extracted with MeOH. A solvent temperature of 55–60 °C has worked quite well, and large-scale extractions are possible (Hasegawa et al. 1989).

Limonoid glucosides can be isolated by conventional open-column chromatography (Bennett et al. 1989; Hasegawa et al. 1989) or by preparative HPLC chromatography (Bennett et al. 1991). Since crude seed extracts contain flavonoid glycosides which coelute with limonoid glucosides during column chromatography, it is helpful to first separate these two types of glycosides before subsequent chromatographic fractionation. This can be accomplished by dissolving the extract in a minimum of H_2O, adjusting the pH to 6.5, and loading onto a DEAE Sephacel column (Cl form). The column is washed several times with H_2O. The negatively charged limonoid glucosides are weakly absorbed, but the contaminating flavonoid glycosides are washed out. The limonoid glucosides are then eluted with a 0.2 M NaCl solution. The pH of the eluate is readjusted to 3.5, and the eluate is loaded onto an XAD-2 column. The column is washed several times with H_2O to remove the NaCl, and the glucosides are eluted with MeOH. After the MeOH is evaporated, the residue is dissolved in a minimum of H_2O and is then used for isolation.

Individual limonoid glucosides are easily isolated by C_{18} reverse-phase preparative HPLC (Bennett et al. 1991). In this method, the column (2.2×25 cm, 10-μm particle size) is eluted at 3 ml/min using a linear gradient starting with 15% and ending with 45% CH_3CN in H_2O at 150 min. Sharper gradients can be used with a faster flow rate. Some glucosides require additional fractionation using MeOH or Me_2CO as the mobile phase.

References

Bennett RD, Hasegawa S (1980) Isolimonic acid, a new citrus limonoid. Phytochemistry 19:2417–2419

Bennett RD, Hasegawa S (1981) Limonoids of calamondin seeds. Tetrahedron 37:17–24

Bennett RD, Hasegawa S (1982) 7-Oxygenated limonoids from the Rutacea. Phytochemistry 21:2349–2354

Bennett RD, Herman Z, Hasegawa S (1988) Ichangensin: a new citrus limonoid. Phytochemistry 27:1543–1545

Bennett RD, Hasegawa S, Herman Z (1989) Glucosides of acidic limonoids of citrus. Phytochemistry 28:2777–2781

Bennett RD, Miyake M, Ozaki Y, Hasegawa S (1991) Limonoid glucosides in *Citrus aurantium*. Phytochemistry 30:3803–3805

Carter BA, Oliver DG, Jang L (1985) A comparison of methods for determining the limonin content of processed California navel orange juice. Food Technol 39:82–86, 97

Chandler BV (1971) Some solubility relationships of limonin. Their importance in orange juice bitterness. CSIRO Food Res Q 31:36–40

Chandler BV, Kefford JF (1966) The chemical assay of limonin, the bitter principle of oranges. J Sci Food Agric 17:193–197

Dekker RFH (1988) De-bittering of citrus fruit juices: specific removal of limonin and other bitter principles. Aust J Biotech 2:65–76

Dreyer DL (1965) Citrus bitter principles. III. Isolation of deacetylnomilin and deoxylimonin. J Org Chem 30:749–751

Dreyer DL (1968) Limonoid bitter principles. In: Zechmeister L (ed) Progress in the chemistry of organic natural products. Springer, Berlin Heidelberg New York, pp 191–244

Emerson OH (1948) The bitter principles of citrus fruit. I. Isolation of nomilin, a new bitter principle from the seeds of oranges and lemons. J Am Chem Soc 70:545–549

Fisher JF (1973) Fluorometric determination of limonin in grapefruit and orange juice. J Agric Food Chem 21:1109–1110

Fong CH, Hasegawa S, Herman Z, Ou P (1989) Limonoid glucosides in commercial citrus juices. J Food Sci 54:1505–1506

Hasegawa S (1989) Biochemistry and biological removal of limonoid bitterness in citrus juices. In: Jen JJ (ed) Quality factors of fruits and vegetables: chemistry and technology. ACS Symp Ser 405. American Chemical Society Press, Washington, DC, pp 84–96

Hasegawa S, Bennett RD, Maier VP, King AD (1972) Limonoate dehydrogenase from *Arthrobacter globiformis*. J Agric Food Chem 20:1031–1034

Hasegawa S, Bennett RD, Verdon CP (1980) Limonoids in citrus seeds: origin and relative concentration. J Agric Food Chem 28:922–925

Hasegawa S, Bennett RD, Herman Z, Fong CH, Ou P (1989) Limonoid glucosides in citrus. Phytochemistry 28:1717–1720

Hashinaga F, Hasegawa S (1989) Limonoids in seeds of sudachi (*Citrus sudachi* Hort. ex Shirai). J Jpn Soc Hortic Sci 58:227–229

Hashinaga F, Herman Z, Hasegawa S (1990) Limonoids in seeds of Yuzu (*Citrus junos* Sieb. ex Tanaka). Nippon Shokuhin Kogyo Gakkaishi 37:380–382

Herman Z, Hasegawa S, Fong CH, Ou P (1989) Limonoids in *Citrus ichangensis*. J Agric Food Chem 37:850–851

Herman Z, Fong CH, Ou P, Hasegawa S (1990) Limonoid glucosides in orange juices by HPLC. J Agric Food Chem 38:1860–1861

Jourdan PS, Mansell RL, Oliver DG, Weiler EW (1984) Competitive solid phase enzyme-linked immunoassay for the quantification of limonin in citrus. Anal Biochem 138:19–24

Kruger AJ, Colter CE (1972) Gas chromatographic identification of limonin in citrus juice. Proc Fla State Hortic Soc 85:206–210

Lam LKT, Li Y, Hasegawa S (1989) Effects of citrus limonoids on glutathione S-transferase activity in mice. J Agric Food Chem 37:878–880

Liu Y, Alford AR, Rajab MS, Bentley MD (1990) Effects and modes of action of citrus limonoids against *Leptinotarsa decemlineata*. Physiol Entomol 15:37–45

Maier VP, Beverly GD (1968) Limonin monolactone, the nonbitter precursor responsible for delayed bitterness in certain citrus juices. J Food Sci 33:488–492

Maier VP, Grant ER (1970) Specific thin-layer chromatography assay of limonin, a citrus bitter principle. J Agric Food Chem 18:250–252

Maier VP, Hasegawa S, Hera E (1969) Limonin D-ring-lactone hydrolase. A new enzyme from *Citrus* seeds. Phytochemistry 8:405–407

Maier VP, Hasegawa S, Bennett RD, Echols LC (1980) Limonin and limonoids: chemistry, biochemistry, and juice bitterness. In: Nagy S, Attaway JA (eds) *Citrus* nutrition and quality. ACS Symp Ser, 143. American Chemical Society Press, Washington, DC, pp 63–82

Mansell RL, Weiler EW (1980) Radioimmunoassay for the determination of limonin in *Citrus*. Phytochemistry 19:1403–1407

Miller EG, Fanous R, Rivera-Hidalgo F, Binnie WH, Hasegawa S, Lam LKT (1989) The effect of citrus limonoids on hamster buccal pouch carcinogenesis. Carcinogenesis 10:1535–1537

Miyake M, Ayano S, Ozaki Y, Maeda H, Ifuku Y, Hasegawa S (1991) Extraction of neutral limonoids from citrus seeds. Nippon Nogeikagaku Kaishi 65:987–992

Ozaki Y, Fong CH, Herman Z, Maeda H, Miyake M, Ifuku Y, Hasegawa S (1991) Limonoid glucosides in citrus seeds. Agric Biol Chem 55:137–141

Ram BP, Jang L, Martins L, Singh P (1988) An improved enzyme immunoassay for limonin. J Food Sci 53:311–312
Rouseff RL, Nagy S (1982) Distribution of limonoids in citrus seeds. Phytochemistry 21:85–90
Van Beek TA, Blaakmeer A (1989) Determination of limonin in grapefruit juice and other citrus juices by high-performance liquid chromatography. J Chromatogr 464:375–386
Weiler EW, Mansell RL (1980) Radioimmunoassay of limonin using a tritiated tracer. J Agric Food Chem 28:543–545
Wilson KW, Crutchfield CA (1968) Spectrophotometric determination of limonin in orange juice. J Agric Food Chem 16:118–124

Subject Index